# Climatology: An Atmospheric Science

# Climatology: An Atmospheric Science

Edited by **Andrew Hyman**

New York

Published by Callisto Reference,
106 Park Avenue, Suite 200,
New York, NY 10016, USA
www.callistoreference.com

**Climatology: An Atmospheric Science**
Edited by Andrew Hyman

International Standard Book Number: 978-1-63239-632-7 (Hardback)

Printed in the United States of America.

# Contents

**Permissions**

**List of Contributors**

# Preface

Climate is the fixed or slightly varying pattern of weather of an area prevailing over a period of time. Climatology attempts to explain the impact of climatic conditions, to assist people in planning and building infrastructure and issuing warnings about potential dangers. Paleoclimatology, historical climatology, and paleotempestology are the different approaches used for determining the past and future global climatic patterns. Since climates are dynamic and complex, they are difficult to deal with. Constant in-depth research and efforts are required on the part of climatologists to deal with these frequently varying patterns. This book presents the research and studies performed by experts across the globe. It strives to provide a fair idea about this discipline and to help develop a better understanding of the latest advances within this field. It will provide comprehensive knowledge to the readers. This text will benefit climatologists, environmentalists, atmospheric scientists, professional and students alike.

All of the data presented henceforth, was collaborated in the wake of recent advancements in the field. The aim of this book is to present the diversified developments from across the globe in a comprehensible manner. The opinions expressed in each chapter belong solely to the contributing authors. Their interpretations of the topics are the integral part of this book, which I have carefully compiled for a better understanding of the readers.

At the end, I would like to thank all those who dedicated their time and efforts for the successful completion of this book. I also wish to convey my gratitude towards my friends and family who supported me at every step.

**Editor**

# Amplified Feedback Mechanism of the Forests-Aerosols-Climate System

**Thomas Hede, Caroline Leck, and Jonas Claesson**

*Department of Meteorology, Stockholm University, 106 91 Stockholm, Sweden*

Correspondence should be addressed to Thomas Hede; hede@misu.su.se

Academic Editor: Alexey V. Eliseev

Climate change very likely has effects on vegetation so that trees grow faster due to carbon dioxide fertilization (a higher partial pressure increases the rate of reactions with Rubisco during photosynthesis) and that trees can be established in new territories in a warmer climate. This has far-reaching significance for the climate system mainly due to a number of feedback mechanisms still under debate. By simulating the vegetation using the Lund-Potsdam-Jena guess dynamic vegetation model, a territory in northern Russia is studied during three different climate protocols assuming a doubling of carbon dioxide levels compared to the year 1975. A back of the envelope calculation is made for the subsequent increased levels of emissions of monoterpenes from spruce and pine forests. The results show that the emissions of monoterpenes at the most northern latitudes were estimated to increase with over 500% for a four-degree centigrade increase protocol. The effect on aerosol and cloud formation is discussed and the cloud optical thickness is estimated to increase more than 2%.

## 1. Introduction

Climate change is a global problem facing humanity and at the same time it is of anthropogenic origin [1]. However, the mechanisms for global climate change are interlinked between all the components of planet Earth, including atmosphere, hydrosphere, and biosphere. Although climate change has begun to become more and more evident, there are still large uncertainties in future projections of the climate. In order to project the evolution of the climate change, we use general circulation models (GCMs), but the accuracy of the projection is dependent on both the input data and the compliance with real processes that affect the climate. One category of processes that is of particular importance concerns feedback mechanisms. A feedback mechanism is a chain of events that in the end has an impact on the initial step, either if it is an enhancing effect, so called positive feedback, or if it has an inhibiting effect, so called negative feedback.

Three examples of a positive feedback mechanism that result from an enhanced greenhouse effect with a subsequent increase in temperature are given by Mencuccini and Grace [2], Swann et al. [3], and Walter et al. [4] Mencuccini and

Grace [2] discuss the leaf to sapwood area ratio in Scots pine as of a changed difference in water vapor pressure deficit in air. Swann et al. [3] propose one example of an enhanced positive feedback mechanism. It consists of two components: (1) broad-leaf deciduous trees assumed to invade the warming tundra and thereby alter the albedo to less reflective land by replacing snow covered land with darker forests and (2) increased release of water vapour through evapotranspiration compared to evergreen boreal needle forests contributing to the greenhouse effect. In a warming climate, the first component indicates that there is a change in type of vegetation with new establishment of broad-leaf deciduous trees in the tundra as well as a replacement of evergreen boreal needle forests. The latter has already been mentioned in component 2. The two components are interlinked through the change in vegetation type. Walter et al. [4] show the importance of emissions of methane from thaw lakes in Siberia through ebullition. The thawing permafrost along the lake margins accounted for the largest emissions, which was estimated to constitute an accelerating effect on climate warming.

One example of a negative feedback mechanism was proposed by Kulmala et al. [5], which links changes in growth of forests and emissions of volatile organic compounds

Negative feedback mechanism (Kulmala et al., 2004)

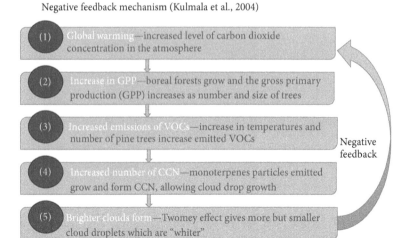

FIGURE 1: Schematic representation of the negative feedback mechanism proposed by Kulmala et al. [5]. GPP is the gross primary production and increases as the boreal forest grows. VOCs are volatile organic compounds. CCN is cloud condensation nuclei.

(VOCs) with subsequent generation of aerosol particles and possible changes in the number of particles available for cloud droplet activation called cloud condensation nuclei (CCN) and cloud optical properties. Figure 1 shows a schematic representation.

Number 1 in Figure 1: global warming causally related to an increase of atmospheric carbon dioxide ($CO_2$) and other so called greenhouse gases [1]. Number 2 in Figure 1: photosynthesis is driven by an increase of atmospheric carbon dioxide with a subsequent increase in the amount of chemical energy as biomass that primary producers create in a given length of time called the gross primary production (GPP) [6]. GPP increases when the boreal forest grows by both tree number and size. Number 3 in Figure 1: boreal forests contribute to the emissions of volatile organic compounds (VOCs). According to O'Dowd et al. [7] the oxidizing products of monoterpenes *cis*-pinonic and pinic acids are abundant VOCs in the organic fraction of newly formed ambient aerosol particles. Number 4 in Figure 1: an increased fraction of surface-active aerosol particles, which due to a decrease of surface tension [8] become efficient CCN, owed to an increase of VOC emissions. The surface-active properties of the VOCs oxidizing products have been discussed in several studies [9–12]. Even though Facchini et al. [9] may have oversimplified their estimates [13–15], surface-active organic material remains to constitute a key component of atmospheric aerosols in cloud droplet activation [16]. A study by Dalirian et al. [17] indicates that the partitioning between water soluble and water insoluble matter in activating CCN is as important as the adsorption properties of the surface. Hings et al. [18] suggest that hydrophobic particles may be coated with slightly soluble material and thereby increase the activation process. The studies of both Dalirian et al. [17] and Hings et al. [18] therefore relate to the activation properties of the slightly water soluble oxidation products of monoterpenes. Number 6 in Figure 1: the increase in number of cloud droplets ultimately could alter cloud optical (microphysical) properties and form brighter clouds [19].

These brighter clouds are proposed to reflect the incoming solar radiation more effectively and thereby enhance surface cooling and by that provide a negative feedback mechanism.

In this study we extend the Kulmala et al. [5] study to also discuss possible implications for the albedo of low-level clouds in the high Arctic as of an increase of VOCs and forest growth in Siberia causally related to global warming due to increase of levels of atmospheric carbon dioxide. Figure 2 gives a schematic representation of the proposed negative feedback mechanisms under study. Main differences from the feedback mechanism proposed by Kulmala et al. [5] are marked with orange.

To estimate changes of GPP upon a doubling of $CO_2$ levels relative to the year 1975, we used the Lund-Potsdam-Jena (LPJ-GUESS) dynamic vegetation model [21, 22] The simulation was started in 1975 and carried out for a period of 200 years over a terrestrial territory in the Siberian inland of northern Russia. During the time of simulation a linear increase in temperature of 2, 3, and 4 degrees centigrade, respectively, resulted after doubling of the atmospheric carbon dioxide level. From the estimates of GPP subsequent emissions of monoterpenes were calculated. At last, the simulations and levels of emissions were discussed in terms of cloud formation and possible climate effects.

## 2. Simulation Method to Estimate GPP

To simulate GPP the dynamic vegetation model LPJ-GUESS was used [20, 22–29]. LPJ-GUESS can be viewed as a framework for ecosystem-modelling which has used some elements from the equilibrium terrestrial biosphere model (BIOME) family of models [29]. The LPJ-GUESS model shows an intermediate response and sensitivity to atmospheric $CO_2$ concentrations and temperatures [27, 28]. Large-scale dynamics of terrestrial vegetation and interactions between the biosphere and atmosphere in terms of exchange of carbon dioxide and water vapour are represented in the model. Processes like photosynthesis and evapotranspiration are simulated daily

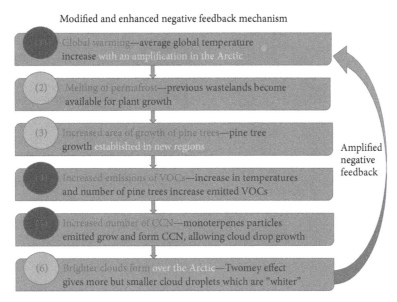

Modified and enhanced negative feedback mechanism

(1) Global warming—average global temperature increase with an amplification in the Arctic

(2) Melting of permafrost—previous wastelands become available for plant growth

(3) Increased area of growth of pine trees—pine tree growth established in new regions

(4) Increased emissions of VOCs—increase in temperatures and number of pine trees increase emitted VOCs

(5) Increased number of CCN—monoterpenes particles emitted grow and form CCN, allowing cloud drop growth

(6) Brighter clouds form over the Arctic—Twomey effect gives more but smaller cloud droplets which are "whiter"

Amplified negative feedback

FIGURE 2: Schematic representation of the negative feedback mechanism proposed in this study. Marked in orange are modifications from the negative feedback mechanism proposed by Kulmala et al. [5], which was schematically presented in Figure 1. VOC is volatile organic compounds; CCN is cloud condensation nuclei.

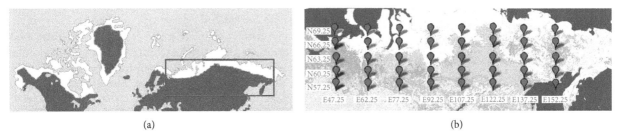

(a)

(b)

FIGURE 3: (a) The permafrost region (white), (b) Siberia and the area under study. Each of the simulated red coordinates does correspond to the coordinates marked out in Figures 6–11.

while slower processes, like biomass growth, are implemented annually. The input data to the model consist of climate parameters, atmospheric $CO_2$ concentrations, and a soil code. The climate parameters are monthly or daily values of mean diurnal air temperature, precipitation, and incoming shortwave radiation (sunshine). The soil code is used for parameters such as hydrology and thermal diffusivity of the soil. Climate forcing, that is mean monthly temperature, precipitation, and cloud fraction, was using the Climate Research Unit (CRU) TS 3.0 data set [26]. Atmospheric $CO_2$ concentrations were taken from observations [20]. Each simulation was started with a 1000-year initialisation to establish vegetation and soil budgets for the equilibrated climate of year 1975. CRU climate data cycled repeatedly to force the model during the initialisation using $CO_2$ concentrations of the year 1975 (330 ppm(v)). After the initialisation, a production simulation was applied. During the production simulation, the temperature increased monotonically with adding an offset of 2–4 degrees centigrade to the baseline climate dataset. The simulations were performed across a $0.5° \times 0.5°$ (latitude–longitude) grid [25]. Vegetation in a grid cell is described in terms of the fractional coverage of populations of different plant functional types (PFTs). Each PFT is assigned bioclimatic limits, which determine whether it can grow and spread under the current condition during the simulation. We used LPJ-GUESS version 2.1, which includes the PFT set and modifications described in [24].

*2.1. Simulated Area and Global Warming Protocols (1 in Figure 2).* The simulated area is located in Siberia in northern Russia. We simulate GPP for a specific geographical coordinate. This is done in a grid-box representing the selected area. Because of the spherical shape of Earth, the grid-box in the model is not rectangular, but rather a trapetzoid as a first simplification.

This study's model domain is marked in Figure 3(b). This is a part of the permafrost region, wasteland with permanently frozen soil; see Figure 3(a). The total area simulated is $8\,424\,510\ km^2$. This can be compared to the total area of Siberia, which is about $13.1 \times 10^6\ km^2$. The simulated area is thus about 64% of the total aerial extent of Siberia. The total simulated area that was forested after the initial equilibration of 1000 years was $6\,115\,464\ km^2$. This value corresponds to the vegetation in the year 1975. Results by

TABLE 1: The emission protocols assumed in this study.

| Emission protocol | Carbon dioxide level increase | Simulated years | Temperature increase |
|---|---|---|---|
| A | Doubling | 200 | 2 degrees centigrade |
| B | Doubling | 200 | 3 degrees centigrade |
| C | Doubling | 200 | 4 degrees centigrade |

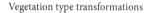

Vegetation type transformations

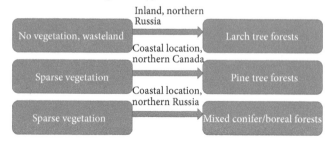

FIGURE 4: Vegetation establishing in permafrost regions after climate change.

Esser et al. [23] show that the historical carbon storage in the boreal forests is located in the western part of Siberia from the years 1860 to 2002.

Three different emission protocols were used to model the response in forest growth. We assumed a doubling of the $CO_2$ level compared to the year 1975 (330 ppm(v)) for all of the protocols after the simulated period of 200 years. The temperature increase corresponds to different levels of climate sensitivity, namely, 2, 3, or 4 degrees centigrade increase. The emission protocols are presented in Table 1. The level of $CO_2$ increase could also be varied, so that higher levels than a doubling can be obtained. It is also possible to keep the $CO_2$ at a constant level during temperature increase and thereby determine how much of the effect is attributed to the $CO_2$ doubling. These types of assumptions will not be considered in this study but could be recommended for future studies.

## 3. Tree Growth Established in New Areas

In the LPJ-GUESS model larch forest is referred to as boreal needle summergreen (BNS) vegetation. In Siberia 46% of the forested area is made up by larch forest [30]. Moreover, spruce and pine trees are named boreal needle evergreen (BNE) vegetation in the model. The grass is not contributing significantly to the emission of monoterpenes and is therefore not considered in this study. In this study also broadleaved summer green trees are not considered as a source of monoterpene emissions, even though there are exceptions such as oak trees [31].

*3.1. Simulations Indicating Melted Permafrost (2 in Figure 2).* To see whether any new vegetation would be established, as a first survey we carried out LPJ-GUESS simulations over the permafrost region using the three emission protocols listed in Table 1. Typical results are shown in Figure 4. The simulations resulted in the fact that new vegetation was established in the wasteland regions as well as in changes of the composition of the forest. We were encouraged by the results from this first survey that a more detailed study could be considered, and that the primary region of interest should be Siberia, located in northern Russia. This was motivated by that the landscape and vegetation were both highly affected by climate change in this region.

*3.2. Simulations of Forest Growth (3 in Figure 2).* Based on the three emission protocols in Table 1 we calculated the relative growth of two types of forests, BNE and BNS, defined as the difference in biomass ($kgC/m^2$) between the simulated year 1975 and after 200 years.

In Figures 6–11, the simulated BNE relative growth in $kgC/m^2$ for northern Russia is listed for each of the protocols given in Table 1. Light shading represents the relative growth as defined; no colour indicates a moderate relative growth (0–3 $kgC/m^2$); yellow colour indicates a substantial relative growth (>3 $kgC/m^2$); green colour indicates a very high relative growth (>5 $kgC/m^2$); and red colour indicates a relative decrease in growth (<0 $kgC/m^2$). Dark shade marks biomass change in percentage; no colour indicates a moderate change in percent (0–200%); orange colour indicates a substantial change in percent (>200%); green colour indicates a very high change in percent (>500%); and red colour indicates a negative change in percent (<100%). The symbol "#" marks locations where there was a growth of forest on land with previously not established forestation. The relative growth in percent is given as the percentage of the biomass in the year 1975; that is, 100% after 200 years equals the biomass level in the year 1975. The background image is the box showed in Figure 3(a).

As can be seen from Figures 6, 8, and 10, the relative growth (%) of BNE forest is located in the northwest region of the investigated area, which is in northern Russia. Further, there is a belt of increased growth of BNE forest stretching from the northwest to southeast of the region, and this is at the same time, at least in the southeast part, and at an expense of BNS forest, which is decreasing in biomass. The BNE forest is decreasing in biomass in the southwest corner of the region. This is according to the model due to a change of vegetation type where conifer forest is growing at an expense of BNE forest. Similar trends are shown in all the three climate protocols: A, B, and C. The trend is however most visible for the emission protocols with the highest temperature increase (C in Table 1).

From Figures 7, 9, and 11 we can conclude that BNS forest is growing in the northeast region and decreasing in biomass in the southwest region as well as in the belt described above. All in all, the general trend is that the boreal forest of northern Russia is growing and moving northward as a result of an increase in temperature and atmospheric $CO_2$. This result is supported by the findings of Xu et al. [32].

TABLE 2: Examples of Pinaceae emissions of monoterpenes [20].

| Examples of Pinaceae | Monoterpene emission $(\mu g\, g_{dw}^{-1}\, h^{-1})$ |
|---|---|
| *Cedrus deodara*, deodar cedar | 0.9 |
| *Picea sitchensis*, Sitka spruce | 1.1 |
| *Pinus halepensis*, Aleppo pine | 1.0 |
| *Pinus pinaster*, maritime pine | 1.0 |
| *Pinus radiata*, Monterey pine | 0.7 |

# 4. Emissions of Monoterpenes and Cloud Droplet Activation

*4.1. Estimated Emissions of VOC and CCN (4 in Figure 2).* In order to perform a back of the envelope calculation of the increase in VOC emissions in the sub-Arctic region that would follow forestation of the tundra, VOC emission rates of spruce, pine, and larch were used. Petersson [33] estimated the fraction of needles to total tree weight to be 9.2% for spruce and 4.2% for pine. An average of these number fractions was used in the simulated BNS and BNE forests.

Lind [34] reported the carbon content of dry mass in spruce and pine to be close to 50%. These numbers are used in this study to convert the biomass carbon into weight of needles per km$^2$. A typical value of 15 kgC/m$^2$ would correspond to 30 kg dry mass per m$^2$, which is $3.0 * 10^7$ kg/km$^2$. This can be compared with the numbers given by Albrektson [35]; for 194 450 trees per km$^2$ with an average dry mass of 172 kg makes $3.3 * 10^7$ kg/km$^2$ dry mass.

Ruuskanen et al. [30] report emissions of monoterpenes at 30°C of $5.2$–$21\,\mu g\, g_{dw}^{-1}\, h^{-1}$ for Siberian Larch (*Larix sibirica*). Emissions from Scots pine (*Pinus sylvestris*) and Norwegian spruce (*Picea abies*) of monoterpenes at 20°C were reported by Janson [36] to be $0.8 \pm 0.4\,\mu g\, g_{dw}^{-1}\, h^{-1}$ and $0.5 \pm 0.7\,\mu g\, g_{dw}^{-1}\, h^{-1}$, respectively. Similar levels of emission are reported for other members of the Pinaceae family; see Table 2 [31].

However, emissions of monoterpenes are temperature dependent. To account for this we used the following equation given by Kesselmeier and Staudt [31]:

$$E(t) = F \times E(20) \times 10^{(0.04 \times \Delta t)}. \tag{1}$$

Here, $\Delta t$ is the temperature difference between ambient temperature ($t$) and 20°C. The factor $F$, which is multiplied with the emission rate at 20°C, varies as a function of temperature.

The factor for calculating the temperature dependency of monoterpene emissions was estimated from the average monthly temperature at longitudes E47.25, E92.25, and E152.25, respectively. The temperature increase according to the protocols A, B, and C (the increase in temperature due to climate change and thereby the deviation from 20°C) is compensated by an increase of the factor $F$ proportionally by 1.20, 1.32, and 1.45, respectively. A typical temperature profile is shown in Figure 5.

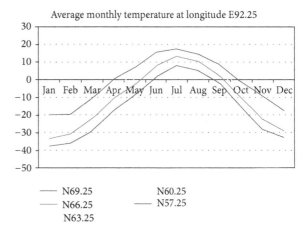

Average monthly temperature at longitude E92.25

| — N69.25 | N60.25 |
| — N66.25 | — N57.25 |
| N63.25 | |

FIGURE 5: Average monthly temperature at longitude E92.25.

The number of days is divided into bins of temperature. Each bin corresponds to a factor ($F$) for which the emission rates are corrected, shown in Table 3.

The number of days accounted for in each temperature bin is based on one month (30 days) for which the temperature is within the interval. On average, the model and the temperature profile are in agreement.

All days when the temperature exceeds 0°C have been assumed to emit monoterpenes. The emission rates by Ruuskanen et al. [30] and Janson [36] given above were used in the calculations. To compensate for lower temperatures in the area the numbers of days were additionally corrected with a factor and done in the same way as for the increased growth listed in Figures 6–11. The resulting estimates of the annual increase of monoterpenes for the three emission protocols are as follows:

for A: 14.5 Tg y$^{-1}$,

for B: 16.4 Tg y$^{-1}$,

for C: 17.0 Tg y$^{-1}$.

As a comparison, Laothawornkitkul et al. [37] estimated the annual global emission of monoterpenes to range from 33 to 480 TgC y$^{-1}$. Corresponding to the year 1975 (330 ppm(v) CO$_2$) the annual total emission of monoterpenes, in the area under study, is calculated to be 26.7 Tg. The relative increase is 55%, 62%, and 64% for the protocols A, B, and C, respectively. These results were compared well with the estimates by Xu et al. [32] that report a relative greening of the boreal region from 34 to 41%. Most striking is the estimated emission of monoterpenes at the most northern latitude investigated (N69.25), which increased from 0.54 Tg y$^{-1}$ with 258% for A protocol, 355% for B protocol, and 496% for C protocol, respectively.

*4.2. CCN Production (5) and Cloud Droplet Activation (6) in Figure 2.* The number of CCN available for cloud droplet activation is hard to estimate because the process of condensational growth of the newly formed particles involving monoterpenes is not fully understood [38]. Given the complexity, Spracklen et al. [39] estimated that a factor of

TABLE 3: Temperature bins and factors for emission rates for different locations in the area under study (Figure 3(b)). For each longitude, shown in paranthesis, the number of days that are applied.

| Temp. bin (°C) | Factor | Latitude (N) | Number of days for longitudes (E47.25, E62.25, E77.25) | Latitude | Number of days for longitudes (E92.25, E107.25, E122.25) | Latitude (N) | Number of days for longitudes (E137.25, E152.25) |
|---|---|---|---|---|---|---|---|
| <0 | 0 | 69.25 | N/A | 69.25 | 270 | 69.25 | 240 |
| | | 66.25 | 210 | 66.25 | 240 | 66.25 | 240 |
| | | 63.25 | 210 | 63.25 | 210 | 63.25 | 210 |
| | | 60.25 | 150 | 60.25 | 210 | 60.25 | 240 |
| | | 57.25 | 150 | 57.25 | 180 | 57.25 | N/A |
| 0–5 | 0.21 | 69.25 | N/A | 69.25 | 30 | 69.25 | 30 |
| | | 66.25 | 30 | 66.25 | 30 | 66.25 | 30 |
| | | 63.25 | 0 | 63.25 | 60 | 63.25 | 60 |
| | | 60.25 | 60 | 60.25 | 30 | 60.25 | 60 |
| | | 57.25 | 30 | 57.25 | 30 | 57.25 | N/A |
| 5–10 | 0.30 | 69.25 | N/A | 69.25 | 60 | 69.25 | 60 |
| | | 66.25 | 60 | 66.25 | 60 | 66.25 | 30 |
| | | 63.25 | 60 | 63.25 | 0 | 63.25 | 0 |
| | | 60.25 | 60 | 60.25 | 30 | 60.25 | 60 |
| | | 57.25 | 30 | 57.25 | 60 | 57.25 | N/A |
| 10–15 | 0.52 | 69.25 | N/A | 69.25 | 0 | 69.25 | 30 |
| | | 66.25 | 60 | 66.25 | 30 | 66.25 | 60 |
| | | 63.25 | 60 | 63.25 | 60 | 63.25 | 90 |
| | | 60.25 | 30 | 60.25 | 60 | 60.25 | 0 |
| | | 57.25 | 60 | 57.25 | 30 | 57.25 | N/A |
| >15 | 0.83 | 69.25 | N/A | 69.25 | 0 | 69.25 | 0 |
| | | 66.25 | 0 | 66.25 | 0 | 66.25 | 0 |
| | | 63.25 | 30 | 63.25 | 30 | 63.25 | 0 |
| | | 60.25 | 60 | 60.25 | 30 | 60.25 | 0 |
| | | 57.25 | 90 | 57.25 | 60 | 57.25 | N/A |

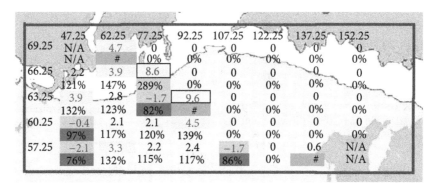

FIGURE 6: BNE relative growth (%) in northern Russia for protocol A after 200-year simulation. The columns show the result for eastern longitudes and the rows for northern latitudes marked in Figure 4(b). Longitudes E47.25, N69.25 and E152.25, N57.25 are not terrestrial and are thus marked with not available (N/A). Light shading represents relative growth and dark shade marks biomass change in percentage. The symbol "#" marks locations where there was a growth of forest on a land area with previously not established forestation.

5 increase in monoterpenes yields up to a factor of 3 in the enhancement of CCN from new particle formation. Again, at present the production of CCN from monoterpenes can only roughly be estimated as the processes involved are not fully understood.

In the boreal forest Kerminen et al. [40] observed nucleating particles, on average $<150\,\mathrm{cm}^{-3}$, followed by subsequent rapid growth. This produced a large number of particles in sizes larger than 50 nm in diameter. 12 hours later, a cloud entered the region and >60% of the particles in the size range 50–100 nm diameter had been activated into cloud droplets. 67% of the cloud droplets were estimated to originate from particles below ca 15 nm in diameter.

|  | 47.25 | 62.25 | 77.25 | 92.25 | 107.25 | 122.25 | 137.25 | 152.25 |
|---|---|---|---|---|---|---|---|---|
| 69.25 | N/A | 0 | 0 | 0 | 3.0 | 5.6 | 4.7 | 5.2 |
|  | N/A | # | 0% | 0% | 895% | 243% | 1522% | 341% |
| 66.25 | −0.1 | 0.1 | 0.1 | 3.9 | 2.6 | 2.0 | 4.4 | 4.5 |
|  | 55% | 492% | # | 449% | 300% | 128% | # | 238% |
| 63.25 | 0.1 | 0 | 0.2 | −4.8 | 1.7 | 0.6 | 6.9 | 2.2 |
|  | 184% | 116% | 370% | 2% | 137% | 57% | 1074% | 156% |
| 60.25 | 0 | 0.1 | 0 | 0 | 2.6 | 0.8 | 5.9 | 0 |
|  | 118% | 189% | 195% | 109% | 262% | 109% | # | 0% |
| 57.25 | −0.2 | −0.1 | 0 | 0.1 | 0.1 | 4.7 | 0 | N/A |
|  | 23% | 60% | 136% | 229% | 171% | 181% | # | N/A |

FIGURE 7: BNS relative growth (%) in northern Russia for protocol A after 200-year simulation. The columns show the result for eastern longitudes and the rows for northern latitudes marked in Figure 4(b). Longitudes E47.25, N69.25 and E152.25, N57.25 are not terrestrial and are thus marked with not available (N/A). Light shading represents relative growth and dark shade marks biomass change in percentage. The symbol "#" marks locations where there was a growth of forest on a land area with previously not established forestation.

|  | 47.25 | 62.25 | 77.25 | 92.25 | 107.25 | 122.25 | 137.25 | 152.25 |
|---|---|---|---|---|---|---|---|---|
| 69.25 | N/A | 7.9 | 1.1 | 0 | 0 | 0 | 0 | 0 |
|  | N/A | # | # | 0% | 0% | 0% | 0% | 0% |
| 66.25 | 3.5 | 3.4 | 6.1 | 0 | 0 | 0 | 0 | 0 |
|  | 133% | 141% | 235% | 0% | 0% | 0% | 0% | 0% |
| 63.25 | 1.8 | 1.9 | 0.7 | 12.1 | 0 | 0 | 0 | 0 |
|  | 115% | 116% | 107% | # | 0% | 0% | 0% | 0% |
| 60.25 | −1.0 | 0.1 | 0.7 | 2.4 | 0.9 | 0 | 0 | 0.1 |
|  | 92% | 101% | 107% | 121% | # | 0% | 0% | # |
| 57.25 | −0.2 | −1.6 | 1.9 | −0.9 | −0.6 | 1.4 | 8.3 | N/A |
|  | 97% | 84% | 113% | 94% | 95% | # | # | N/A |

FIGURE 8: BNE relative growth (%) in northern Russia for protocol B after 200-year simulation. The columns show the result for eastern longitudes and the rows for northern latitudes marked in Figure 4(b). Longitudes E47.25, N69.25 and E152.25, N57.25 are not terrestrial and are thus marked with not available (N/A). Light shading represents relative growth and dark shade marks biomass change in percentage. The symbol "#" marks locations where there was a growth of forest on a land area with previously not established forestation.

|  | 47.25 | 62.25 | 77.25 | 92.25 | 107.25 | 122.25 | 137.25 | 152.25 |
|---|---|---|---|---|---|---|---|---|
| 69.25 | N/A | 0 | 0 | 0 | 3.3 | 4.7 | 6.8 | 5.3 |
|  | N/A | # | # | 0% | 959% | 222% | 2143% | 345% |
| 66.25 | −0.0 | 0.1 | 0 | 3.6 | 3.3 | 1.3 | 7.7 | 5.3 |
|  | 58% | 615% | # | 432% | 356% | 118% | # | 265% |
| 63.25 | 0.1 | 0 | 0.1 | −4.9 | 1.4 | 1.7 | 6.2 | 3.9 |
|  | 178% | 67% | 241% | 2% | 130% | 121% | 975% | 199% |
| 60.25 | −0.1 | 0.1 | 0.1 | 0.0 | 1.9 | 1.6 | 7.8 | 0 |
|  | 58% | 230% | 422% | 144% | 141% | 119% | # | 0% |
| 57.25 | −0.1 | 0 | 0 | 0.1 | 0.1 | 2.7 | 0 | N/A |
|  | 37% | 95% | 137% | 211% | 146% | 146% | # | N/A |

FIGURE 9: BNS relative growth (%) in northern Russia for protocol B after 200-year simulation. The columns show the result for eastern longitudes and the rows for northern latitudes marked in Figure 4(b). Longitudes E47.25, N69.25 and E152.25, N57.25 are not terrestrial and are thus marked with not available (N/A). Light shading represents relative growth and dark shade marks biomass change in percentage. The symbol "#" marks locations where there was a growth of forest on a land area with previously not established forestation.

## 5. Discussion of the Negative Feedback Mechanisms Postulated in Figure 2

*5.1. The Feedback Mechanism.* Kulmala et al. [5] proposed that a global fertilization of atmospheric $CO_2$ with subsequent temperature increase would favour photosynthesis and the growth of Scots pine trees. The increased biomass would emit larger amounts of VOCs adding to particle formation and growth into CCN-sizes. A higher aerosol load would contribute to net cooling of the land surface both due to reflection and scattering and to the indirect effect of aerosol particles [19]. There are however four issues regarding the

| | 47.25 | 62.25 | 77.25 | 92.25 | 107.25 | 122.25 | 137.25 | 152.25 |
|---|---|---|---|---|---|---|---|---|
| 69.25 | N/A | 8.6 | 8.1 | 0 | 0 | 0 | 0 | 0 |
| | N/A | # | # | 0% | 0% | 0% | 0% | 0% |
| 66.25 | 2.6 | 4.2 | 8.2 | 3.4 | 0 | 0 | 0 | 0 |
| | 124% | 151% | 282% | # | 0% | 0% | 0% | 0% |
| 63.25 | 1.8 | −0.1 | 3.7 | 12.8 | 0 | 0 | 0 | 0 |
| | 115% | 99% | 137% | # | 0% | 0% | 0% | 0% |
| 60.25 | −2.1 | 3.2 | 3.8 | −0.3 | 4.7 | 0 | 0 | 3.3 |
| | 83% | 126% | 135% | 97% | # | 0% | 0% | # |
| 57.25 | −3.6 | −0.6 | 1.6 | −1.1 | −0.4 | 5.7 | 10.2 | N/A |
| | 59% | 94% | 111% | 92% | 97% | # | # | N/A |

FIGURE 10: BNE relative growth (%) in northern Russia for protocol C after 200-year simulation. The columns show the result for eastern longitudes and the rows for northern latitudes marked in Figure 4(b). Longitudes E47.25, N69.25 and E152.25, N57.25 are not terrestrial and are thus marked with not available (N/A). Light shading represents relative growth and dark shade marks biomass change in percentage. The symbol "#" marks locations where there was a growth of forest on a land area with previously not established forestation.

| | 47.25 | 62.25 | 77.25 | 92.25 | 107.25 | 122.25 | 137.25 | 152.25 |
|---|---|---|---|---|---|---|---|---|
| 69.25 | N/A | 0.1 | 0.1 | 0 | 3.8 | 4.4 | 8.1 | 5.6 |
| | N/A | # | # | 0% | 1106% | 212% | 2519% | 359% |
| 66.25 | 0.0 | 0.1 | 0.1 | 2.0 | 4.4 | −0.2 | 7.6 | 3.7 |
| | 125% | 554% | # | 281% | 436% | 98% | # | 214% |
| 63.25 | 0 | 0 | 0 | −4.9 | 1.8 | 1.6 | 8.5 | 5.0 |
| | 159% | 118% | 130% | 2% | 138% | 120% | 1307% | 227% |
| 60.25 | 0 | 0 | 0.1 | 0.1 | −1.1 | 2.9 | 8.0 | 0.1 |
| | 95% | 129% | 230% | 228% | 77% | 132% | # | # |
| 57.25 | 0 | 0 | 0 | 0.1 | 0 | −1.0 | 0.1 | N/A |
| | 78% | 100% | 98% | 171% | 140% | 83% | # | N/A |

FIGURE 11: BNS relative growth (%) in northern Russia for protocol C after 200-year simulation. The columns show the result for eastern longitudes and the rows for northern latitudes marked in Figure 4(b). Longitudes E47.25, N69.25 and E152.25, N57.25 are not terrestrial and are thus marked with not available (N/A). Light shading represents relative growth and dark shade marks biomass change in percentage. The symbol "#" marks locations where there was a growth of forest on a land area with previously not established forestation.

feedback mechanism proposed that we believe they are worth discussing in view of the results obtained in this study.

Firstly, the increase in average temperature and the increase of $CO_2$ in the atmosphere as a result of climate change not only contribute to an increase of numbers of trees in the forests, as stated by Kulmala et al. [5], but it would also be possible for new forests to establish in former wastelands such as the subpolar permafrost region in the north of Russia, Scandinavia, Greenland, Canada, and Alaska. Further, if the permafrost will melt in the region, vegetation can start to grow. This effect would also be amplifying the effect described by Kulmala et al. [5], since more forests would emit more VOCs contributing to the population of aerosol particles.

Secondly, the effect on surface tension and thereby the lowering of supersaturation of water vapour needed for condensational growth by water vapour of droplets according to Köhler theory [8] is questioned by Sorjamaa et al. [13] and Kokkola et al. [14]. They state that the partitioning between surface and bulk reduces the effect of surface tension reduction. At a first assumption Li et al. [41] show that the natural surfactant cis-pinonic acid (an oxidation product of monoterpenes) greatly can reduce the surface tension of nanosized aerosol particles by accumulating at the surface;

however, Hede et al. [11, 12] showed that cis-pinonic acid can form micelle-like aggregates inside the nanoaerosol, but the surface tension reduction is not limited by this behaviour in accordance with experimental results by Schwier et al. [42]. Therefore, the effect of surface tension depression is not irrelevant for cloud formation processes, implying that a greater number of cloud droplets can form as a result of an increase in the number of available particles consisting of surface tension reducing surfactants. The clouds that are formed are thus "whiter" (as of changed microphysical properties) with a higher albedo and reflecting more of the incoming solar radiation. This effect is amplifying the negative feedback process proposed by Kulmala et al. [5].

Thirdly, monoterpenes evaporated from boreal trees react quickly with oxidizing reactants like ozone and hydroxyl radicals producing low-volatile substances such as cis-pinonic acid that take place in gas-to-particle conversion processes [43, 44]. According to Boy et al. [45] the growth rate of nucleation mode particles is from 30 to 50% explained by hydroxyl radical oxidation. According to Riipinen et al. [46] the growth of nanoparticles is governed by the presence organics. Given time in the atmosphere newly formed particles may grow to form CCN. Depending on the synoptical scale meteorology, CCN generated and grown from VOCs originating over Siberia could potentially reach the high Arctic pack ice region.

In the high Arctic summer there are few of available particles for cloud droplet formation, since the air is clean with limited influences from manmade activities [47–49]. An increase in particles due to regional transport would have an impact on the microphysical properties of the clouds being formed. Generally, clouds over the pack ice constitute a warming factor [50, 51], but adding particles when CCN > $10 \, cm^{-3}$ will have a net cooling effect [52]. This is again an amplification of the negative feedback mechanism.

The fourth issue is the most important. Kulmala et al. [5] use an assumed moderate estimate of the increase of VOC emissions of 10%. This value of VOCs emissions increase is corresponding to an increase of optical thickness of the clouds by 1 to 2%. Kulmala et al. [5] also state that a doubling (100%) of increase of VOCs emissions corresponds to 20% increase in optical thickness of the clouds. In the present study we have found that the increase of emissions of VOCs is higher than 10%. This result is not based on a moderate estimate but on simulations using the LPJ-GUESS vegetation model and calculations of emission rates reported in literature. Even though there are approximations included in both simulations as well as in the calculations, we hope that the results derived from this study are more accurate than a moderate estimation. That our estimates of the relative increase of VOCs are higher than the estimates by Kulmala et al. [5] would work in the direction towards an amplification of the feedback mechanism under study.

*5.2. Simplifications and Uncertainties.* In any kind of attempt to simulate feedback mechanisms there are both simplifications made and uncertainties to consider. For Step 1 in Figure 2, we do not know the magnitude of the global warming. The Intergovernmental Panel on Climate Change [1] concludes that there likely is an increase in average global temperature, and that the increase is somewhere between $0.3°C$ and $0.7°C$ for the period of 2016–2035 compared to the period 1986–2005. For the period 2081–2100, different climate model scenarios (Representative Concentration Pathways, RCPs) project the average global temperature increase as $0.3°C$–$1.7°C$ (RCP2.6), $1.1°C$–$2.6°C$ (RCP4.5), $1.4°C$–$3.1°C$ (RCP6.0), and $2.6°C$–$4.8°C$ (RCP8.5) [1]. In future studies, the RCPs can be used to model the temperature increase during the simulations. For Step 2, the present knowledge is insufficient to know how much warmer the tundra region must be for the permafrost to melt permanently. For Step 3, we rely on the model simulations to be correct, and as mentioned the model is simplified and the result comes accompanied with uncertainties. For Step 4, the rate of emissions of VOCs is temperature dependent and also different for different types of vegetation. For Step 5, even if we knew the number of newly formed organic particles from emissions of VOCs, the number of particles available as CCN is hard to predict because the process of condensational growth of the newly formed particles is not fully understood [38]. We have seen that the conventional theory used, Köhler theory [8], lacks in

the description of organic compounds for sizes below 5 nm and complex morphology as the CCN. However, we have previously shown that there are methods for better describing such systems [11, 53]. For Step 6, the aerosol-cloud-radiation-albedo relationship over the Arctic pack ice is discussed by Leck and Bigg [49] and was concluded to be more complex than elsewhere.

For the calculations of increase in emissions of monoterpenes from forest growth, the resolution of the simulated area is of vital importance. The more accurate the resolution of the calculated locations is, the higher the degree of information would be. To meet the need for highly resolved simulations we use the 50 km × 50 km horizontal grid resolution provided by the LPJ-GUESS model. By this we increase the resolution by 5 to 10 times relative to using atmospheric GCMs. However, the uncertainties in the assumptions used in the simulations described above limit the level of accuracy in the calculations. An even higher resolution would therefore not favor the final results.

The temperature increase that we set to 2, 3, and 4 degrees centigrade is corresponding to an increase globally in the LPJ-GUESS model evenly distributed. This is, however, not the case since we can see an amplification of the warming in the Arctic region and a global mean of 2-degree centigrade increase could actually correspond to a higher temperature in the Arctic region. This effect is not considered in our simulations. This suggests that the enhancement of the forests-aerosol-climate negative feedback mechanism is most likely even more pronounced at a lower global temperature increase.

## 6. Summary and Conclusions

In this study model simulations of the vegetation in northern Russia were carried out for the climate of the year 1975 as a reference and for three possible future projections: two-, three-, and four-degree centigrade increase of mean global temperature and with a doubling of the atmospheric carbon dioxide concentration over 200 years. The simulations were carried out using the LPJ-GUESS model. The growth of the forest was converted to emissions of monoterpenes based on literature values and compensating for the number of days of sunlight and to a temperature factor.

The results show the following.

(i) There was a growth of the boreal forest both for spruce and pine (in the central and western parts of the forest area) and larch (in the north-eastern part of the forest), as well as new forestation of previously wastelands of the tundra. Based on the simulations the boreal forest will move northward as a result of global warming.

(ii) The emissions of monoterpenes at the most northern latitudes were estimated to increase with close to 500% for a four-degree centigrade increase protocol.

(iii) The likelihood for organic vapours and particles to be transported northward over the central Arctic Ocean is highest for this northern latitude at the same time as the projected increase in precursors for particle formation is huge. Kulmala et al. [5] suggested that the cloud optical thickness would increase 1 to 2% for an increase in VOC emissions by 10% and the cloud optical thickness would increase 20% for an increase in VOC emissions by 100%. As we report an overall increase in VOC emissions by 64%, the cloud optical thickness would then increase more than 2%. However, a back of the envelope calculation indicates that the optical thickness may be up to 9%. For more accurate calculations of cloud optical thickness, a future study could incorporate a box model for radiation and convection. The preliminary results however indicate that the cloud deck to a greater extent would reflect solar radiation back to space and thereby cool the surface of the Arctic region more than what was previously expected. Kurtén et al. [54] estimate the global contribution from boreal aerosol formation to radiative forcing to be from $-0.03$ to $-1.1\,\mathrm{Wm}^{-2}$ and therefore an increase of VOC emissions would be expected to increase the radiative forcing of the same magnitude. This can be compared to the radiative forcing from surface albedo changes which has a global mean value of $-0.20\,\mathrm{Wm}^{-2}$ or $0.35\,\mathrm{Wm}^{-2}$ radiative forcing caused by an atmospheric $CO_2$ increase of 19 ppm(v) since 800 AD [55].

## Conflict of Interests

The authors have no conflict of interests to declare.

## References

[1] IPCC, "Climate change 2013: the physical science basis," in *Contribution of Working Group I to the Fifth Assessment Report of the Intergovernmental Panel on Climate Change*, T. F. Stocker, D. Qin, G.-K. Plattner et al., Eds., p. 1535, Cambridge University Press, Cambridge, UK, 2013.

[2] M. Mencuccini and J. Grace, "Climate influences the leaf area/sapwood area ratio in Scots pine," *Tree Physiology*, vol. 15, no. 1, pp. 1–10, 1995.

[3] A. L. Swann, I. Y. Fung, S. Levis, G. B. Bonan, and S. C. Doney, "Changes in arctic vegetation amplify high-latitude warming through the greenhouse effect," *Proceedings of the National Academy of Sciences of the United States of America*, vol. 107, no. 4, pp. 1295–1300, 2010.

[4] K. M. Walter, S. A. Zimov, J. P. Chanton, D. Verbyla, and F. S. Chapin III, "Methane bubbling from Siberian thaw lakes as a positive feedback to climate warming," *Nature*, vol. 443, no. 7107, pp. 71–75, 2006.

[5] M. Kulmala, T. Suni, K. E. J. Lehtinen et al., "A new feedback mechanism linking forests, aerosols, and climate," *Atmospheric Chemistry and Physics*, vol. 4, no. 2, pp. 557–562, 2004.

[6] R. Valentini, G. Matteucci, A. J. Dolman et al., "Respiration as the main determinant of carbon balance in European forests," *Nature*, vol. 404, no. 6780, pp. 861–865, 2000.

[7] C. D. O'Dowd, P. Aalto, K. Hämeri, M. Kulmala, and T. Hoffmann, "Atmospheric particles from organic vapours," *Nature*, vol. 416, no. 6880, pp. 497–498, 2002.

[8] H. Köhler, "The nucleus in and the growth of hygroscopic droplets," *Transactions of the Faraday Society*, vol. 32, pp. 1152–1161, 1936.

[9] M. C. Facchini, S. Decesari, M. Mircea, S. Fuzzi, and G. Loglio, "Surface tension of atmospheric wet aerosol and cloud/fog droplets in relation to their organic carbon content and chemical composition," *Atmospheric Environment*, vol. 34, no. 28, pp. 4853–4857, 2000.

[10] H. Rodhe, "Clouds and climate," *Nature*, vol. 401, no. 6750, pp. 223–225, 1999.

[11] T. Hede, X. Li, C. Leck, Y. Tu, and H. Ågren, "Model HULIS compounds in nanoaerosol clusters—investigations of surface tension and aggregate formation using molecular dynamics simulations," *Atmospheric Chemistry and Physics*, vol. 11, no. 13, pp. 6549–6557, 2011.

[12] T. Hede, C. Leck, L. Sun, Y. Tu, and H. Ågren, "A theoretical study revealing the promotion of light-absorbing carbon particles solubilization by natural surfactants in nanosized water droplets," *Atmospheric Science Letters*, vol. 14, no. 2, pp. 86–90, 2013.

[13] R. Sorjamaa, B. Svenningsson, T. Raatikainen, S. Henning, M. Bilde, and A. Laaksonen, "The role of surfactants in Köhler theory reconsidered," *Atmospheric Chemistry and Physics*, vol. 4, no. 8, pp. 2107–2117, 2004.

[14] H. Kokkola, R. Sorjamaa, A. Peräniemi, T. Raatikainen, and A. Laaksonen, "Cloud formation of particles containing humic-like substances," *Geophysical Research Letters*, vol. 33, no. 10, Article ID L10816, 2006.

[15] R. Sorjamaa and A. Laaksonen, "The influence of surfactant properties on critical supersaturations of cloud condensation nuclei," *Journal of Aerosol Science*, vol. 37, no. 12, pp. 1730–1736, 2006.

[16] V. F. McNeill, N. Sareen, and A. N. Schwier, "Surface-active organics in atmospheric aerosols," *Topics in Current Chemistry*, vol. 339, pp. 201–259, 2013.

[17] M. Dalirian, H. Keskinen, L. Ahlm et al., "CCN activation of fumed silica aerosols mixed with soluble pollutants," *Atmospheric Chemistry and Physics Discussions*, vol. 14, no. 16, pp. 23161–23200, 2014.

[18] S. S. Hings, W. C. Wrobel, E. S. Cross, D. R. Worsnop, P. Davidovits, and T. B. Onasch, "CCN activation experiments with adipic acid: effect of particle phase and adipic acid coatings on soluble and insoluble particles," *Atmospheric Chemistry and Physics*, vol. 8, no. 14, pp. 3735–3748, 2008.

[19] S. Twomey, "The influence of pollution on the shortwave albedo of clouds," *Journal of the Atmospheric Sciences*, vol. 34, no. 7, pp. 1149–1152, 1977.

[20] A. D. McGuire, S. Sitch, J. S. Clein et al., "Carbon balance of the terrestrial biosphere in the twentieth century: analyses of $CO_2$, climate and land use effects with four process-based ecosytem models," *Global Biogeochemical Cycles*, vol. 15, no. 1, pp. 183–206, 2001.

[21] B. Smith, I. C. Prentice, and M. T. Sykes, "Representation of vegetation dynamics in the modelling of terrestrial ecosystems: comparing two contrasting approaches within European climate space," *Global Ecology and Biogeography*, vol. 10, no. 6, pp. 621–637, 2001.

[22] S. Sitch, B. Smith, I. C. Prentice et al., "Evaluation of ecosystem dynamics, plant geography and terrestrial carbon cycling in the LPJ dynamic global vegetation model," *Global Change Biology*, vol. 9, no. 2, pp. 161–185, 2003.

[23] G. Esser, J. Kattge, and A. Sakalli, "Feedback of carbon and nitrogen cycles enhances carbon sequestration in the terrestrial biosphere," *Global Change Biology*, vol. 17, no. 2, pp. 819–842, 2011.

[24] A. Ahlström, G. Schurgers, A. Arneth, and B. Smith, "Robustness and uncertainty in terrestrial ecosystem carbon response to CMIP5 climate change projections," *Environmental Research Letters*, vol. 7, no. 4, Article ID 044008, 2012.

[25] B. Smith, D. Wärlind, A. Arneth et al., "Implications of incorporating N cycling and N limitations on primary production in an individual-based dynamic vegetation model," *Biogeosciences*, vol. 11, no. 7, pp. 2027–2054, 2014.

[26] T. D. Mitchell and P. D. Jones, "An improved method of constructing a database of monthly climate observations and associated high-resolution grids," *International Journal of Climatology*, vol. 25, no. 6, pp. 693–712, 2005.

[27] W. Cramer, A. Bondeau, F. I. Woodward et al., "Global response of terrestrial ecosystem structure and function to $CO_2$ and climate change: results from six dynamic global vegetation models," *Global Change Biology*, vol. 7, no. 4, pp. 357–373, 2001.

[28] P. Friedlingstein, P. Cox, R. Betts et al., "Climate-carbon cycle feedback analysis: results from the C4MIP model intercomparison," *Journal of Climate*, vol. 19, no. 14, pp. 3337–3353, 2006.

[29] A. Haxeltine and I. C. Prentice, "BIOME3: an equilibrium terrestrial biosphere model based on ecophysiological constraints, resource availability, and competition among plant functional types," *Global Biogeochemical Cycles*, vol. 10, no. 4, pp. 693–709, 1996.

[30] T. M. Ruuskanen, H. Hakola, M. K. Kajos, H. Hellén, V. Tarvainen, and J. Rinne, "Volatile organic compound emissions from Siberian larch," *Atmospheric Environment*, vol. 41, no. 27, pp. 5807–5812, 2007.

[31] J. Kesselmeier and M. Staudt, "Biogenic volatile organic compounds (VOC): an overview on emission, physiology and ecology," *Journal of Atmospheric Chemistry*, vol. 33, no. 1, pp. 23–88, 1999.

[32] L. Xu, R. B. Myneni, F. S. Chapin et al., "Temperature and vegetation seasonality diminishment over northern lands," *Nature Climate Change*, vol. 3, no. 6, pp. 581–586, 2013.

[33] H. Petersson, "Biomassafunktioner för trädfaktorer av tall, gran och björk i Sverige," SLU Arbetsrapport 59, SLU, Umeå, Sweden, 1999.

[34] T. Lind, "Kolinnehåll i skog och mark i Sverige—Baserat på Riksskogstaxeringens data," SLU Arbetsrapport 86, SLU, Umeå, Sweden, 2001.

[35] A. Albrektson, "Sapwood basal area and needle mass of scots pine (*Pinus sylvestris* L.) trees in central Sweden," *Forestry*, vol. 57, no. 1, pp. 35–43, 1984.

[36] R. W. Janson, "Monoterpene emissions from Scots pine and Norwegian spruce," *Journal of Geophysical Research*, vol. 98, no. 2, pp. 2839–2850, 1993.

[37] J. Laothawornkitkul, J. E. Taylor, N. D. Paul, and C. N. Hewitt, "Biogenic volatile organic compounds in the Earth system," *New Phytologist*, vol. 183, no. 1, pp. 27–51, 2009.

[38] J. Julin, M. Shiraiwa, R. E. H. Miles, J. P. Reid, U. Pöschl, and I. Riipinen, "Mass accommodation of water: Bridging the gap between molecular dynamics simulations and kinetic condensation models," *Journal of Physical Chemistry A*, vol. 117, no. 2, pp. 410–420, 2013.

[39] D. V. Spracklen, K. S. Carslaw, M. Kulmala et al., "Contribution of particle formation to global cloud condensation nuclei concentrations," *Geophysical Research Letters*, vol. 35, no. 6, Article ID L06808, 2008.

[40] V.-M. Kerminen, H. Lihavainen, M. Komppula, Y. Viisanen, and M. Kulmala, "Direct observational evidence linking atmospheric aerosol formation and cloud droplet activation," *Geophysical Research Letters*, vol. 32, no. 14, Article ID L14803, pp. 1–4, 2005.

[41] X. Li, T. Hede, Y. Tu, C. Leck, and H. Ågren, "Surface-active *cis*-pinonic acid in atmospheric droplets: a molecular dynamics study," *The Journal of Physical Chemistry Letters*, vol. 1, no. 4, pp. 769–773, 2010.

[42] A. Schwier, D. Mitroo, and V. F. McNeill, "Surface tension depression by low-solubility organic material in aqueous aerosol mimics," *Atmospheric Environment*, vol. 54, pp. 490–495, 2012.

[43] R. Atkinson, "Atmospheric chemistry of VOCs and $NO_x$," *Atmospheric Environment*, vol. 34, no. 12–14, pp. 2063–2101, 2000.

[44] A. Calogirou, B. R. Larsen, and D. Kotzias, "Gas-phase terpene oxidation products: a review," *Atmospheric Environment*, vol. 33, no. 9, pp. 1423–1439, 1999.

[45] M. Boy, Ü. Rannik, K. E. J. Lehtinen, V. Tarvainen, H. Hakola, and M. Kulmala, "Nucleation events in the continental boundary layer: long-term statistical analyses of aerosol relevant characteristics," *Journal of Geophysical Research D: Atmospheres*, vol. 108, no. 21, article 4667, 2003.

[46] I. Riipinen, T. Yli-Juuti, J. R. Pierce et al., "The contribution of organics to atmospheric nanoparticle growth," *Nature Geoscience*, vol. 5, no. 7, pp. 453–458, 2012.

[47] E. K. Bigg and C. Leck, "Cloud-active particles over the central Arctic Ocean," *Journal of Geophysical Research D: Atmospheres*, vol. 106, no. 23, pp. 32155–32166, 2001.

[48] C. Leck, M. Norman, E. K. Bigg, and R. Hillamo, "Chemical composition and sources of the high Arctic aerosol relevant for cloud formation," *Journal of Geophysical Research*, vol. 107, no. D12, p. 4135, 2002.

[49] C. Leck and E. K. Bigg, "A modified aerosol-cloud-climate feedback hypothesis," *Environmental Chemistry*, vol. 4, no. 6, pp. 400–403, 2007.

[50] J. M. Intrieri, C. W. Fairall, M. D. Shupe et al., "An annual cycle of Arctic surface cloud forcing at SHEBA," *Journal of Geophysical Research C: Oceans*, vol. 107, no. 10, pp. 1–14, 2002.

[51] M. Tjernström, "The summer arctic boundary layer during the arctic ocean experiment 2001 (AOE-2001)," *Boundary-Layer Meteorology*, vol. 117, no. 1, pp. 5–36, 2005.

[52] T. Mauritsen, J. Sedlar, M. Tjernström et al., "An arctic CCN-limited cloud-aerosol regime," *Atmospheric Chemistry and Physics*, vol. 11, no. 1, pp. 165–173, 2011.

[53] X. Li, T. Hede, Y. Tu, C. Leck, and H. Ågren, "Glycine in aerosol water droplets: a critical assessment of Köhler theory by predicting surface tension from molecular dynamics simulations," *Atmospheric Chemistry and Physics*, vol. 11, no. 2, pp. 519–527, 2011.

[54] T. Kurtén, M. Kulmala, M. Dal Maso et al., "Estimation of different forest-related contributions to the radiative balance using observations in southern Finland," *Boreal Environment Research*, vol. 8, no. 4, pp. 275–285, 2003.

[55] J. Pongratz, C. H. Reick, T. Raddatz, K. Caldeira, and M. Claussen, "Past land use decisions have increased mitigation potential of reforestation," *Geophysical Research Letters*, vol. 38, no. 15, Article ID L15701, 2011.

# Time Series Analysis: A New Methodology for Comparing the Temporal Variability of Air Temperature

**Piia Post[1] and Olavi Kärner[2]**

[1] *Institute of Physics, University of Tartu, 50090 Tartu, Estonia*
[2] *Tartu Observatory, 61602 Tõravere, Estonia*

Correspondence should be addressed to Piia Post; piia.post@ut.ee

Academic Editors: E. Paoletti, A. Rutgersson, and A. P. Trishchenko

Temporal variability of three different temperature time series was compared by the use of statistical modeling of time series. The three temperature time series represent the same physical process, but are at different levels of spatial averaging: temperatures from point measurements, from regional Baltan65+, and from global ERA-40 reanalyses. The first order integrated average model IMA(0, 1, 1) is used to compare the temporal variability of the time series. The applied IMA(0, 1, 1) model is divisible into a sum of random walk and white noise component, where the variances for both white noises (one of them serving as a generator of the random walk) are computable from the parameters of the fitted model. This approach enables us to compare the models fitted independently to the original and restored series using two new parameters. This operation adds a certain new method to the analysis of nonstationary series.

## 1. Introduction

Atmospheric reanalyses are widely used in meteorological and climatological research, as it makes available long time series of gridded meteorological variables, to explore climatic trends and low-frequency variations. Traditional analysis of air temperature time series, focused on trend detection and fitting of statistical models, has shown itself a useful tool (e.g., [1, 2] and references wherein). A use of structural time series models (i.e., linear trend plus red noise) has been popular in recent years (e.g., [3, 4]).

Our paper differs from cited above approach because no a priori model structure is supposed. We use the structure and correlation functions of temperature time series to select an applicable model type according to the scheme introduced by [5] inside the family of autoregressive integrated moving average (ARIMA) models. The scheme has been tested by means of various daily series before by [6, 7], and the earlier analysis shows that the first order integrated moving average model IMA(0, 1, 1) is applicable for daily temperature anomaly time series.

Nonstationary nature of the fitted IMA(0, 1, 1) model indicates that the traditional characterization of temperature variability on the basis of sample moments is unjustified. Due to nonstationarity, the moments do not converge if the sample size increases. This means that in order to compare temporal variability of air temperatures from different sources (e.g., measured and reanalysed), some other characteristic parameters are needed.

A nonstationary IMA(0, 1, 1) model is divisible into a sum of white noise (WN) and random walk (RW) component [5]. The variances for both white noises (the other of them serving as the random walk generator) are computable from two parameters of the fitted model. This approach enables us to compare the coincidence of original and restored series by means of two separated components. One of them is stationary (WN) and the other nonstationary (RW).

The aim of our paper is to show that time series analysis offers a tool with which it is possible to estimate the accuracy of reanalysis method to restore statistical characteristics of the initial time series. Our characterization also introduces different parameters for the comparison of the variability in time series. Two climate scale characteristics are the mean annual cycle and the tolerance. Two proposed weather scale characteristics are the ranges for lower and upper outliers, respectively.

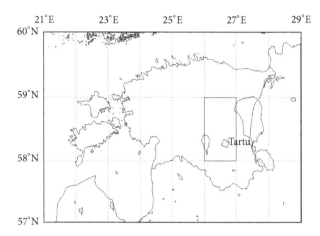

FIGURE 1: Positions of the used reanalysis' grids near Tartu: the small rectangle shows Baltan65+ and the larger ERA-40 grid-cell.

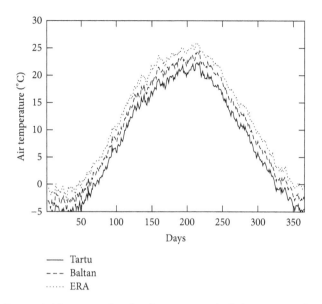

FIGURE 2: Mean annual cycle of temperature (in °C) at 12 GMT for 3 datasets: Tartu - original position, Baltan +2°C, ERA-40 +4°C.

Applicability of the idea is demonstrated on the basis of three daily air temperature time series representing the same territory but produced using different procedures. The results enable us to distinguish between recoverable and nonrecoverable parts of the IMA(0, 1, 1) model. This operation introduces a new method for the analysis of nonstationary series.

## 2. Initial Data

The temperatures we used were measured at Tartu-Ülenurme station (58°18′N, 26°41′E), 7 km out of Tartu, where the station was moved in 1950. Since 1997, it is not official SYNOP station for Tartu any more, nevertheless due to the airport the meteorological measurements are still carried on. We have analyzed the temperatures measured at 6, 12, and 18 GMT for the period from 1966 to 2005.

We used two different reanalysis outputs: one of them is well-known ERA-40 (the European Centre for Medium-Range Weather Forecasts (ECMWF) Reanalysis) [8] and the other Baltan65+—a regional reanalysis for the Baltic Sea region [9]. ERA-40 has a spatial resolution of 1 degree and Baltan65+ of 11 km. Data used to represent Tartu were from the closest grid point (see Figure 1). In Baltan65+ database, the development of the climate system is calculated with High Resolution Limited Area Model (HIRLAM) (version 7.1.4.), for the period 01.01.1965–31.12.2005. Every 6 h modeling cycle begins with providing new observational data and boundary conditions. Atmospheric pressure, air temperature, wind speed, specific humidity, and sea surface temperature are applied as boundary fields. These fields are created by interpolation of corresponding ERA-40 fields to the HIRLAM grid. As ERA-40 is calculated for the period 1957–2002, then after 2002, the ECMWF operational model is used for boundary data. Data assimilation contains also quality control for both observations and background fields. Observational data for Baltan65+ project is acquired from the ERA-40 database where possible and from Finnish Meteorological Institute's operational HIRLAM observation archive for the period after

2002. Measurements from Tartu SYNOP station are also used in the data assimilation.

We denote the different temperatures in the following way: $T_{T,k}$ for Tartu, $T_{E,k}$ for ERA-40, and $T_{B,k}$ for Baltan. Here, $k = 6, 12, 18$ stands for GMT observation time. From ERA-40 temperatures we used 12 GMT ones only. Unfortunately, not all three time-series are of the same length and for exactly the same period. The lengths of the original temperature time series are 14610 days for the measured, 14975 for Baltan65+, and 16436 for ERA-40.

## 3. Day to Day Comparison

We fit an acceptable model to anomalies with respect to annual mean cycle. Therefore, the mean annual cycles for available periods are computed for $T_{T,k}$, $T_{B,k}$, and $T_{E,k}$ series. One or two year difference in the length of initial series has no considerable influence for that characteristic. An example showing all three cycles corresponding to the noon observations (12 GMT) is shown in Figure 2. The curves are raw, that is, averaged over each calendar day (February 29 included), that avoids generation of approximation errors for the anomalies. Since the cycles are very similar to each other, then the curves are shifted for better overview. The maximum difference between any two cycles does not exceed 0.5 centigrades.

The daily anomalies are compared over the common interval 1966–2005 in order to describe the distribution of restoration accuracy. The frequencies of temperature difference are collected in three groups and shown in Table 1. The maximum frequency corresponds to the group containing the occasions with temperature difference less than one degree. About one percent of observations shows the occasions with the corresponding difference exceeding 3 degrees.

TABLE 1: Cumulative frequencies of temperature differences between Tartu and Baltan65+ anomalies.

| $(X_T - X_B)$ | 6 GMT | 12 GMT | 18 GMT |
|---|---|---|---|
| ±1 degree | 0.848 | 0.869 | 0.856 |
| ±2 degrees | 0.966 | 0.975 | 0.970 |
| ±3 degrees | 0.987 | 0.990 | 0.990 |

The annual cycles are similar but still show a systematic shift towards warming for the mean values as a result of reproduction. Forty years mean shift for the measurements at 6, 12, and 18 GMT is 0.037, 0.035, and 0.028, respectively. This difference is considered to be unimportant for future analysis.

Anomalies (deviations) with respect to the annual cycles of measured $X_{T,k}$, Baltan65+$X_{B,k}$, and ERA-40 $X_{E,k}$ values are further used for modeling the temporal variability.

## 4. Model Fitting Summary

Before fitting any statistical model to experimental time series, an initial exploration of the series temporal variability is necessary. Experience shows that several local daily air temperature series appeared to be the series with stationary daily increments [10]. This means that instead of the customary assumption about stationarity, it is reasonable to apply more general theory that is capable of analyzing the series with stationary increments. This leads to the use of structure function (instead of the correlation function) to examine the series variability.

The structure function for a series $X(t)$, where $t = 1, \ldots, n$, is the second moment of time series increments as a function of the increment interval $\tau$ (see [12] for details):

$$D(\tau) = \frac{1}{n-\tau} \sum_{t=1}^{n-\tau} [X(t+\tau) - X(t)]^2. \qquad (1)$$

This function enables us to distinguish between short and long range variabilities in the initial series by means of its different scaling exponent, depending on the increment interval (e.g. [6, 7]). The growth of $D(\tau)$ for three temperature series (corresponding to the observation time 12 GMT) over the increment range from 1 to 4096 days is shown in Figure 3(a).

There appears to be a remarkable growth of the structure function as the increment interval grows up to two weeks. The growth rate decreases considerably as the increment interval grows beyond. The further growth is very slow for daily temperature anomaly series. The same behaviour is well-known on the basis of several earlier works (e.g., [10, 11]).

If $\tau$ grows beyond a month, the $D(\tau)$ growth becomes proportional to $\tau^\alpha$ where $\alpha$ is a small number (approximately 0.1 or less) for these series. The property is called scaling [12]. For the current analysis, this means that the correlation between consecutive increments as a function of the increment interval $\tau$ is capable of suggesting an applicable autoregressive integrated moving average (ARIMA) model to represent the series long range temporal variability. More

precisely, the named correlation is the lag one autocorrelation of the series (e.g., [6]),

$$x_j(t) = X((t+1)\tau + j) - X(t\tau + j). \qquad (2)$$

Here, $j = 1, 2, \ldots, \tau$ because the original series $X(t)$ where $t = 1, 2, \ldots n$ is divisible into just $\tau$ different increment series. The necessary correlation $r_\tau(1)$ between the consecutive increments is computed as the mean over $\tau$ values of the corresponding lag one autocorrelations for time series (2) at each $j$. The results as functions of $\tau$ for three temperature anomaly series are shown in Figure 3(b).

Figure 3(b) shows that the correlation between consecutive increments for these series saturates at a negative level near −0.5 as the increment interval grows. This indicates that the first order moving average MA(1) model is applicable to represent temporal variability of the subseries of long range increments [6, 11]. The saturation means that the applicable model type does not change if the increment interval stays larger than 30 days.

In this case, the increment interval $\tau = 56$ days will be used for modeling. The interval is chosen earlier [6] in order to match with total solar irradiance variability. This means that the model has been fitted to 56 subseries.

Let us consider subseries $X(t)$ from the initial daily series over the time interval $\tau = 56$ days. Fixing this time interval, we can write the MA(1) model for the corresponding increments formally as

$$X(t) - X(t-1) = a(t) - \Theta_1 a(t-1), \qquad (3)$$

where $a(t)$ is white noise (WN) and $\Theta_1$ is a fitted coefficient. Here, we use (3) to stand for the mean model for original series $X(t)$ obtained by averaging the fitted coefficients over 56 subseries models.

For each subseries model, a diagnostic test has been carried out using the modified portmanteau statistic Ljung-Box test statistic [13]:

$$Q = N(N+2) \sum_{j=1}^{K} \frac{r_a^2(j)}{(N-j)}. \qquad (4)$$

Here, $N$ is the length of subseries, $K = 24$, and $r_a(j)$ is the autocorrelation of residual series $a(t)$. Critical values for $Q$ in our case (for 23 degrees of freedom) are 35.2 and 41.6 at 95% and 99% significance levels, respectively. The test is generally passed at 95% level. In a few occasions the value of $Q$ appeared to be between 35.2 and 41.6. (not shown).

Adding both sides of (3) over $t, t-1, \ldots, -\infty$, we obtain the model for temperature deviations $X(t)$ (i.e., IMA(0, 1, 1)):

$$X(t) = \Lambda \sum_{i=1}^{\infty} a(t-i) + a(t), \qquad (5)$$

where $\Lambda = 1 - \Theta_1$.

As a result, we obtained a simple model which describes long range temperature anomaly variability by means of two parameters, $\Lambda$ and the variance of residuals $\sigma_a^2$.

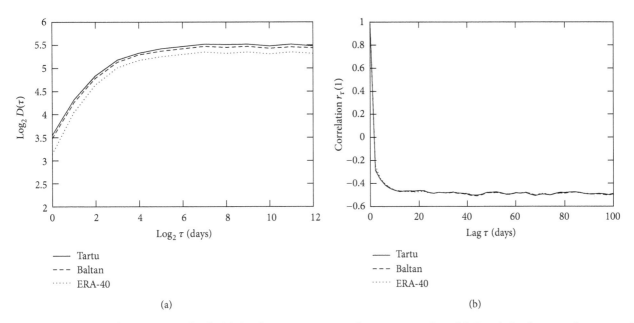

FIGURE 3: Exploratory analysis. (a) Growth of $D(\tau)$ for the increment range from 1 to 4096 days. (b) Correlation between the consecutive increments as a function of the increment interval.

## 5. Climatological Interpretation of the Model

Integrated moving average model IMA(0, 1, 1) can be presented as a sum of two components (see [5] for details):

$$X(t) = Y(t) + b(t), \tag{6}$$

where $Y(t) = \sum_{i=1}^{\infty} u(t-i)$ is a nonstationary random walk (RW) with the generator $u(t)$ which is white noise. The other component $b(t)$ is also white noise (WN), but independent of $u(t)$. It is possible to compute the variances for both components knowing $\Lambda$ and $\sigma_a^2$ [5]:

$$\sigma_b^2 = (1 - \Lambda) \sigma_a^2$$
$$\sigma_u^2 = \Lambda^2 \sigma_a^2. \tag{7}$$

Now, we have four parameters: $\Lambda$, $\sigma_a$, $\sigma_b$, and $\sigma_u$ to compare the reproduced Baltan65+ and ERA-40 series with the original one.

*5.1. Revealed Difference in Model Parameters.* Every subseries model depends on one fitted parameter $\Lambda$, that in our case fits all into a narrow interval from 0.04 to 0.14. The average values for both variables $\Lambda$ and $\sigma_a^2$ are taken to represent the model in the following analysis. These values of $\Lambda$ at three observation times are shown in the second, third, and fourth columns of Table 2. The other nine columns show mean standard deviations for the series of $a(t)$, $b(t)$, and $u(t)$, respectively.

The most significant difference between the produced model parameters occurs in $\Lambda$ and $\sigma_u$ values. They both indicate weakening of the influence of RW component as the modeling object changes from the original $X_T$ to the reproduced variable of $X_B$ or $X_E$. This means that the reproducing process essentially smooths the temporal variability

of temperature. Using $(|\sigma_T - \sigma_B|)/\sigma_T$ as a measure of relative restoration error, two of them appear to be reasonably small 0.03 (for the series $a(t)$) and 0.01 (for $b(t)$) but the third is huge 0.7 (for $u(t)$). This is an indication of the fact that a random walk path length is not predictable as well as reproducible (see [5] for details). But the mean $\sigma_b$ over 56 models appears to be approximately equal in the models for the original and reproduced variables.

## 6. About the Air Temperature Tolerance

The standard deviation $\sigma_b$ is about ten times larger than $\sigma_u$ as shown in Table 2. This means that the fitted model appears to be nearly stationary. It is natural to describe the range of its variability by the aid of standard deviation of the stationary component of the fitted model (i.e., $\sigma_b$). In case of normal stationary series, the interval $0 \pm 2\sigma_b$ would contain approximately 95% of the observed anomalies over the available sample. Without loss of generality, we can use the same yardstick to compute the frequency of outliers to compare the behaviour of the current empirical series. Since the temperature anomaly fluctuations are generally asymmetric, thus the interval should be shifted a bit towards lower values in order to reconcile it with the 0.025 and 0.975 percentile values for the whole anomaly histogram from the available time period. We call the interval between 0.025 and 0.975 percentiles of the total histogram of the temperature anomalies the local temperature (anomaly) tolerance. Its width is initiated by $\pm 2\sigma_b$ due to the general tendency that 95% of the anomaly observations fit into it. Individual deviations due to asymmetry lead to specification of the interval by means of the percentiles. The results in centigrades for two variables are shown in Table 3.

Table 2: IMA(0, 1, 1) model results for three variables. See text for details.

| Time | $\Lambda_T$ | $\Lambda_B$ | $\Lambda_E$ | $\sigma_{aT}$ | $\sigma_{aB}$ | $\sigma_{aE}$ | $\sigma_{bT}$ | $\sigma_{bB}$ | $\sigma_{bE}$ | $\sigma_{uT}$ | $\sigma_{uB}$ | $\sigma_{uE}$ |
|------|------|------|------|------|------|------|------|------|------|------|------|------|
| 6 GMT | .107 | .062 | — | 5.144 | 4.978 | — | 4.861 | 4.821 | — | .549 | .310 | — |
| 12 GMT | .096 | .058 | .042 | 4.950 | 4.783 | 4.552 | 4.705 | 4.641 | 4.455 | .477 | .279 | .192 |
| 18 GMT | .104 | .063 | — | 4.902 | 4.764 | — | 4.640 | 4.612 | — | .511 | .300 | — |

Table 3: Estimated boundaries (in centigrades) for local temperature anomaly tolerance for two variables.

| Time | Lower for $X_T$ | Lower for $X_B$ | Upper for $X_T$ | Upper for $X_B$ |
|------|------|------|------|------|
| 6 GMT | −11.47 | −11.16 | 8.46 | 8.40 |
| 12 GMT | −9.69 | −9.42 | 8.64 | 8.48 |
| 18 GMT | −10.41 | −10.19 | 8.15 | 8.14 |

Using the tolerance for comparison of time series behaviour is reasonable. It helps to get rid of the traditional statistical way to describe the air temperature variability by means of sample moments whereas they are useless due to nonstationarity. The new approach partly satisfies the assumption by North and Cahalan [14] about stationarity of the marginal distribution. The only concession is the use of selected 95% of the sample instead of the whole 100% sample. The use of 100% of the sample would be justified only if $\Lambda = 0$ in the fitted model. Then, the IMA(0, 1, 1) model would be reduced to IMA(0, 1, 0) model, which shows that the temperature anomaly series behaves as WN. In our model there is a small but still nonzero $\Lambda$ indicating evident difference from any pure WN case. This means that the RW component is important in the description of the temperature series variability. But due to the appeared small frequency of the outliers from the tolerance, their influence can be treated as a result of extremal weather events not influential to the climate.

Table 3 shows that the reproducing is connected to some narrowing down of the regions. This is in good accord with the difference between the variances $\sigma_b^2$. More informative action is to compute the change of outliers frequency if the reproduced tolerance is used in order to handle the original temperature anomalies. The new frequencies are expected to be slightly higher than 0.025 for both edges. For lower/higher edge, they are 0.027/0.026 (for 6 GMT), 0.028/0.028 (for 12 GMT), and 0.027/0.026 (for 18 GMT). This means that the summary number of outliers increased less than one percent.

Random walk sample paths are unpredictable, and their influence to our climate scale characteristic (tolerance) is small. Thus, an analysis of outliers ranges is not important for our main problem of estimating the climate scale influence of time series reproduction by means of HIRLAM-based reanalysis.

But this analysis becomes more important if a determination of climate variability and/or change signals arise [7].

## 7. Conclusions

We fitted statistical models to three 40-year long temperature anomalies time series; one was measured at Tartu, and two others were from different reanalyses—ERA-40 and Baltan65+. Modeling of daily air temperature series showed that a nonstationary first order integrated average IMA(0, 1, 1) model was applicable.

Despite that the daily time series themselves may differ essentially, the comparison of the three series models shows that the white noise (climatological) component of the series barely differ. This means that the traditional comparison by means of statistical moments is inapplicable, as they remain because of nonstationarity sample dependent and, thus, are unable to produce a reliable basis for any comparison. Applying IMA(0, 1, 1) enabled us to overcome this obstacle and organize the comparison of measured and reanalysed datasets on the basis of stationary components picked out from the series.

(1) The calculations with anomaly series show that the separation of white noise component is crucial for successful comparison—only the white noise component variance is reproducible by means of a reanalysis (the mean values remain zero during both operations).

(2) The comparison shows that a nonstationary IMA(0, 1, 1) model enables us to extract the standard deviation ($\sigma_b$) belonging to its stationary white noise component. This standard deviation enables us to produce a quantitative scheme to estimate fitness of the reanalyzed variables.

(3) The separated random walk components of measured and reanalysed data differ essentially. This is self-evident because random walk is not predictable (and thus, also not reproducible). Its presence in a stochastic process enhances its nonstationarity. This emphasizes the necessity of using the white noise components variance in climate analysis.

(4) The described small comparison shows that the regional Baltan65+ reanalysis is capable of producing a reliable dataset for future climate studies provided that a separation of stationary component will be extracted.

(5) The same procedure is applicable if one needs to compare temperature variability between two closely situated stations or between outputs of numerical models which pretend to describe the temperature variability over the same area.

## Acknowledgments

The authors would like to thank Andres Luhamaa for providing Baltan65+ temperature data, Marili Kangur for preparing the measured temperature data, and COST733 colleagues for preparing ERA-40 data. The study was supported by the Estonian Ministry of Education and Research (Grant SF0180038s08 and Grant ETF9134) and by the EU Regional Development Foundation, Environmental Conservation and Environmental Technology R & D Program Project no. 3.2.0801.12-0044.

# References

[1] W. A. Woodward and H. L. Gray, "Global warming and the problem of testing for trend in time series data," *Journal of Climate*, vol. 6, no. 5, pp. 953–962, 1993.

[2] X. Zheng and R. E. Basher, "Structural time series models and trend detection in global and regional temperature series," *Journal of Climate*, vol. 12, no. 8, pp. 2347–2358, 1999.

[3] S. T. J. Grieser and C. Schönwiese, "Statistical time series decomposition into significant components andapplication to European temperatures," *Theoretical and Applied Climatology*, vol. 71, pp. 171–183, 2002.

[4] M. Mudelsee, *Climate Time Series Analysis*, Springer, New York, NY, USA, 2010.

[5] G. E. Box, G. M. Jenkins, and G. C. Reinsel, *Time Analysis, Forecasting and Control*, Prentice-Hall, Englewood Cliffs, NJ, USA, 1994.

[6] O. Kärner, "ARIMA representation for daily solar irradiance and surface air temperature time series," *Journal of Atmospheric and Solar-Terrestrial Physics*, vol. 71, no. 8-9, pp. 841–847, 2009.

[7] O. Kärner and C. R. de Freitas, "Modelling long-term variability in daily air temperature time series for southern hemisphere stations," *Environmental Modeling and Assessment*, vol. 17, no. 3, pp. 221–229, 2012.

[8] S. M. Uppala, P. W. Kållberg, A. J. Simmons et al., "The ERA-40 re-analysis," *Quarterly Journal of the Royal Meteorological Society*, vol. 131, no. 612, pp. 2961–3012, 2005.

[9] A. Luhamaa, K. Kimmel, A. Männik, and R. Rõõm, "High resolution re-analysis for the Baltic Sea region during 1965–2005 period," *Climate Dynamics*, vol. 36, no. 3, pp. 727–738, 2011.

[10] S. Lovejoy and D. Schertzer, "Scale invariance in climatological temperatures and the local spectral plateau," *Annales Geophysicae B*, vol. 4, no. 4, pp. 401–410, 1986.

[11] O. Kärner, "Some examples of negative feedback in the earth climate system," *Central European Journal of Physics*, vol. 3, no. 2, pp. 190–208, 2005.

[12] A. M. Yaglom, *Correlation Theory of Stationary and Related Random Functions: Volume I: Basic Results*, Springer, New York, NY, USA, 1987.

[13] G. M. Ljung and G. E. P. Box, "On a measure of lack of fit in time series models," *Biometrika*, vol. 65, no. 2, pp. 297–303, 1978.

[14] G. R. North and R. F. Cahalan, "Predictability in a solvable stochastic climate model," *Journal of the Atmospheric Sciences*, vol. 38, no. 3, pp. 504–513, 1981.

# 3

# Periods of Excess Energy in Extreme Weather Events

## Igor G. Zurbenko[1] and Amy L. Potrzeba-Macrina[2]

[1] *Department of Epidemiology & Biostatistics, University at Albany, 1 University Place, Rensselaer, NY 12144, USA*
[2] *Department of Mathematics, Northern Virginia Community College, Annandale Campus, 8333 Little River Turnpike, Annandale, VA 22003, USA*

Correspondence should be addressed to Amy L. Potrzeba-Macrina; amacrina@nvcc.edu

Academic Editors: I. Alvarez, P. Canziani, L. Makra, and E. Paoletti

The reconstruction of periodic signals that are embedded in noise is a very important task in many applications. This already difficult task is even more complex when some observations are missed or some are presented irregularly in time. Kolmogorov-Zurbenko (KZ) filtration, a well-developed method, offers a solution to this problem. One section of this paper provides examples of very precise reconstructions of multiple periodic signals covered with high level noise, noise levels that make those signals invisible within the original data. The ability to reconstruct signals from noisy data is applied to the numerical reconstruction of tidal waves in atmospheric pressure. The existence of such waves was proved by well-known naturalist Chapman, but due to the high synoptic fluctuation in atmospheric pressure he was unable to numerically reproduce the waves. Reconstruction of the atmospheric tidal waves reveals a potential intensification on wind speed during hurricanes, which could increase the danger imposed by hurricanes. Due to the periodic structure of the atmospheric tidal wave, it is predictable in time and space, which is important information for the prediction of excess force in developing hurricanes.

## 1. Introduction

The spectral analysis of longitudinal data is the method of analyzing natural periodicities of data that are observed over long periods of time. However, the problem with longitudinal studies is that data is not collected at regular (or equally spaced) time intervals. Studies have been conducted as to how to handle the problem presented by having missing values. Examples of such studies include separating the data into the missing part and the nonmissing part called the EM (Expectation Maximization) algorithm by Shumway and Stoffer [1] and the CLEAN method formulated by Baish and Bokelmann [2], which they tested on simulated noisy data and due to their results applied this method to seismological data. Other researchers have also considered estimating the missing data [3–5].

The Kolmogorov-Zurbenko (KZ) filter is an efficient means of analyzing data even when datasets are considered to have missing values ([6–8], *Kolmogorov-Zurbenko filters*). Several computer algorithms are available that will investigate a time series with irregularly spaced data, including the nonequispaced Fast Fourier Transform (NFFT) developed by Keiner et al. and the Mathematics Faculty at Chemnitz University of Technology (offered in their free software) and the KZA algorithm provided in R-software by Close and Zurbenko [9].

This paper highlights the strength of a method of periodic signal reconstruction that can be used for longitudinal data or more specifically a time series with missing values, namely, the Kolmogorov-Zurbenko Fourier Transform (KZFT) filter. The strength of the filter in determining spectra of datasets with irregularly spaced (or missing) values and its ability to reconstruct the periodic signal after determining such dominant frequencies are presented. This method is then used to reconstruct the atmospheric tidal wave at locations during times of major meteorological events. The goal of this paper is to discuss the effects of the atmospheric tidal waves on the intensification of major hurricanes/storms such as Super Storm Sandy and Hurricane Katrina.

## 2. The KZFT Algorithm and Irregularly Spaced Data

KZ algorithms were developed for use in a highly noisy environment and are able to provide a numerical result that is very close to the actual hidden signals. With high accuracy the KZ algorithms are able to detect the frequency characteristics of the hidden signals as well as provide signal reconstructions. More details can be found in Zurbenko [6], Zurbenko and Porter [7], and Yang and Zurbenko [10]. All KZ algorithms are currently available in R-software.

The spectra are important diagnostic tools for the reconstruction of hidden signals. The spectrum displays the dominant frequencies of the data. Dominant frequencies can be detected and then used to reconstruct the periodic signal inherent to the noisy dataset. A highly efficient example of such reconstruction for equally spaced data is provided in Potrzeba and Zurbenko [11]. The authors will investigate the efficiency of such reconstructions by simulations of complicated hidden signals in a very noisy environment. Through simulation it will be demonstrated that this filter is extremely capable of determining spectra and reconstructing periodic signals from noisy datasets that have irregularly-spaced data (or data that is a time series with missing values).

A signal that contains a couple of periodic sine waves is complex; therefore it can be difficult to recognize its periodic components. The total amplitude of the summation may oscillate with low difference frequency between basic components. When such a signal is buried in noise the outcome appears absolutely chaotic. Unequally spaced data or data with missing observations can also provide extra difficulties for the recovery of the original signal. The authors will estimate the accuracy of the reconstruction of such an original signal through simulations. For the simulations, the authors formed a wave of 17,000 consecutive time observations. In order to consider a longitudinal dataset with irregularly-spaced data a percentage of observations were randomly "removed" from the wave formed from 17,000 consecutive time observations. The parameter for the percentage of missing observations is denoted by $p$. Close and Zurbenko's [9] KZFT algorithm in the KZA package of R-software was used. This algorithm has three main parameters. When using this algorithm one must specify a vector of observations, a vector of the observations corresponding time, $m$ = window size of the filter, $k$ = number of iterations of the filter, and $f$ = frequency. When using this tool to estimate spectra the frequency parameter, $f$, should be set to NULL. An adaptive algorithm, DZ [12] for spectral estimation, is also available in the same package. The DZ parameter in the KZFT algorithm defaults to 5%; it is used to address the energy levels of interest and may be adjusted by users.

The aforementioned simulated signal was used to test the KZFT algorithm's ability to accurately determine spectra of datasets with missing values. Several simulations were performed, first with low percentage of missing values, $p$ = 20%, with KZFT parameters $m$ = 150 and $k$ = 1. The dominant frequencies were accurately displayed in the spectrum even with heavier noise added to the signal (noise $\sim N(0, 32)$ and again with noise $\sim N(0, 64)$). The percentage of missing

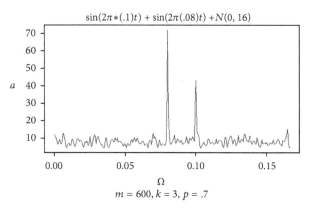

FIGURE 1: The spectrum of the signal that is the summation of two sine waves about frequencies 0.08 and 0.10 cycles per unit time plus noise $\sim N(0, 16)$ with $p$ = 70% missing values. The KZFT filter was used with parameters $m$ = 600, $k$ = 3.

information, $p$, was increased following these successful results. Values of hidden frequencies were reconstructed with the help of the DZ algorithm (the smoothing parameter that is at 5%) and used for further reconstruction. The reconstructed signal was compared with the original signal over the entire timeframe and randomly chosen small intervals of time to observe quality of reconstruction.

For practical considerations, the percentage of missing values was increased to $p$ = 60%. The KZFT algorithm was applied with parameters $m$ = 300 and $k$ = 3 to determine the spectrum. The result of this simulation was that the spectrum again accurately depicted the dominant known frequencies. To further test the use of the KZFT algorithm, the signal with $p$ = 60% was embedded within a noisier dataset by increasing noise to $N(0, 32)$ and then repeated again with noise $\sim N(0, 64)$ added to the signal. For each simulation, the filter was able to precisely capture the hidden frequencies detected by the DZ algorithm in the noisy signal.

Due to the abovementioned successful results, the missing rate of observations was further increased to $p$ = 70% to determine if the spectrum could continue to capture the dominant frequencies. Using a wider window length of $m$ = 600 and $k$ = 3 iterations, the KZFT algorithm was used to determine the spectrum for the simulated longitudinal dataset. It is apparent in Figure 1 that the dominant frequencies of 0.08 and 0.10 cycles per unit time are identifiable as signal's inherent frequencies.

The results of these aforementioned simulations demonstrate the effectiveness of the KZFT filter to estimate spectra for longitudinal data where time is randomly, uniformly, i.i.d. over an interval. The parameters ($m$ and $k$) were first selected to keep the amount of available time points in a window equal and then increased to improve the results. A shorter window length (smaller $m$) is possible, but it will yield a much wider statistical confidence. The current window sizes for the meteorological data, in Section 3, were chosen from the consideration that within $\sim$1 lunar cycle its astronomic influences will remain the same. This assumption is realistic for practical purposes where observations are considered to be made at arbitrary time points. Computer readings always provide a limited accuracy (days, hours, etc.), therefore a discrete scale is used in practice, with the exception that those time points can be missed. After using the spectra to determine any periodic signal inherent to data, researchers can reconstruct those periodic signals.

Following the determination of spectra for the simulated longitudinal datasets where time points were missed with probabilities 20%, 60%, and 70%, the KZFT algorithm was then applied to reconstruct each signal. With the dominant frequencies determined from each spectrum, the KZFT filter was then applied by setting the frequency parameter ($f$) to each of the determined frequencies; varying window sizes and number of iterations have been used. For multiple peaks in a spectrum, the KZFT filter was applied multiple times, adjusting for each desired frequency parameter. The resulting periodic signal is the summation of these filters. The accuracy is approximately $1/\sqrt{N}$, where $N$ is the number of available observations within the window length; for more information regarding the variance of the KZFT filter and the resolution of wavelets see Yang and Zurbenko [10] and Zurbenko and Porter [7]. Accuracy may be increased by two ways; by increasing the window size parameter, $m$ or through repeated simulations. The former may have some computer limitations.

The authors demonstrate the ability of the KZFT algorithm's reconstructive properties for longitudinal data that was the sum of two sine waves about periods 10 and 12.5 with noise $\sim N(0, 16)$ with $p = 60\%$. Given the corresponding frequencies found from the spectrum, the KZFT filter was applied to the data to reconstruct the signal. The KZFT filter in the KZA package of R-software has a parameter $f$ = frequency. By defining this parameter for each of the known dominant frequencies found in the spectrum, the authors applied this filter with parameters $m = 300$ and $k = 3$ to reconstruct the signal about each frequency (0.08 and 0.10 cycles per unit time). The reconstructed signal was determined by applying the KZFT filter twice (once about each dominant frequency) and then by summing the results of each filter. The correlation between the true signal and the reconstructed signal was 96.4% displayed in Figure 2. The original observations provide no guess of the complex, hidden periodicity, which was perfectly reconstructed by the algorithm. If necessary those already high correlations can be improved essentially with wider time windows in use. The high quality of reconstruction does not affect the major study of effects of hidden periodicities.

Since the correlation between the reconstructed signal and the true signal was high with 60% missing observations, the percentage of missing values was increased to $p = 70\%$. The KZFT filter was applied to the simulated signal that was the sum of two sine waves about frequencies 0.10 and 0.08 cycles per unit time with noise $\sim N(0, 16)$, with $p = 70\%$. The KZFT parameters used were $m = 800$ and $k = 3$ (a higher window size was used for the higher rate of missing observations) to reconstruct the signal about each frequency determined from the spectrum. The correlation between the true signal and the reconstructed signal was 92.1%.

The ability to determine spectra and to reconstruct periodic signals allows researchers to perform important

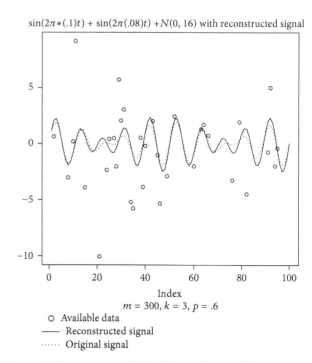

$\sin(2\pi*(.1)t) + \sin(2\pi(.08)t) + N(0, 16)$ with reconstructed signal

Index
$m = 300, k = 3, p = .6$

o    Available data
—    Reconstructed signal
......    Original signal

FIGURE 2: The reconstructed signal that is the sum of two sine waves about frequencies 0.08 and 0.10 cycles per unit time from an original signal with added noise $\sim N(0, 16)$ and where 60% of the values were not available.

diagnostics on any longitudinal dataset or time series with missing data. The figures in this section clearly indicate the strong ability of the algorithm to accurately restore small hidden periodic signals from a very noisy original dataset that is absolutely, chaotically observed in time. This method has a wide range of practical uses.

## 3. Application of Tidal Wave Reconstruction in Determination of Most Hazardous Moments of Hurricanes

Tidal waves in the five oceans greatly influence all aspects of life in the oceans and on their shores. Schedules of those tides are available and their predictions are widely distributed as a highly important factor affecting surrounding life. Tides in the oceans can be measured directly and it is a continual procedure to do so in the network over the globe. Existence of similar tides in atmospheres around the globe has been discovered by distinguished scientist Chapman [13, 14].

Atmospheric tides can be noticeable for us only by changing atmospheric pressure. Measurements of atmospheric pressure are available in the network over the Globe, but these measurements are strongly affected by much larger synoptic fluctuations; thus making the possibility of the direct observation of tides in the atmosphere almost impossible. Chapman proved the existence of those tides, but could not reconstruct those waves in a numerical way. The development of Kolmogorov-Zurbenko (KZ) filtration ([6, 7, 10], *Kolmogorov-Zurbenko filters*) made that solution possible [15,

16]. Tidal waves in the atmosphere are generated by lunar gravitational forces and solar thermal influences, having strict astronomic frequencies $12^{-1} = 0.0833$, $12.42^{-1} = 0.080515$, $24^{-1} = 0.04167$, and $24.84^{-1} = 0.04026$ cycles per hour that correspond to 1 and 1/2 sun and moon days on Earth. Existence of those strict lines in the spectrum of atmospheric pressure and the method of the reconstruction has been developed by [15, 16].

One goal of this paper is to examine the effects of atmospheric tidal waves on the intensification of hurricanes, such as Super Storm Sandy of 2012, Hurricane Katrina of 2005, and Hurricane Rita of 2005. Any developed hurricane has strictly synoptic reasons for its development and trajectory in time and space. The strength of hurricanes is usually described by a level of low pressure in the center that is characteristic of the hurricane itself. Outside, high pressure is a feeding force of the hurricane, thus creating circular descending winds proportionally to the difference in pressure. Periodic fluctuations in outside pressure due to the tidal wave will create periodic fluctuations in the strength of the wind. The difference in pressure is the original force yielding acceleration, which is the derivative of speed. The 24-25 hour periodic wave in pressure will be approximately a one-quarter period ahead of the resulting periodic wind speed, by connection between the periodic sine wave and its derivative. Thus, one may expect a 4–7 hour delay in wind speed compared with the periodic wave in the outside pressure due to the atmospheric tidal wave. These general notions are very well supported by data analysis of wind speed and pressure in the aforementioned hurricanes. The fluctuations in strength of hurricanes are organized in time when the hurricane itself is moving due to the synoptic conditions in the area. Spatial scale of tidal waves in the atmosphere is about 6000 miles and those spatial scales are approximately periodic with a period of approximately 25 hours. Hence, for each hurricane, every 25 hours there are potentially more dangerous moments. During hurricane trajectory in space there exists more dangerous times of expected force. In this paper the authors are not addressing the details of dynamic forces that are responsible for hurricane constructions; the scope of this paper is statistical evidence of periodic reinforcements of developed hurricanes due to the effect of the atmospheric tidal wave.

Data collection during hurricanes is frequently interrupted by their destructive forces; therefore the authors will use the methods described in the second section of this paper to process data with many of the missing values. As it was shown in the previous section, the power of our methods of reconstruction of a periodic wave is not diminished with those missing observations. The data used in this paper is from the ISD-Lite (Integrated Surface Database) from the National Climatic Data Center (NCDC) (http://www.ncdc.noaa.gov/oa/climate/isd/index.php?name=isd-lite).

Figure 3 displays raw atmospheric pressure data along with the reconstructed tidal wave during Super Storm Sandy at Atlantic City International, New Jersey, during October 2012. With the approaching storm the atmospheric pressure

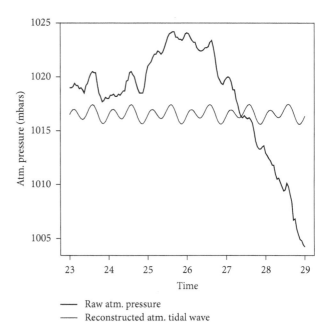

FIGURE 3: Raw atmospheric pressure versus the reconstructed atmospheric tidal wave at Atlantic City International, New Jersey. The horizontal axis represents points of midnight UTC during October 2012. It is seen that the tidal wave and raw pressure data are in-phase.

rapidly declined; measurements were interrupted during the center of the storm's arrival, but it did not affect the ability of the reconstruction of the tidal wave. It is shown that the raw pressure and reconstructed periodic wave were in-phase. Figure 4 displays the tidal wave shifted forward 5 hours and sustained wind speed at that area; the periodic influence of the tidal wave on wind speed was visually and statistically evident. It was determined that the highest correlation between wind speed and the tidal wave was achieved with a shift of 5 hours; therefore, the reconstructed tidal wave was shifted 5 hours ahead in time. The recorded landing time was a few hours later than the highest effect of the tidal wave, so if an actual hurricane arrived at a point a few hours earlier its effect could have been substantially stronger.

Figure 5 displays the tidal wave reconstruction shifted 5 hours ahead and wind speed during Hurricane Katrina in New Orleans, Louisiana, in August 2005. It displays undeniable evidence of periodic force of hurricane intensified by the atmospheric tidal wave. The wind speed practically repeats periodic fluctuations in tidal wave in the approaching hurricane. There was no record of wind speed at the landing time due to the destruction of the hurricane. Nevertheless, the authors were able to reconstruct the tidal wave during the entire event. Shown in the figure are high wind speeds with high tide at the recorded area before landing time. Sustained wind speed as a 3-hour moving average of wind speed and the reconstructed tidal wave are observed to be in-phase. There was a short relief of wind speed due to the low tide hours before landing time. The landing time coincides with a secondary peak in the tidal wave; hours later the peak in the tide came to the area together with enormous destruction. In

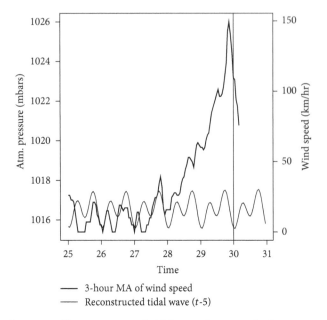

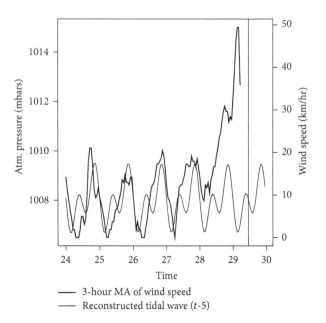

FIGURE 4: The reconstructed tidal wave in atmospheric pressure wind speed, and landfall of Super Storm Sandy. The vertical line represents landfall of Super Storm Sandy in New Jersey. The horizontal scale uses dates in October 2012 at points of midnight UTC. The 3-hour moving average of wind speed and the shifted reconstructed tidal wave are seen to be in-phase.

FIGURE 5: A 3-hour moving average of wind speed and the reconstructed atmospheric tidal wave around the time Hurricane Katrina made landfall in New Orleans, LA. The vertical line represents the landfall of Hurricane Katrina. The horizontal scale is time in August 2005 with points at midnight UTC.

many cases, data collections, unfortunately, are interrupted in and around the landing time of hurricanes; therefore it is difficult to analyze correlation during these events. It is the authors' consideration that pattern influences with the tidal wave that are clearly displayed before the event will remain, regardless of the fact that formal correlation analysis is impossible within the rapidly changing wind speeds during the event.

Figure 6 displays a reconstructed tidal wave shifted forward 7 hours and sustained wind speed during Hurricane Rita at Houston, Texas, during September 2005. Again there is an obvious correlation of wind speed with the amplitude of the atmospheric tidal wave before the arrival of the hurricane. Hurricane Rita was expected to be very strong and cities were evacuated prior to its arrival. It is seen in Figure 6 that the landing time of the hurricane coincides with lowest effect of tidal wave in the area. Destruction caused by that hurricane was sufficiently low. It might have been a very different story if the hurricane had arrived to the area 12 hours earlier or 12 hours later.

## 4. Conclusions and Future Developments

Restorations of periodic signals in a very noisy environment, with the possibility of substantially missed observations, recently became possible and provided the opportunity to look at the same events with new technological opportunities of time and space data analysis. This opportunity enabled the authors to better understand time and space descriptions for extreme weather effects and climate fluctuations [16–18]. It permits one to see fluctuations and effects in time and space

that are very well hidden in high level noise or events of different natural lows.

This paper provides a few examples of evidence of periodic reinforcements of extreme weather phenomena. Weather fluctuations are regularly described by synoptic forces, which permit predictions up to a few weeks in advance. Tidal waves in the atmosphere are smaller than synoptic fluctuations of pressure, but they are organized by astronomic forces of gravity of the moon and sun daily thermal effect that are periodic by nature. The periodic nature of the atmospheric tidal waves makes them predictable for a long time into the future. Thus, for weather prediction analysis, atmospheric tidal waves are creating an opportunity to improve total prediction by using highly, well predictable values of tidal waves. Hurricane trajectories and landing times are synoptically, very well predicted about 1-2 days in advance. Tidal waves in the atmosphere can be reproduced up to one month in advance. Extra security precautions should be made in areas where a hurricane is scheduled to arrive, together with the high tide effect. Schedules of the tides in atmosphere should be reconstructed everywhere just as these schedules exist with tides in the oceans. The atmosphere has no shore bounds so atmospheric tides are more continuously constructed over the globe and can be easily reproduced over it with a prediction opportunity of at least one month ahead. The amplitude of tidal waves in the atmosphere has slow fluctuations due to differences in astronomic forces, which may also be examined to make predictions of tides for a long time into the future.

Tables of tidal waves in the atmosphere should be created in time and space over the planet and these new schedules

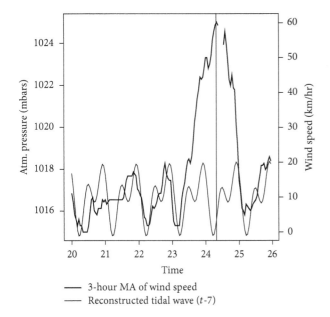

—— 3-hour MA of wind speed
—— Reconstructed tidal wave (*t*-7)

FIGURE 6: A 3-hour moving average of wind speed, atmospheric pressure, and the reconstructed atmospheric tidal wave around the time Hurricane Rita made landfall in Texas. The vertical line represents the landfall of Hurricane Rita. The horizontal scale is time in September 2005 at points of midnight UTC.

will be just as valuable as the schedules of tides on the shorelines of the oceans. This information provides awareness for extreme weather events, such as hurricanes, that will arrive at a vulnerable time at specific locations. It will be very interesting to examine the time record of the pressure in the center of developed hurricanes as long as such data will be available. It will also be interesting to detect the periodic component of tidal waves in the center and compare it with tidal wave outside the hurricane. There is a possibility of phase shift between periodic components, which may contribute to further reinforcement of hurricane. In proper synoptic conditions it is likely that atmospheric tides are affecting the strength and probability of appearance of tornados. The appearance of extreme events is controlled by synoptic conditions, but in the examples examined in this paper, tidal waves are making the difference in their strength in time and space due to their periodic structure.

Gravity forces are applicable to any substance, including the plasma of the sun; so gravity of the planets must create tidal waves on the sun. Tidal forces are coherent for big distances and can easily create currents. Coriolis forces will cause cyclones in those currents. Cyclonic turbulences in such currents can be strongly reinforced by the electromagnetic structure to create strong eruptions. The frequency of such eruptions is dictated by periodicities of tidal waves amplitudes, which for example, have close to an 11-year cycle due to the periodicities in planets' movements. Examination of those so-called difference frequencies caused by planets periodicities around the Sun was provided in Zurbenko and Potrzeba [16]. The difference frequency of 11 years came from the examination of a couple of planets with the most influential gravity on the sun. All of the planets together

may cause a range of several difference frequencies, including some very long periods. Restoration of those periodicities can be run as a big scale computer project, which may reproduce entire difference frequencies for a very long period of time. The restoration of those periodicities may help to explain and predict long term changes of the climate on Earth. The climate on Earth changed dramatically within a substantial timeframe. Those changes were partially due to the changing Sun activities. The reproduction of periods in those activities may provide partial explanation to the long scale climate changes on Earth. Sun activity is affecting all aspects of human life. It might cause magnetic storms and it might affect rate of some diseases; see Valachovic and Zurbenko [19]. Any periodic events can be predicted for an extended time. For signals that are embedded in high levels of noise, it is truly a treasure that the periodic signals can be detected and forecasted for extended time.

## Acknowledgment

The authors express their gratitude to Dr. Robert Henry from New York State Department of Environmental Conservation for providing data that was used in this paper and for the many discussions regarding the results.

## References

[1] R. H. Shumway and D. S. Stoffer, *Time Series Analysis and Its Applications*, Springer, New York, NY, USA, 2006.

[2] S. Baisch and G. H. R. Bokelmann, "Spectral analysis with incomplete time series: an example from seismology," *Computers and Geosciences*, vol. 25, no. 7, pp. 739–750, 1999.

[3] A. C. Harvey and R. G. Pierse, "Estimating missing observations in economic time series," *Journal of the American Statistical Association*, vol. 79, no. 385, pp. 125–131, 1984.

[4] R. J. A. Little and D. B. Rubin, *Statistical Analysis with Missing Data*, Wiley Series in Probability and Mathematical Statistics, John Wiley & Sons, New York, NY, USA, 1986.

[5] E. Parzen, "Mathematical considerations in the estimation of spectra," *Technometrics*, vol. 3, no. 2, pp. 167–190, 1961.

[6] I. G. Zurbenko, *The Spectral Analysis of Time Series*, North Holland Series in Statistics and Probability, 1986.

[7] I. G. Zurbenko and P. S. Porter, "Construction of high-resolution wavelets," *Signal Processing*, vol. 65, no. 2, pp. 315–327, 1998.

[8] S. T. Rao, I. G. Zurbenko, R. Neagu, P. S. Porter, J. Y. Ku, and R. F. Henry, "Space and time scales in ambient ozone data," *Bulletin of the American Meteorological Society*, vol. 78, no. 10, pp. 2153–2166, 1997.

[9] B. Close and I. Zurbenko, KZFT in KZA package, R-software first published version kza_0. 4 (27 Apr 2004) and latest published version (2013), Package Sources, 2010, http://cran.us.r-project.org/.

[10] W. Yang and I. Zurbenko, "Kolmogorov-Zurbenko filters," *Wiley Interdisciplinary Reviews: Computational Statistics*, vol. 2, no. 3, pp. 340–351, 2010.

[11] A. L. Potrzeba and I. G. Zurbenko, "Algorithm of periodic signal reconstruction with applications to tidal waves in atmosphere," in *Poster Topic Engineering and Physical Sciences, Chemometrics*,

JSM Proceedings, American Statistical Association, Alexandria, Va, USA, 2008.

[12] A. G. DiRienzo and I. G. Zurbenko, "Semi-adaptive non-parametric spectral estimation," *Journal of Computational and Graphical Statistics*, vol. 8, no. 1, pp. 41–59, 1999.

[13] S. Chapman, "The lunar tide in the earth's atmosphere," *Proceedings of the Royal Society of London A*, vol. 151, no. 872, pp. 105–117, 1935.

[14] S. Chapman and R. S. Lindzen, *Atmospheric Tides*, Gordon and Breach, New York, NY, USA, 1970.

[15] I. G. Zurbenko and A. L. Potrzeba, "Tidal waves in the atmosphere and their effects," *Acta Geophysica*, vol. 58, no. 2, pp. 356–373, 2010.

[16] I. G. Zurbenko and A. L. Potrzeba, "Tides in the Atmosphere," *Air Quality, Atmosphere & Health*, vol. 6, no. 1, pp. 39–46, 2013.

[17] I. G. Zurbenko and D. D. Cyr, "Climate fluctuations in time and space," *Climate Research*, vol. 57, no. 1, pp. 93–94, Addendum in Climate Research, vol. 46, no. 1, pp. 67–76, 2011.

[18] I. Zurbenko and M. Luo, "Restoration of time-spatial scales in global temperature data," *American Journal of Climate Change*, vol. 1, no. 3, pp. 154–163, 2012.

[19] E. Valachovic and I. Zurbenko, *Skin Cancer and the Solar Cycle: An Application of Kolmogorov-Zurbenko Filters*, JSM Proceedings, 2013.

# 4

# An Efficient Prediction Model for Water Discharge in Schoharie Creek, NY

**Katerina G. Tsakiri,**[1,2] **Antonios E. Marsellos,**[3,4] **and Igor G. Zurbenko**[5]

[1] *School of Computing, Engineering and Mathematics, University of Brighton, Brighton BN2 4GJ, UK*

[2] *Division of Math, Science and Technology, Nova Southeastern University, 3301 College Avenue, Fort Lauderdale-Davie, FL 33314, USA*

[3] *School of Environment and Technology, University of Brighton, Brighton BN2 4GJ, UK*

[4] *Department of Geological Sciences, University of Florida, 241 Williamson Hall, P.O. Box 112120, Gainesville, FL 32611, USA*

[5] *Department of Biometry and Statistics, State University of New York at Albany, One University Place, Rensselaer, NY 12144, USA*

Correspondence should be addressed to Katerina G. Tsakiri; ktsakiri@gmail.com

Academic Editor: Maite deCastro

Flooding normally occurs during periods of excessive precipitation or thawing in the winter period (ice jam). Flooding is typically accompanied by an increase in river discharge. This paper presents a statistical model for the prediction and explanation of the water discharge time series using an example from the Schoharie Creek, New York (one of the principal tributaries of the Mohawk River). It is developed with a view to wider application in similar water basins. In this study a statistical methodology for the decomposition of the time series is used. The Kolmogorov-Zurbenko filter is used for the decomposition of the hydrological and climatic time series into the seasonal and the long and the short term component. We analyze the time series of the water discharge by using a summer and a winter model. The explanation of the water discharge has been improved up to 81%. The results show that as water discharge increases in the long term then the water table replenishes, and in the seasonal term it depletes. In the short term, the groundwater drops during the winter period, and it rises during the summer period. This methodology can be applied for the prediction of the water discharge at multiple sites.

## 1. Introduction

It has been shown that there is a connection between increased river flooding and climate change [1–3]. In addition to this increased flooding activity, it has also been proven that global atmospheric, surface, and troposphere temperatures are rising [4, 5], coincident with increasing amounts of water vapor in the atmosphere [6]. Consequently, storms with increasing supplies of moisture produce more intense precipitation events therefore increasing the risk of flooding [7]. Temperature that increases during the winter period is known to break up the river's icy surface and hence initiate "ice jams." An ice jam is a localized accumulation of ice that can create blockages of river flow and increase the probability of subsequent upstream flooding [8].

Over the last decade, Schoharie Creek has experienced a significant increase in water discharge [9–11]. This increase can be explained by the increase of the winter melt events associated with warming temperatures [9].

Water discharge depends on the watershed area and the tributaries that drain into the stream. Schoharie Creek (hydrologic unit: 02020005 with a drainage basin of approximately 886 mi$^2$ (2300 km$^2$) in New York, USA, flows north 93 miles (150 km) from the foot of Indian Head Mountain in the Catskill Mountains through the Schoharie Valley to the Mohawk River. It is one of the two principal tributaries of the Mohawk River, with the other being West Canada Creek. Eight further tributaries drain into the Schoharie Creek. It is critical to model flooding in order to help local governments and communities to take precautionary steps to minimize or to prevent damage on property from flooding.

Several studies on flooding in Schoharie or in different locations (e.g., in Germany, Indus River in India, or in

Elbe River in Czech [1–3, 12]) show a moderate explanation of the water discharge by the climatic variables. The main reason is the presence of uncertainties in all the variables of the time series. In particular, it has been shown that uncertainties in basic variables or from the effects of scales related with different frequencies of the variables can change the inferences of the analysis between the variables [13–15]. Moreover, it has been shown that separation of time series into different components is necessary to avoid interference from different covariance structures existing between the components of the time series. An absence of separation of scales may lead to erroneous estimated parameters of the linear regression model or other multivariate models (principal components, canonical pairs). For this reason, the time series decomposition is essential before performing analysis since absence of separation of scales can lead to misinterpretation [16].

In previous studies regarding water discharge, the time series of the water discharge variable has rarely been decomposed into different components [1–3, 12]. Specifically, Kotlarski et al. investigate the Elbe River flooding in August 2002, using the precipitation variable [3]. In their model, the authors determined a correlation pattern of 0.75 yielding an $R^2$ value of 0.56. A moderate interpretation of the water discharge by the climatic variables might be due to a mixed interference of scales of the time series. The explanation of the water discharge can be improved by using the decomposition of the time series.

The main purpose of this paper is to present a novel methodology to explain and improve the predictive capability of river water discharge time series by using climatic and hydrological variables. We introduce a new model which incorporates separate treatments of frequency scales. Those scales have been determined based on spectral and cospectral analysis, and they provide a physical explanation of the hydrology of this area. In particular, we separate the data into seasonal and long and short term components by using the Kolmogorov-Zurbenko (KZ) filter [17]. The KZ filter provides effective separation of frequencies and its properties are described in the following section. For this paper, we use two models for the prediction of the water discharge time series to avoid mixed interferences. This is especially necessary in high latitudes where winter and summer seasons are prolonged. Prolonged seasons, such as winter (extended by a month or more) in this latitude, may provide a mixed interference due to their different correlation structure, and they must be separated as evidenced in standard statistical approaches.

One model is designed for the winter period and another model for the summer period. This is because flooding in Schoharie Creek may be caused by heavy rainfall during the summer or rapid snowmelt or "ice jam" during the prolonged winter period [18]. For both models, we first separate the different scales of the time series of water discharge, the depth of water level below land surface (referred as groundwater level), and the climatic variables by different frequency bands. We then design different multivariate models for each scale of the water discharge time series. Different frequencies

are always uncorrelated and as a result multivariate models can be designed for each scale of the water discharge time series, separately. The long term component and the seasonal component are the same for the winter and summer models. However, the short term component has been designed using two different models for the prediction of the water discharge time series. Physical phenomena (such as an ice jam or flooding) are a result of contributions from all components (seasonal and long and short term) and not simply short term variations. We prove that by using the KZ filter, we have improved the explanation of the water discharge time series up to 81%.

## 2. Methods

*2.1. Data.* Daily time series of water discharge, groundwater level, tides, and climatic variables have been obtained through the online public database of National Oceanic and Atmospheric Administration (NOAA; http://www.ncdc .noaa.gov/data-access/land-based-station-data/land-based-datasets/quality-controlled-local-climatological-data-qclcd/; Quality Controlled Local Climatological Data, Albany, Hudson River, NY) and the U.S. Geological Survey (USGS; http://waterdata.usgs.gov/nwis/, Schoharie Creek at Burtonsville, NY, for water discharge and for depth of water level data below surface). All monitoring stations are located in an area nearby the Schoharie Watershed. In particular, daily time series of air temperature (°F), wind speed (m/sec), total rainfall and snowfall precipitation (mm/hr), and tide (feet) data from a nearby station of Schoharie Creek have been obtained by the National Oceanic and Atmospheric Administration for January 2005–February 2013 period. Moreover, daily time series of water discharge and groundwater level have been analyzed for the same area and time period.

*2.2. Decomposition of Time Series.* In several studies, it has been shown that the separation of a time series into different components is essential in order to avoid contributions from different covariance structures between the components of the time series [13–16, 19]. The time series of a variable can be expressed by

$$X(t) = L(t) + \text{Se}(t) + \text{Sh}(t), \tag{1}$$

where $X(t)$ represents the original time series of a variable, $L(t)$ is the long term trend component, $\text{Se}(t)$ is the seasonal component, and $\text{Sh}(t)$ is the short term component. The long term component describes the fluctuations of a time series defined as being longer than a given threshold, the seasonal component describes the year-to-year fluctuations, and the short term component describes the short term variations. The long term trend component incorporates information regarding the trend component together with the cyclical component [20]. The cyclical component of a time series describes the fluctuations around the trend. The cyclical component can be viewed as those fluctuations in a time series which are longer than a given threshold, for example, one and a half years but shorter than those attributed to the trend.

The KZ filter, which separates long term variations from short term variations in a time series [17], provides a simple design and the smallest level of interferences between scales (seasonal and long and short term components) of a time series. It allows a physical interpretation of the scales [21, 22]. Furthermore, the KZ filter provides effective separation of frequencies for application directly to datasets containing missing observations [23–25]. KZ filtration is also known to provide the best and closest results to the optimal mean square of error [17, 22, 26]. As examples of its use, the KZ filter has been applied effectively for the explanation and prediction of the ozone problem and for the explanation of the water use time series in Gainesville, Florida [19, 24, 27–30].

Specifically, the KZ filter is a low pass filter, defined by $p$ iterations of a simple moving average of $m$ points. The moving average of the KZ can be expressed by

$$Y_t = \left(\frac{1}{m}\right) \sum_{j=-k}^{k} X_{t+j}, \qquad (2)$$

where $m = 2k + 1$. The output of the first iteration becomes the input for the second iteration, and so on. The time series produced by $p$ iterations of the filter described in expression (2) is denoted by

$$Y_t = \mathrm{KZ}_{m,p}(X_t). \qquad (3)$$

The parameter $m$ in expression (3) has been determined to provide the maximum explanation of the water discharge time series by the climatic variables. In this study, first, we examine the periodograms of all the time series. By using the periodograms we verify that the decomposition of the time series is essential for the prediction of the water discharge time series using the climatic variables. We design different multivariate models corresponding to each scale of the time series. For the decomposition of the time series, we use the KZ filter. The multivariate models describe the projection of the water discharge time series in the space defined by the groundwater variable and the climatic variables. This projection can be expressed through linear regression between the water discharge time series and the hydrological and climatic variables. Therefore, we decompose all the time series of the variables using the KZ filter. After the application of the KZ filter, we need to predict each component of the water discharge time series, separately. For this reason, we use multivariate models for the prediction of each component of the water discharge time series. For the application of the multivariate model, we select different climatic variables to explain the components. Finally, we estimate the water discharge time series using the climatic variables for the winter and summer period.

In the following sections, we describe the method for prediction of the seasonal and long and short term components of the water discharge time series using Schoharie Creek in New York as an example of application area.

*2.2.1. Raw Data Analysis of the Water Discharge Time Series.* Here, we apply the previous methodology for the decomposition of the time series to explain and predict the water

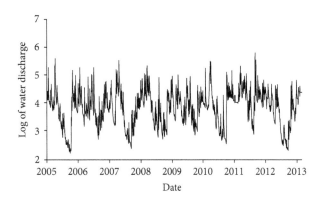

FIGURE 1: Time series of logarithm of water discharge data (ft$^3$/sec) during the period January 2005–February 2013.

discharge time series of Schoharie Creek. For the explanation and prediction of the water discharge, we use the climatic variables and the groundwater level. The logarithm of the water discharge from the study area was measured from January 2005 to February 2013, and it has been presented in Figure 1. We use the logarithm transform for the water discharge time series in order to stabilize the variance.

Table 1 presents the correlation matrix between the raw data of the variables. It can be noted that based on the significance level, the correlations between the water discharge and the climatic variables are statistically significant. To explain the water discharge time series using the raw data of the climatic variables, a linear regression is performed with regression coefficient, $R^2$, equal to 0.590. This model yields a moderate $R^2$ for the water discharge time series (expression (4)). Specifically, the time series of the water discharge raw data can be expressed through a linear regression as follows:

$$\mathrm{WD}(t) = -0.378\mathrm{T}(t) + 0.113\mathrm{TD}(t) + 0.049\mathrm{WS}(t)$$
$$- 0.551\mathrm{GWL}(t) + 0.118\mathrm{PR}(t) + \varepsilon(t), \qquad (4)$$
$$R^2 = 0.590,$$

where WD($t$), T($t$), TD($t$), WS($t$), GWL($t$), and PR($t$) denote the raw data of the logarithm of water discharge, temperature, tide, wind speed, groundwater level, and precipitation, respectively. Furthermore, in expression (4), $\varepsilon(t)$ represents the residuals of this relationship and $R^2$ is the square of the correlation coefficient.

The relationship of the water discharge with the climatic variables can be strengthened by separating the seasonal and long and short term variations in all the time series. The separation of scales in the time series can also be verified through examination of the coherence between the variables. Table 2 shows the main periods of all the variables derived by the periodograms of the variables. It can be observed that almost all the variables consist of a 365-day period describing the seasonal component of the variables. Some variables consist of short periods (7 days, 4 days, etc.) which are related to the short term components of the variables. In particular, the precipitation variable consists mostly of short periods and for this reason this variable contributes

TABLE 1: Correlation matrix of the raw data between the water discharge, the climatic variables, and the groundwater level.

| Variables | Water discharge | Groundwater level | Temperature | Tide | Wind speed | Precipitation |
|---|---|---|---|---|---|---|
| Water discharge | 1 | −0.65 | −0.481 | 0.254 | 0.203 | 0.109 |
| Groundwater level | −0.65 | 1 | 0.178 | −0.277 | −0.094 | 0.033 |
| Temperature | −0.481 | 0.178 | 1 | 0.041 | −0.193 | −0.007 |
| Tide | 0.254 | −0.277 | 0.041 | 1 | 0.105 | −0.008 |
| Wind speed | 0.203 | −0.094 | −0.193 | 0.105 | 1 | 0.139 |
| Precipitation | 0.109 | 0.033 | −0.007 | −0.008 | 0.139 | 1 |

TABLE 2: Periods derived by the graph of periodograms of the water discharge variable and the climatic variables.

| Variable | Period (days) | | |
|---|---|---|---|
| | First peak of period | Second peak of period | Third peak of period |
| Water discharge | 365 | 594 | |
| Groundwater level | 365 | 991 | 198 |
| Tide | 365 | 185 | 14 |
| Precipitation | 27 | 7 | 4 |
| Wind speed | 365 | | |
| Temperature | 365 | | |

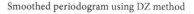

Smoothed periodogram using DZ method

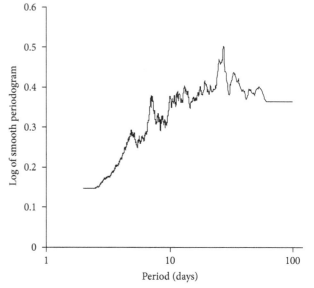

FIGURE 3: Periodogram of the precipitation variable. The $y$-axis describes the spectral density using the KZFT (Kolmogorov Zurbenko Fourier Transform) method.

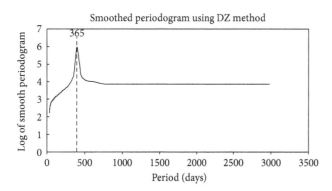

FIGURE 2: Periodogram of the temperature variable. The $y$-axis describes the spectral density using the KZFT (Kolmogorov Zurbenko Fourier Transform) method.

mostly to the short term component of the water discharge. Figures 2 and 3 show two examples of the periodograms for the temperature and the precipitation variable using the DZ (DiRienzo and Zurbenko [31]) algorithm in R software. From Figures 2 and 3 and Table 2, we can conclude that it is essential to decompose the time series into different components to avoid the contribution of different frequencies between the scales of the time series.

For the decomposition of the time series, we use the $KZ_{33,3}$ filter (length is 33 with 3 iterations) to provide a physical based explanation of the water discharge time series. Following Rao et al. [24], the parameters of the KZ filter are chosen to provide the optimal solution for our study. In particular, the parameters have been estimated in order to sustain the properties of each time series component and

reduce the short term variations displayed in the long and seasonal component of the time series.

The KZ filter is applied to the logarithm of daily water discharge and produces a time series devoid of short term variations and consisting only of the long term variations of the time series ($L(t)$). The same filter is applied to the variables of daily temperature, tide, wind speed, precipitation, and groundwater level.

### 2.2.2. Prediction of the Long Term Component of Water Discharge.
To explain the long term component of water discharge and its relationship with groundwater level and climatic variables, we examine the filtered daily temperature, tide, wind speed, and precipitation of the sum of four continuous days. For the analysis, we denote the long term components of the water discharge, temperature, tide, wind speed, precipitation, and groundwater level time series with $WD_{KZ}(t)$, $T_{KZ}(t)$, $TD_{KZ}(t)$, $WS_{KZ}(t)$, $PR_{KZ}(t)$, and $GWL_{KZ}(t)$, respectively.

The correlation between the long term component of the water discharge and the precipitation is weak (Table 3). This can be verified by the peaks in the periodogram

TABLE 3: Correlation matrix between the long term components of the water discharge, the climatic variables, and the groundwater level.

| Variable | Water discharge | Groundwater level | Temperature | Tide | Wind speed | Precipitation (4 days) |
|---|---|---|---|---|---|---|
| Water discharge | 1 | −0.744 | −0.641 | 0.39 | 0.596 | 0.175 |
| Groundwater level | −0.744 | 1 | 0.26 | −0.412 | −0.43 | −0.224 |
| Temperature | −0.641 | 0.26 | 1 | 0.033 | −0.691 | 0.103 |
| Tide | 0.39 | −0.412 | 0.033 | 1 | 0.462 | −0.114 |
| Wind speed | 0.596 | −0.43 | −0.691 | 0.462 | 1 | −0.131 |
| Precipitation (4 days) | 0.175 | −0.224 | 0.103 | −0.114 | −0.131 | 1 |

of the precipitation variable which occur over short periods (Table 2). Consequently, the precipitation variable contributes mostly to the short term component of the water discharge while groundwater level contributes mostly to the long term component of the water discharge (Table 3). The groundwater level is the only variable that consists of time periods greater than a year (991 days; Table 2). This period may be explained by larger time scale atmospheric episodes or from river flow changes due to anthropogenic impacts (e.g., dams, power projects, lock stations, bridges, etc.).

For the prediction of the long term component of water discharge, a linear regression is performed using the filtered logarithm of the water discharge, the filtered climatic variables, and the groundwater level time series with an $R^2$ value of 0.833. The long term component of the logarithm of the water discharge time series can then be expressed by

$$WD_{KZ}(t) = -0.704T_{KZ}(t) + 0.337TD_{KZ}(t) - 0.241WS_{KZ}(t)$$

$$- 0.494GWL_{KZ}(t) + 0.143PR_{KZ}(t) + \varepsilon(t),$$

$$R^2 = 0.833.$$

$$(5)$$

It can be noted that the $P$ values for the coefficients of the variables of the linear regression model are equal to zero. Thus, the coefficients of the linear regression are statistically significant. The scale of $\varepsilon(t)$ corresponds to percent change in the long term component of water discharge data due to effects other than the climatic variables.

From expression (5), we can conclude that 83.3% of the long term fluctuations of water discharge can be explained by long term groundwater level fluctuations and major climatic variables. If we were to consider additional climatic variables such as sea level, wind direction, or relative humidity, the coefficients related to those variables would not be statistically significant (the $P$ value associated with the $t$-test is greater than the significance level). For this reason, we did not consider additional climatic variables in the model.

Because expression (5) is given for the natural logarithm of water discharge, the additive term $\varepsilon(t)$ has a multiplicative effect, $\exp(\varepsilon(t))$, in the original data. Since the term $\varepsilon(t)$ is sufficiently small, we can obtain that

$$\exp(\varepsilon(t)) \approx 1 + \varepsilon(t); \text{ that is,}$$
$$\ln(1 + \varepsilon(t)) \approx \varepsilon(t). \tag{6}$$

Therefore, $\varepsilon(t) * 100$ corresponds to percent changes in the long term of water discharge unexplained by the groundwater level and the climatic variables.

*2.2.3. Prediction of the Seasonal Component of Water Discharge.* The seasonal component of a time series represents the year-to-year fluctuations of the corresponding variable. In order to predict the seasonal component of the water discharge time series, we use seasonal components of the climatic variables and the groundwater level. In particular, the seasonal component of the water discharge, $WD_{Se}(d)$, can be defined by

$$WD_{Se}(d) = \frac{1}{Y} \sum_{y=1}^{Y} \left( WD(t) - WD_{KZ}(t) \right), \tag{7}$$

$d = 1, \ldots, 365, d = 1$ represents January, and so forth

$$y = 1, \ldots, Y, \tag{8}$$

where $d$ represents the days of a year, $y$ is the number of years of the observed values, $WD(t)$ is the raw data of the time series of the water discharge, and $WD_{KZ}(t)$ is the long term component of the water discharge time series derived by the application of the KZ filter. Similarly, we define the seasonal components of the climatic variables and the groundwater level. In particular, the seasonal components of the temperature, tide, wind speed, the sum of four continuous days of precipitation, and groundwater level variables are denoted by $T_{Se}(d)$, $TD_{Se}(d)$, $WS_{Se}(d)$, $PR_{Se}(d)$, and $GWL_{Se}(d)$, respectively. The maximum correlation between the water discharge and the precipitation variable occurs when we consider the sum of the precipitation variable for four days.

To investigate the relationship between the seasonal component of the water discharge and the climatic variables, we estimate their correlation matrix (Table 4). The correlation between the seasonal component of the water discharge and two of the climatic variables (temperature and tide) is strong as is the case with the groundwater level. Moreover, all examined variables have a period of 365 days (Table 2). The most correlated variable with the seasonal component of the water discharge is the seasonal component of tide (Table 4).

To predict the seasonal component of the water discharge, we perform linear regression using the seasonal components of the climatic variables and the groundwater level. The coefficient of determination, $R^2$, is equal to 0.912.

TABLE 4: Correlation matrix between the seasonal components of the water discharge, the climatic variables, and the groundwater level.

| Variable | Water discharge | Groundwater level | Temperature | Tide | Wind speed | Precipitation (4 days) |
|---|---|---|---|---|---|---|
| Water discharge | 1 | 0.63 | 0.579 | 0.893 | 0.418 | 0.27 |
| Groundwater level | 0.63 | 1 | 0.492 | 0.831 | 0.406 | 0.012 |
| Temperature | 0.579 | 0.492 | 1 | 0.627 | 0.3 | −0.046 |
| Tide | 0.893 | 0.831 | 0.627 | 1 | 0.428 | 0.089 |
| Wind speed | 0.418 | 0.406 | 0.3 | 0.428 | 1 | 0.044 |
| Precipitation (4 days) | 0.27 | 0.012 | −0.046 | 0.089 | 0.044 | 1 |

In particular, the seasonal component of the logarithm of the water discharge can be estimated by

$$\mathrm{WD}_{\mathrm{Se}}(d) = 0.042\mathrm{T}_{\mathrm{Se}}(d) + 1.103\mathrm{TD}_{\mathrm{Se}}(d) + 0.061\mathrm{WS}_{\mathrm{Se}}(d)$$
$$- 0.335\mathrm{GWL}_{\mathrm{Se}}(d) + 0.174\mathrm{PR}_{\mathrm{Se}}(d) + \varepsilon(t),$$
$$R^2 = 0.912. \tag{9}$$

Thus, 91.2% of the variability of the seasonal component of the water discharge can be explained by the climatic variables and the groundwater level as described in expression (9). The $P$ values for the coefficients of the variables of the linear regression are equal to zero. Thus, the coefficients of the linear regression are statistically significant. The addition of extra climatic variables such as sea level, wind direction, and relative humidity in expression (9) does not change the value of $R^2$.

### 2.2.4. Prediction of the Short Term Component of Water Discharge.
For the prediction of the short term component of the water discharge, we consider the short term components of the climatic variables and groundwater level. The short term component of the water discharge time series, $\mathrm{WD}_{\mathrm{Sh}}(t)$, can be defined as follows:

$$\mathrm{WD}_{\mathrm{Sh}}(t) = \mathrm{WD}(t) - \mathrm{WD}_{\mathrm{KZ}}(t) - \mathrm{WD}_{\mathrm{Se}}(d), \tag{10}$$

where $\mathrm{WD}(t)$ represents the raw data of the water discharge time series, $\mathrm{WD}_{\mathrm{KZ}}(t)$ represents the long term component of water discharge, and $\mathrm{WD}_{\mathrm{Se}}(d)$ is the seasonal component of the water discharge time series. Similarly, we can define the short term components of the remaining variables.

For the short term component of the water discharge, we consider two different models. One describes the prediction of the short term component during the summer period (May through September) and the other during winter (December through March). We consider two different models because flooding in the rivers in New York State is caused by extensive precipitation (e.g., extensive rainfall or tropical storms during the summer period) or by rapid snowmelt (e.g., ice jams during the prolonged winter period). The overall explanation of the winter model is maximum during the prolonged winter period (December to March), while the summer model shows maximum explanation during the prolonged summer period (May to September).

### 2.2.5. Prediction of the Short Term Component of Summer Water Discharge.
To predict the short term component of the water discharge time series, we perform a linear regression by using the short term components of daily temperature, precipitation, the sum of four continuous days of daily precipitation, and daily groundwater level. The short term components of the above variables are denoted by $\mathrm{T}_{\mathrm{Sh}}(t)$, $\mathrm{PR}_{\mathrm{Sh}}(t)$, $\mathrm{PR4}_{\mathrm{Sh}}(t)$, and $\mathrm{GWL}_{\mathrm{Sh}}(t)$, respectively. Table 5 shows the correlation matrix between the short term components of the variables for the summer period (May through September).

For the prediction of the short term component of the water discharge for the summer period, a linear regression is performed by using the short term components of the above climatic variables and the groundwater level with coefficient of determination, $R^2$, equal to 0.447. Specifically, the short term component of the water discharge can be expressed through a linear regression as follows:

$$\mathrm{WD}_{\mathrm{Sh}}(t) = -0.097\mathrm{T}_{\mathrm{Sh}}(t) - 0.377\mathrm{GWL}_{\mathrm{Sh}}(t) + 0.199\mathrm{PR}_{\mathrm{Sh}}(t)$$
$$+ 0.453\mathrm{PR4}_{\mathrm{Sh}}(t) + \varepsilon(t),$$
$$R^2 = 0.447. \tag{11}$$

The $P$ values for the coefficients of the variables of the linear regression model are equal to zero.

### 2.2.6. Prediction of the Short Term Component of Winter Water Discharge.
Flooding in the rivers during the winter period occurs through rapid snowmelt due to the increase of air temperature. For this reason, the average temperature is used for the prediction model in the linear regression. The maximum correlation between water discharge and average temperature occurs for the average temperature over four days. Moreover, for the prediction model, we consider the variables: tide, wind speed, groundwater level, and the sum of four days of precipitation. The short term components of the above variables are denoted by $\mathrm{T4}_{\mathrm{Sh}}(t)$, $\mathrm{WS}_{\mathrm{Sh}}(t)$, $\mathrm{PR4}_{\mathrm{Sh}}(t)$, and $\mathrm{GWL}_{\mathrm{Sh}}(t)$, respectively. Table 6 shows the correlation matrix for the short term components of the variables.

To predict the short term component of the water discharge time series, we perform a linear regression by using the short term components of the above climatic variables with resultant coefficient of determination, $R^2$, equal to 0.719.

TABLE 5: Correlation matrix between the short term components of the water discharge, the climatic variables, and the groundwater level for the summer period.

| | Water discharge | Groundwater level | Temperature average (4 days) | Precipitation | Precipitation (4 days) |
|---|---|---|---|---|---|
| Water discharge | 1 | −0.379 | −0.247 | 0.124 | 0.502 |
| Groundwater level | −0.379 | 1 | 0.099 | 0.194 | −0.069 |
| Temperature average (4 days) | −0.247 | 0.099 | 1 | 0.004 | −0.249 |
| Precipitation | 0.124 | 0.194 | 0.004 | 1 | −0.004 |
| Precipitation (4 days) | 0.502 | −0.069 | −0.249 | −0.004 | 1 |

TABLE 6: Correlation matrix between the short term components of the water discharge, the climatic variables, and the groundwater level for the winter period.

| | Water discharge | Groundwater level | Temperature average (4 days) | Tide | Wind speed | Precipitation (4 days) |
|---|---|---|---|---|---|---|
| Water discharge | 1 | 0.398 | 0.57 | 0.732 | 0.25 | 0.229 |
| Groundwater level | 0.398 | 1 | 0.408 | 0.597 | 0.12 | −0.005 |
| Temperature average (4 days) | 0.57 | 0.408 | 1 | 0.444 | 0.136 | 0.038 |
| Tide | 0.732 | 0.597 | 0.444 | 1 | 0.181 | 0.117 |
| Wind speed | 0.25 | 0.12 | 0.136 | 0.181 | 1 | 0.155 |
| Precipitation (4 days) | 0.229 | −0.005 | 0.038 | 0.117 | 0.155 | 1 |

The short term component of water discharge can be expressed as follows:

$$WD_{Sh}(t) = 0.323T4_{Sh}(t) + 0.629TD_{Sh}(t) + 0.087WS_{Sh}(t)$$

$$- 0.117GWL_{Sh}(t) + 0.130PR4_{Sh}(t) + \varepsilon(t),$$

$$R^2 = 0.719.$$

(12)

The variables considered for the prediction of the short term component of water discharge during the winter period are different to those explaining the summer period. Therefore, the separation of scales of the time series is essential since different components are related to different physical phenomena (rapid snowmelt and "ice jam" for the winter period while extensive rainfall or tropical storms for the summer period). The $P$ values for the coefficients of the variables of the linear regression model are equal to zero.

## 3. Results

A river can act as a gaining stream, receiving water from the groundwater system, or as a losing stream, losing water to the groundwater system. The water table's height reflects a balance between the rate of replenishment, through precipitation, and removal through discharge and withdrawal. Any imbalance either raises or lowers the water table, acting with the opposite effect to the groundwater level value (because groundwater level value is measured as the distance of the water level depth below surface). In the summer period, the water river discharge of the Schoharie Creek increases as it flows downstream as tributaries and groundwater contribute additional water (recalling that eight tributaries contribute to Schoharie Creek before it reaches Mohawk River).

During summer, groundwater resources appear to increase to ample levels (recharge). Natural replenishment will decrease the measured value of the groundwater level and enhance the river water discharge. Hence, it produces a negative association.

In regions where there is a prolonged winter period, the rainfall to replenish the water table is often scarce, and the rate of water table recharge will be less than the river's water discharge. The groundwater will drop and may result in very low water level (depletion). Groundwater level value will increase (water level drops), and water discharge in the stream increases (primarily through snow melt), which produces a positive association.

Due to the separation of scales in the time series, the long term component of water discharge shows a negative correlation with the long term component of the groundwater level and temperature (Table 3), while the correlation between the seasonal components of those variables is positive (Table 4). A positive correlation has also been observed between the short term components of those variables during the winter (Table 6), while a negative correlation exists during the summer period (Table 5).

The negative correlation between the long-term components of the water discharge and the groundwater level is due to the increased rainfall during the last decade in Schoharie Creek area. This phenomenon has also been observed for the short term components during the summer period when precipitation is intense. A positive correlation between the water discharge and the groundwater level takes place during the prolonged winter period.

The correlation between the short term component of the water discharge and precipitation during the winter is lower than the short term components of those variables during the summer. This is because in areas that experience prolonged

TABLE 7: Summary of variances and the coefficient of determinations for the seasonal and long and short term component for the summer and winter period.

|  | Long term | Seasonal term | Short term |
|---|---|---|---|
| Summer period | | | |
| Variance | 0.588 | 0.178 | 0.232 |
| $R^2$ | 0.833 | 0.912 | 0.447 |
| Winter period | | | |
| Variance | 0.531 | 0.162 | 0.307 |
| $R^2$ | 0.833 | 0.912 | 0.719 |

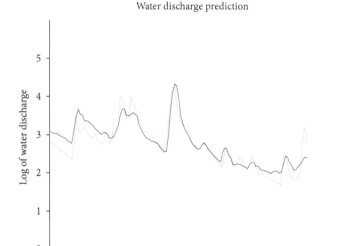

2006
Water discharge prediction

— Predicted water discharge
— Water discharge (USGS data)

FIGURE 4: Raw water discharge data (blue line) along with the prediction model (purple line) for the summer period (May through September 2006).

winter seasons, rainfall does not contribute to water table in the same rate as it would replenish the water table during the summer period.

To estimate the total explanation of the model for the water discharge of the summer period, we combine expressions (5), (9), and (11) to represent the seasonal and long and short term components (Table 7). The contribution of the long term component of the climatic variables and the groundwater level in the time series of the water discharge is 49% ($0.588 \times 0.833$) from expression (5). Furthermore, 16.2% ($0.178 \times 0.912$) is the contribution from the seasonal component using expression (9), while 10.4% ($0.232 \times 0.447$) is the contribution of the short term component from expression (11). By combining expressions (5), (9), and (11), in a similar way to expression (1), we can then explain 75.6% of the total variance of the water discharge time series using the climatic variables and groundwater level during the summer period. Figure 4 shows the raw data of the time series of the water discharge data (blue line) along with the prediction model (purple line) derived by expressions (5), (9), and (11) for the year 2006 (summer flood in June 29; widespread flooding in the Mohawk and Hudson basins, and Catskills was observed). For other years, similar graphs can be presented for the remaining years during the summer period.

By combining expressions (5), (9), and (12), we can estimate the total explanation of the water discharge time series from climatic variables for the winter period (Table 7). In particular, 44.2% ($0.531 \times 0.833$) is the contribution of the long term components of the climatic variables and groundwater level to the time series of the water discharge as described in expression (5). 14.8% ($0.162 \times 0.912$) is the contribution of the seasonal components of the climatic variables as described in expression (9), while 22.1% ($0.307 \times 0.719$) is the contribution of the short term component by expression (12). Consequently, 81.1% is the total explanation of the water discharge time series using the climatic variables and the groundwater level during the winter period. As an example, Figure 5 shows the raw water discharge time series data (blue line) for the year 2010 along with the prediction model (purple line) described by expressions (5), (9), and (12). Our model accounts for the "ice jam" event of 2010 (January 25-26; widespread flooding occurred across east central New York and adjacent western New England from a combination of rain, snowmelt, and frozen ground).

As a consequence of summer and winter model results, we prove that the decomposition of the time series improves our ability to describe and predict the time series variations of water discharge by approximately two times. In particular, the unexplained variance derived by the raw data described in expression (4) is 41%, while the unexplained variance derived from the decomposition method of the winter water discharge time series is 19%. Similar results can be derived for the summer period providing an unexplained variance of 24.4%. This method can be applied in other locations as well. In such cases, the coefficients associated with the above expressions will be different.

## 4. Discussion

This study focuses on predicting the daily water discharge time series using available climatic variables and the groundwater level. We prove that the decomposition of the time series is essential due to the presence of short term variations in the time series. We use the KZ filter to decompose the time series into the seasonal and long and short term components to provide a physical based explanation for the time series of the water discharge. The long term component is associated with long term changes, the seasonal component with year-to-year fluctuations, and the short term component with short term variations.

In this study, we prove that the seasonal and long and short term components of water discharge can be explained through consideration of climatic variables and groundwater level. The results show that as water discharge increases in the long term the water table replenishes, while in the seasonal

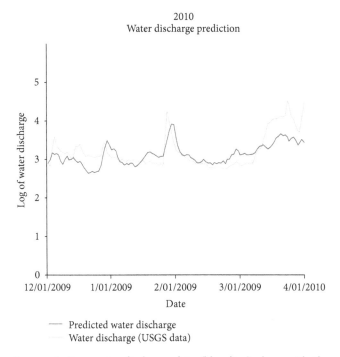

2010
Water discharge prediction

— Predicted water discharge
— Water discharge (USGS data)

FIGURE 5: Raw water discharge data (blue line) along with the prediction model (purple line) for the winter period (December of 2009 through March of 2010).

term it depletes. In the short term, the groundwater drops during the winter period and rises during the summer period. The short term component of the water discharge is related to synoptic weather fluctuations and short term effects of rain, storms, and cyclones during the summer period and the rapid increase in temperature as well as storms during the winter period. As it is described in expressions (5), (9), (11), and (12), the selection of variables and coefficients of the multivariate models is different for the explanation of the seasonal and long and short term components of the water discharge time series. This requires the separation of the different scales of the water discharge time series in order to avoid erroneous results [14–16].

After the application of the KZ filter and the decomposition of all time series, we apply different multivariate models to each component of the water discharge time series. Different scales of time series are associated with different correlation structures (Tables 3–6) and provide different coefficients for the multivariate models [14–16]. Furthermore, for the explanation of the short term component of water discharge, we design two different models. We use different variables to explain the short term component during the winter and summer period and as a result, we need to use two different multivariate models for the explanation of the short term component of water discharge (Tables 5 and 6). This is because, in the specific location, the local climate is controlled mainly by two prolonged seasons instead of four as a typical temperate climate represents.

In our paper, we show that the accuracy of the water discharge time series prediction can be increased up to 81.1% for the winter period and 75.6% for the summer period by

incorporating the KZ filter in a separation of the scales of the time series [17]. Moreover, the coefficient of determination of the water discharge time series is very strong for all scales of water discharge time series, and it exceeds those of the raw data. This model can be used for the prediction of critical levels of water discharge ahead of time as long as other variables will be received prior to the event and are applicable to other locations. Due to the variability of local hydrological and climatic characteristics, it would be likely that other locations would produce variable numeric values but maintain the relative improvements in prediction accuracy.

## 5. Conclusions

The prediction of the daily water discharge time series using the climatic variables and groundwater level can be substantially improved through the decomposition of the time series. The decomposition of the different components (scales), which are the seasonal and long and short term components, avoids erroneous results and approximately doubles the prediction accuracy of the water discharge time series relative to raw data. The resulting isolation of the short term variations by the decomposition of the time series shows a summer period with a water table replenishment and a winter period with a water table depletion. The design of multivariate models (winter and summer) can improve the prediction of flooding caused by storms, rapid snowmelt, and ice jams.

## Conflict of Interests

The authors declare that there is no conflict of interests regarding the publication of this paper.

## References

[1] A. Bronstert, "River flooding in Germany: influenced by climate change?" *Physics and Chemistry of the Earth*, vol. 20, no. 5-6, pp. 445–450, 1995.

[2] H. Hartmann and L. Andresky, "Flooding in the Indus river basin-A spatiotemporal analysis of precipitation records," *Global and Planetary Changes*, vol. 107, pp. 25–35, 2013.

[3] S. Kotlarski, S. Hagamann, P. Krahe, R. Podzun, and D. Jacob, "The Elbe river flooding 2002 as seen by an extended regional climate model," *Journal of Hydrology*, vol. 472-473, pp. 169–183, 2012.

[4] P. Brohan, J. Kennedy, I. Harris, S. F. B. Tett, and P. D. Jones, "Uncertainty estimates in regional and global observed temperature changes: a new dataset from 1850," *Journal of Geophysical Research*, vol. 11, no. 12, 2006.

[5] T. M. Smith and R. W. Reynolds, "A global merged land-air-sea surface temperature reconstruction based on historical observations (1880–1997)," *Journal of Climate*, vol. 18, no. 12, pp. 2021–2036, 2005.

[6] K. E. Trenberth, J. Fasullo, and L. Smith, "Trends and variability in column-integrated atmospheric water vapor," *Climate Dynamics*, vol. 24, no. 7-8, pp. 741–758, 2005.

[7] K. E. Trenberth, "Changes in precipitation with climate change," *Climate Research*, vol. 47, no. 1-2, pp. 123–138, 2011.

[8] S. Beltaos, "Onset of river ice breakup," *Cold Regions Science and Technology*, vol. 25, no. 3, pp. 183–196, 1997.

[9] A. L. Kern, *Study of 20th century trends in stream flow for West Canada and schoharie creeks of the Mohawk-Hudson Rivers watershed [Ph.D. thesis]*, Department of Geology, Union College, Schenectady, NY, USA, 2008.

[10] A. E. Marsellos, J. I. Garver, J. M. H. Cockburn, and K. Tsakiri, "Determination of historical channel changes and meander cut-off points using LiDAR and GIS in Schoharie creek, NY," in *Proceedings of the Mohawk Symposium*, pp. 28–33, Schenectady, NY, USA, 2010.

[11] A. E. Marsellos and J. I. Garver, "Channel incision and landslides identified by LiDAR in the lower reaches of Schoharie creek, New York," *Geological Society of America Abstracts with Programs*, vol. 42, no. 1, p. 73, 2010.

[12] S. A. Johnston and J. I. Garver, "Record of flooding on the Mohawk River from 1634 to 2000 based on historical archives," *Geological Society of America Abstracts with Programs*, vol. 33, no. 1, 2001.

[13] I. G. Zurbenko and M. Sowizral, "Resolution of the destructive effect of noise on linear regression of two time series," *Far East Journal of Theoretical Statistics*, vol. 3, pp. 139–157, 1999.

[14] K. G. Tsakiri and I. G. Zurbenko, "Destructive effect of the noise in principal component analysis with application to ozone pollution," in *Proceedings of the Joint Statistical Meetings*, pp. 3054–3068, Statistical Computing Section, American Statistical Association, Alexandria, Va, USA, 2008.

[15] K. G. Tsakiri and I. G. Zurbenko, "Model prediction of ambient ozone concentrations," in *Proceedings of the Joint Statistical Meetings*, pp. 3054–3068, Statistical Computing Section, American Statistical Association, Alexandria, Va, USA, 2009.

[16] K. Tsakiri and I. G. Zurbenko, "Effect of noise in principal component analysis," *Journal of Statistics and Mathematics*, vol. 2, no. 2, pp. 40–48, 2011.

[17] I. G. Zurbenko, *The Spectral Analysis of Time Series*, Statistics and Probability, North Holland, Amsterdam, The Netherlands, 1986.

[18] J. I. Garver and J. M. H. Cockburn, "A historical perspective of ice jams on the lower Mohawk River," in *Proceedings of the Mohawk Symposium*, pp. 25–29, 2009.

[19] K. Tsakiri and I. G. Zurbenko, "Explanation of fluctuation in water use time series," *Journal of Environmental and Ecological Statistics*, vol. 20, no. 3, pp. 399–412, 2013.

[20] OECD, "Section 4: guidelines for the reporting of different forms of data," in *Data and Metadata Reporting and Presentation Handbook*, OECD, Paris, France, 2005.

[21] Wikipedia, "Kolmogorov-Zurbenko Filters," http://en.wikipedia.org/wiki/Kolmogorov%E2%80%93Zurbenko_filter.

[22] W. Yang and I. Zurbenko, "Kolmogorov-Zurbenko filters," *Wiley Interdisciplinary Reviews*, vol. 2, no. 3, pp. 340–351, 2010.

[23] R. E. Eskridge, J. Y. Ku, S. Trivikrama Rao, P. Steven Porter, and I. G. Zurbenko, "Separating different scales of motion in time series of meteorological variables," *Bulletin of the American Meteorological Society*, vol. 78, no. 7, pp. 1473–1483, 1997.

[24] S. T. Rao, I. G. Zurbenko, R. Neagu, P. S. Porter, J. Y. Ku, and R. F. Henry, "Space and time scales in ambient ozone data," *Bulletin of the American Meteorological Society*, vol. 78, no. 10, pp. 2153–2166, 1997.

[25] W. Yang and I. Zurbenko, "Nonstationarity," *Wiley Interdisciplinary Reviews*, vol. 2, no. 1, pp. 107–115, 2010.

[26] B. Close and I. Zurbenko, "Kolmogorov-Zurbenko adaptive filters," version 3, R-software, 2013, http://cran.r-project.org/web/packages/kza/index.html.

[27] S. T. Rao and I. G. Zurbenko, "Detecting and tracking changes in ozone air quality," *Journal of the Air and Waste Management Association*, vol. 44, no. 9, pp. 1089–1092, 1994.

[28] C. Hogrefe, S. T. Rao, I. G. Zurbenko, and P. S. Porter, "Interpreting the information in ozone observations and model predictions relevant to regulatory policies in the Eastern United States," *Bulletin of the American Meteorological Society*, vol. 81, no. 9, pp. 2083–2106, 2000.

[29] K. G. Tsakiri and I. G. Zurbenko, "Prediction of ozone concentrations using atmospheric variables," *Journal of Air Quality, Atmosphere, and Health*, vol. 4, no. 2, pp. 111–120, 2011.

[30] K. G. Tsakiri and I. G. Zurbenko, "Determining the main atmospheric factor on ozone concentrations," *Journal of Meteorology and Atmospheric Physics*, vol. 109, no. 3-4, pp. 129–137, 2010.

[31] A. G. DiRienzo and I. G. Zurbenko, "Semi-adaptive nonparametric spectral estimation," *Journal of Computational and Graphical Statistics*, vol. 8, no. 1, pp. 41–59, 1999.

# 5

# Influence of Northwest Cloudbands on Southwest Australian Rainfall

**Nicola Telcik and Charitha Pattiaratchi**

*School of Civil, Environmental and Mining Engineering, The UWA Oceans Institute, The University of Western Australia, 35 Stirling Highway, Crawley, WA 6009, Australia*

Correspondence should be addressed to Charitha Pattiaratchi; chari.pattiaratchi@uwa.edu.au

Academic Editor: Silvio Gualdi

Northwest cloudbands are tropical-extratropical feature that crosses the Australian continent originating from Australia's northwest coast and develops in a NW-SE orientation. In paper, atmospheric and oceanic reanalysis data (NCEP) and Reynolds reconstructed sea surface temperature data were used to examine northwest cloudband activity across the Australian mainland. An index that reflected the monthly, seasonal, and interannual activity of northwest cloudbands between 1950 and 1999 was then created. Outgoing longwave radiation, total cloud cover, and latent heat flux data were used to determine the number of days when a mature northwest cloudband covered part of the Australian continent between April and October. Regional indices were created for site-specific investigations, especially of cloudband-related rainfall. High and low cloudband activity can affect the distribution of cloudbands and their related rainfall. In low cloudband activity seasons, cloudbands were mostly limited to the south and west Australian coasts. In high cloudband activity seasons, cloudbands penetrated farther inland, which increased the inland rainfall. A case study of the southwest Australian region demonstrated that, in a below average rainfall year, cloudband-related rainfall was limited to the coast. In an above average rainfall year, cloudband-related rainfall occurred further inland.

## 1. Introduction

The northwest cloudband (NWCB) is one of three tropical-extratropical cloudbands that crosses the Australian continent [1]. It originates from Australia's northwest coast (Figure 1(a)) and develops a NW-SE orientation, with no immediate connection to southern frontal systems [2]. Extratropical cloudbands are not unique to Australia; they also occur over Southern Africa [3], over South America, and in the Northern Hemisphere [4]. The Northern Hemisphere equivalent, the southwest (SW) cloudband, extends from Central America to the North American east coast and from Indochina to Japan and is most pronounced in winter [1]. Kuhnel [4] defined 14 tropical-extratropical cloudbands around the globe (seven in each hemisphere). The NWCB is the third most frequently occurring cloudband globally and the fourth most frequently occurring cloudband in the Southern Hemisphere.

The NWCB is a large cloud structure, which can reach a length of 8000 km from northwest (NW) to southeast (SE)

[5]. It can connect a cold front at its SE end to the intertropical convergence zone at its NW end [6]. If the SE end reaches the Pacific Ocean, the NWCB can also connect to the South Pacific convergence zone [6].

Cloudbands are prominent in Australia during autumn and early winter when the ocean waters around Australia are warm and the subtropical jet stream is strong [1]. The decline in cloudbands in late winter and early spring has been attributed to the cooling of the surrounding oceans [5]. An interannual signal is also present with cloudbands, which may be related to the oceanic conditions in the cloudband formation region [5]. NWCBs are formed through the broad-scale ascent of moist, poleward, and tropical air by a midlatitude baroclinic and could transport latent heat and moisture into higher latitudes [1].

It has been estimated [1] that cloudband-related rainfall provides up to 80% of the annual rainfall for NW Australia and up to 40% of the annual rainfall for SW Australia. Recent positive trends in winter rainfall over northwest Western Australia and Central Australia have been attributed to increased

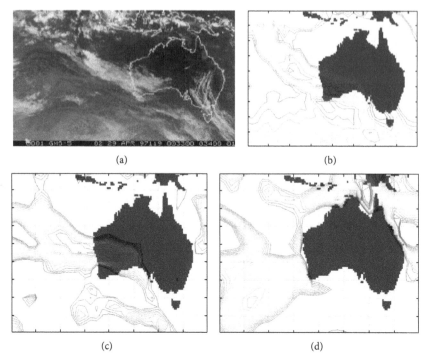

FIGURE 1: (a) Satellite image of a NWCB on 29 April 1996 (image source: Australian Bureau of Meteorology) and corresponding (b) OLR contours ($140$–$250\ \text{W/m}^2$), (c) TCC contours ($70$–$100\%$), and (d) latent heat flux contours ($110$–$300\ \text{W/m}^2$).

northwest cloudband activity [7]. Southern Australia receives most of its cloudband-related rainfall through convective activity, such as cloudbands interacting with cold fronts [1]. Frontal systems produce most of the rainfall that occurs over SW Australia; however, the interaction of these systems with NWCBs increases the rainfall resulting from the frontal system [1]. Thus knowledge of NWCBs and their interactions with other weather phenomena is important for seasonal rainfall forecasting in Australia.

Wright [1] defined "interaction" as the "capture of low-latitude moisture (north of $23°$S), as represented by cloud masses in the circulation of a mid and low-latitude front or depression, and the subsequent amalgamation of mid and low-latitude cloud systems." An interaction is not defined until amalgamation of the mid- and low-latitude cloud systems has occurred or if there is evidence of cyclonic curvature at the SW end of the cloudband, where there are suspected cloudband-associated interactions [1].

*1.1. Climate Influences on Northwest Cloudbands.* The Indian Ocean is not an isolated water body. Teleconnections with other oceans through atmospheric and oceanic circulations affect the Indian Ocean (e.g., [8, 9]). Meehl [10] studied the precipitation, sea level pressure, and sea surface temperature (SST) in the Indo-Pacific to reveal a dynamically coupled ocean-atmosphere system. Variations in this system can affect NWCB formation; however, Webster et al. [11] found no connection between SST and the outgoing longwave radiation (OLR) to the NW of the NWCB formation region in the Indian Ocean. Webster et al.'s findings suggest that

atmospheric dynamical processes or instabilities, not ocean-atmosphere interactions with warm SSTs, mainly produce the convection that occurs in this region.

Two direct influences on NWCB activity are the Madden-Julian oscillation (MJO) and the Indian Ocean dipole. Both phenomena have been observed in the NWCB formation region in the Eastern Indian Ocean. The El Niño southern oscillation also affects NWCBs through tropical forcing.

The MJO, or intraseasonal oscillation, occurs during fluctuations in tropical rainfall on the scale of 30–60 days [12, 13]. These fluctuations are visible in the wind, SST, precipitation, and cloudiness fields [3, 14]. The MJO moves east across the tropics towards the Australian continent every 30–50 days and can set up the synoptic conditions (winds, pressure, convection, and cloudiness) conducive to NWCB development. The MJO also has an influence on cloudbands across Southern Africa [3].

The Indian Ocean dipole (IOD), which occurs every four to five years, is an oscillation of opposing, anomalous SSTs off Sumatra and in the Western Indian Ocean, with accompanying wind and precipitation anomalies [15]. Strong coupling between the air-sea processes is unique to the Indian Ocean; however, this coupling is not completely independent of the El Niño southern oscillation[16]. Ummenhofer et al. [8] found that major droughts in southeast Australia were related to periods when the IOD remained persistently "positive" or "neutral" whilst during a negative phase of the IOD, unusually wet conditions dominated across southern regions of Australia. The dipole mode index [15] is based on monthly SST anomalies between the Western Indian Ocean ($50$–$70°$E, $10$–$10°$N) and the Eastern Indian Ocean ($90$–$110°$E, $10$–$0°$S).

## 2. Methodology

*2.1. Data.* Several data sets were used in this study of NWCBs and their effects on Australian rainfall:

(1) satellite images for 1997 and 1999 from the Australian Bureau of Meteorology,

(2) National Centers for Environmental Prediction (NCEP) reanalysis data for 1950–1999 from the Climate Diagnostics Center, Colorado, including:

  (i) daily total cloud cover (TCC) and latent heat flux on a $2 \times 2°$ grid,

  (ii) six-hour mean sea level pressure; wind fields at 250, 300, 400, 500, 600, and 700 hPa; geopotential heights at 500 and 700 hPa on a $2.5 \times 2.5°$ grid,

(3) daily interpolated OLR on a $2.5 \times 2.5°$ grid for 1974–1998 from the Climate Diagnostics Center, Colorado,

(4) Australian daily rainfall data for 1950–1999 from the Australian Bureau of Meteorology,

(5) one-degree Australian daily rainfall data for 1950–1999 from the Australian Bureau of Meteorology Research Centre (Drosdowsky, pers. comm.); the gridded rainfall was based on the interpolation of the data obtained with the rainfall gauges,

(6) Reynolds reconstructed mean monthly SST for 1950–1999 from the Climate Diagnostics Center, Colorado.

*2.2. Northwest Cloudband Activity.* Satellite data from April to October in 1997 and 1999 were used to analyse NWCBs over Australia. All the days between April and October that had some portion of a NWCB were collated. Cloudband activity days were isolated to determine the atmospheric conditions conducive to cloudband formation (see also [3]). Fifty-three days of cloudband activity were recorded and used for the composite of NWCB activity.

The wind fields, mean sea level pressure, and geopotential heights at 500 hPa over the Eastern Indian Ocean were used to identify the NWCBs; however, it was hard to identify NWCBs using these parameters because NWCBs have similar baroclinic features to a strong frontal system that crosses SW Australia, although the NWCBs did not appear to be connected to this southern frontal system [2]. This preexisting baroclinicity, which the NWCB moves through, is reinforced by the passage of a second cold front through SE Australia [2]. Bell [17] found that this baroclinic instability over Australia caused NWCBs. The OLR distributions were more useful for identifying NWCBs because NWCBs were visible in the OLR fields and distinguishable from other cloud features within the range of 140–250 W/m². 

When the OLR distributions satisfied the prescribed NWCB criteria and some portion was located over the Australian continent the corresponding date was recorded as a "cloudband day" and the NWCB was defined as "mature." The NWCB index is a representation of cloudband days influencing the study region. Surges in tropical energy can reinforce

NWCBs before the NWCBs dissipate [4], which makes it hard to discern whether a cloudband has dissipated and a new one has formed or whether the original cloudband persists. A cloudband's lifetime can vary considerably; thus it was more useful to include the number of days in which a NWCB was over part of Australia than how many cloudbands crossed over the continent.

The criteria for a cloudband day were defined from the studies of Tapp and Barrell [5], Kuhnel [18], and Wright [1] and included the dimensions, orientation, composition, and the formation region as follows:

(1) a cloudband in a NW-SE orientation with a length greater than 2000 km and a width of at least 450 km; this orientation could range from NNW-SSE to WNW-ESE;

(2) a generating or initiation region over the Eastern Indian Ocean (or Northern Australian waters east to 130° E);

(3) a continuous cloud structure, the grid field was $2.5 \times 2.5°$, but the NWCB structure was much larger than the OLR or TCC grid fields and could be located as a continuity in these fields;

(4) a section of the cloudband was located over the Australian continent.

Contour lines of equal OLR or TCC were used to detect NWCBs. These contour lines also had to meet the above criteria before the day was recorded as a cloudband day. The confirmed cloudband days were used to produce a NWCB index for 1974–1999, which was verified with satellite images from 1997 and 1999. To extend the index to 1950, the TCC was used as a proxy for the OLR.

A satellite image of a NWCB on 29 April 1996 and the corresponding OLR contours are shown in Figures 1(a) and 1(b), respectively. The TCC is presented as a percentage between 0 and 100 and treats the atmosphere as one layer; thus NWCBs can be difficult to locate because the cloudband's coherent structure can be obscured. More cloud activity was revealed over the Southern Indian Ocean with the TCC data than with the OLR data, which made it hard to distinguish a clear cloudband perimeter; however, when the latent heat flux data for the Southern Indian Ocean were used, most of this cloud "noise" was screened. The screen was based on a latent heat flux of >100 W/m² in the region of 80–105°E, 22.5–35°S; thus the TCC data for these grid points were removed.

Kuhnel [4] defined the cloudband formation region as 90–145°E; however, the analysis of satellite and OLR data showed that this region did not capture all the mature NWCBs present over Australia. The east end of this region (145°E) captured cloudbands that did not have the characteristic NW-SE orientation of a NWCB. These cloudbands were most likely the continental cloudbands that Wright [1] described. Therefore this study focused on NWCBs in the region of 80–130°E.

*2.3. Northwest Cloudband Activity and Rainfall.* Transects of the cloudband starting at 110°E and moving east were extracted to identify the NWCBs over Australia. The centre of a NWCB was defined as the local minimum in the OLR field and the local maximum in the TCC field and defining the OLR and TCC to the north and south of this point. When the values were within the OLR range (i.e., 140–250 W/m$^2$) and TCC range (60–100%) for a NWCB and generally decreased (OLR) or increased (TCC) from the local minimum, frontal disturbances that were not interacting with the cloudband showed a decrease in OLR and an increase in TCC. The OLR field (with NWCB coverage shaded) for 10 June 1998 is shown in Figure 2(a), and the corresponding land mask of cloudband activity is shown in Figure 2(b). In Figure 2(b), "1" denotes that a NWCB was recorded in the 2.5° grid, and "0" denotes that no NWCB was recorded in that grid.

Two data sets were obtained from the Australian Bureau of Meteorology to analyse the cloudband-related rainfall. The first data set was a daily gridded (1°) rainfall data set (Drosdowsky, pers. comm.), which was used for the large-scale analysis of the cloudband-related rainfall. The second data set consisted of daily rainfall station data. If part of a NWCB was detected in a particular grid, the rainfall that occurred in the corresponding grids was recorded as cloudband-related rainfall. The rainfall gauge data and cloudband-related rainfall data were collated for each month and each NWCB season (April–October) to study the spatial and temporal trends of the NWCBs. If NWCBs were detected in the 2.5° grid containing the gauge, any rainfall recorded that day was classified as cloudband-related. Cloudband-related rainfall data for the NWCB season were collated for each year and compared with the total April–October rainfall data for that gauge.

## 3. Results and Discussion

*3.1. Northwest Cloudband Climatology.* The synoptic conditions associated with a NWCB across Australia generally showed a high-pressure cell over Central and Southern Australia and another pressure cell over the Southern Indian Ocean at around 80°E. Downey et al. [2] also observed these high-pressure features. Between the two pressure cells, a deepening trough was located over the west Australian coast, which connected with a midlatitude depression off the SW coast (see Figure 3(a), which is based on a composite of 327 events of NWCBs between 1979 and 1999). High-pressure systems over the Indian Ocean and Central/SE Australia framed this tropical-extratropical connection of low pressures. A concentration of surface latent heat fluxes was also observed in the NWCB formation region (80–100°E, 10–20°S) (Figure 3(b)).

NWCBs are also associated with an upper-level trough in the Southern Indian Ocean at 80–100°E. This trough is not apparent in the composite of geopotential heights at 500 hPa in Figure 3(c) because of the longwave trough's differing amplitudes. The upper-level wind fields at 500 hPa showed a NW airstream over Australia from the NWCB formation region (Figure 3(d)), which was related to the upper-level trough. The upper-level winds formed a NW-SE "jet" near the NWCB formation region, along which the cloudband followed.

NWCB development is associated with the leading edge of a midlatitude upper-level trough. Downey et al. [2], Bell [17], and Tapp and Barrell [5] found that this upper-level trough, or cut-off low, was generally located off SW Australia. Wright [19] found that the trough's amplitude was correlated with the southern oscillation index (an indicator of the El Niño southern oscillation), an average winter Eastern Indian Ocean cloudiness, and average NW Victorian rainfall. The amplitude and position of this upper-level trough can vary; hence it is not well defined in Figure 3(d). Kuhnel [18] stated that not all upper-level troughs were well defined and that two out of ten upper-level troughs might have small amplitudes. We found that four out of ten upper-level troughs had a NW-SE orientation, which is characteristic of NWCBs.

*3.2. A Northwest Cloudband Index.* The OLR, TCC, and latent heat flux data were used to create a NWCB index for 1950–1999. The seasonal variability of the NWCB index is shown in Figure 4. The five-year, centred, weighted, running mean is included in the figure. Between April and October of 1950–1999, a mature cloudband was recorded over Australia on about 30 days. These NWCBs lasted from a day to more than a week (see Figure 5).

Several extreme NWCB events shown in Figure 4 were related to the El Niño southern oscillation. For example, NWCB activity was low in 1982 and 1986, which corresponded to El Niño years (1982-1983 and 1986-1987); however, NWCB activity was high in 1998, which corresponded to La Niña years (1998-1999). Indonesian throughflow-induced SST anomalies [20] in the NWCB formation region may have contributed to this relationship between the El Niño southern oscillation and NWCB activity.

Kuhnel [18] showed that Western Australia experienced about 15 cloudband days a year and Central and South Australia experienced about nine cloudbands a year, which showed that cloudband activity over Australia decreased from west to east. The mean seasonal NWCB frequencies over Australia for 1979–1999 are shown in Figure 6. The figure shows that the NWCB activity had a NW-SE bias. The state average in this study was slightly higher than that in Kuhnel's [18] study. The high cloudband activities in 1992, 1993, and 1998 caused this increase in the average.

Hirst and Godfrey [21] showed that the SSTs in the Eastern Indian Ocean were linked to the Pacific Ocean via subsurface baroclinic waves. They found regions of convective mixing or strong upwelling that were susceptible to small changes in the throughflow. Variations in the throughflow affect the pattern and response time of the subsurface (and therefore surface) temperature. During El Niño, cool SSTs feed through the Indonesian throughflow into the Eastern Indian Ocean [22], whereas, during La Niña, warm SSTs feed into the Eastern Indian Ocean [22]. These SST anomalies were observed in the NWCB formation region.

The warmer SSTs during La Niña may increase cloud-band activity through enhanced convection; conversely, the cooler SSTs during El Niño may decrease cloudband

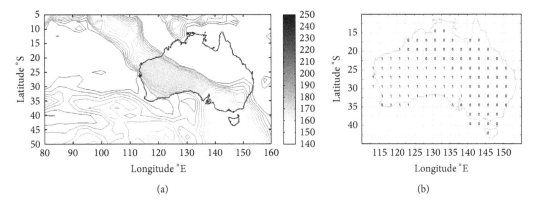

(a)  (b)

FIGURE 2: (a) The OLR field (with NWCB coverage shaded) for 10 June 1998 (OLR in W/m$^2$). (b) The land mask of NWCB activity across Australia used to create the regional NWCB indices. "1" denotes NWCB coverage for that grid.

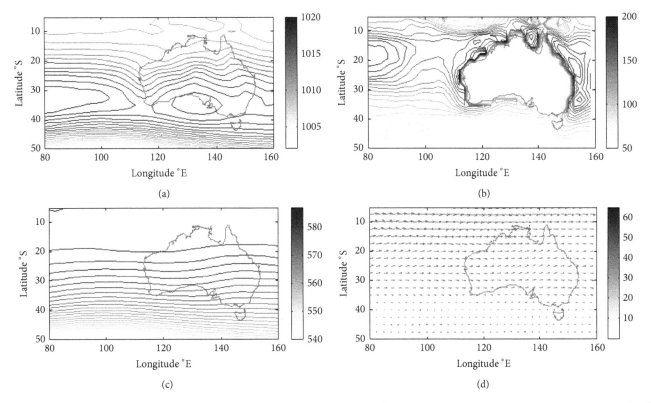

(a)  (b)

(c)  (d)

FIGURE 3: Composites of 327 NWCB events between 1979 and 1998 showing conditions associated with NWCB activity: (a) mean sea level pressure in hPa; (b) latent heat fluxes in W/m$^2$; (c) geopotential height at 500 hPa in dm; (d) upper-level wind fields at 500 hPa.

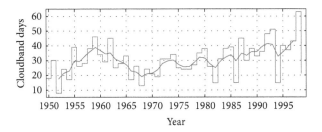

FIGURE 4: The NWCB index for Australia. The index represents the number of days in a season (April–October) in which a NWCB was located over part of Australia. A five-year, weighted, running mean was fitted to this index.

activity through reduced convection. Wright [1] showed that enhanced cloudband activity was associated with warm SST anomalies in the NWCB formation region and cool SST anomalies directly to the SW of this region. This reflects similar regions in an Indian Ocean dipole defined by Nicholls et al. [25].

*3.3. Factors Influencing Northwest Cloudband Activity.* Meehl [23] showed that SST anomalies on the scale of a season, associated with wind anomalies, affected air-sea interactions (especially evaporation) and therefore the latent heat release. Strong winds and warm SSTs increase convection

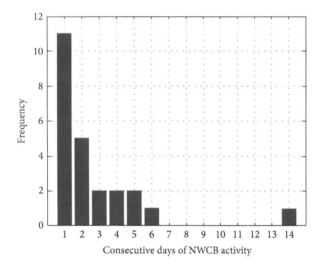

FIGURE 5: The frequency of individual NWCBs over Australia in 1998.

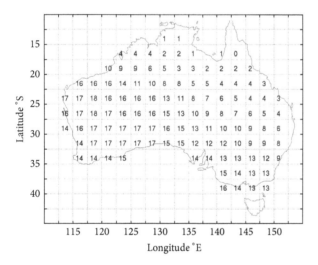

FIGURE 6: The mean seasonal (April–October) NWCB frequencies for 1979–1999.

TABLE 1: Mean monthly (April–October) NWCB days over Australia for 1950–99.

| | Mar. | Apr. | May | June | July | Aug. | Sep. | Oct. |
|---|---|---|---|---|---|---|---|---|
| NWCB days | 5.5 | 7.9 | 7.9 | 5.9 | 3.7 | 2.3 | 1.0 | 1.8 |

TABLE 2: The maximum monthly cross-correlations between NWCB activity and the southern oscillation index (SOI), GEN, the dipole mode index (DMI), the Madden-Julian oscillation (MJO), and Nino12.

| Index with NWCB | Correlation (r) | Correlation month |
|---|---|---|
| SOI | −0.52 | Apr. |
| GEN | 0.61 | Apr. |
| DMI | −0.60 | July |
| MJO | −0.48 | July |
| Nino12 | 0.66 | Apr. |

Kuhnel [4] found a correlation of −0.36 between NWCB activity and the southern oscillation index (SOI); however, this correlation was based on a slightly different study region and a single year of cloudband activity. The April–October NWCB activity accounted for about two-thirds of the cloudband activity that Kuhnel [4] described. The correlation between NWCB activity and the southern oscillation index in April was −0.52 (Table 2). This inverse relationship suggested that a positive southern oscillation index in April was linked to less NWCB activity and a negative southern oscillation index was linked to more NWCB activity. The SST anomalies in the eastern equatorial Pacific Ocean (Nino12: 80–100°W, 5°S–5°N) in April were also related to NWCB activity (correlation of 0.66). These results showed that the El Niño southern oscillation in April was linked to cloudband activity in the eastern Indian Ocean.

Kuhnel [4] defined cloudband activity as a function of the southern oscillation index, which determines the strength and position of the main convective regions. Fluctuations in convective regions affect NWCB formation. The NWCB formation region is located where the 40–50 day oscillation (i.e., the MJO) has been observed.

Tapp and Barrell [5] and Kuhnel [4] stated that the NWCB season was between April and October. The results showed more cloudbands were usually present in the first half of the season (April–July). The mean monthly NWCB days for 1950–1999 are shown in Table 1. Cloudband activity generally peaked in May and then decreased over the following months. These results agreed with those of Kuhnel [4] and Wright [1]. Cloudband activity occurred in March but was more irregular and not usually associated with rainfall. Cloudbands present in the warmer months (November–March) are associated with monsoonal depressions or tropical cyclones NW of Australia [4] and are not considered NWCBs.

The duration of a NWCB over Australia varies but can sometimes be up to two weeks (e.g., the NWCB that occurred in 1998, which appeared on 30 March and lasted until 12 April; see Figure 4). If the synoptic conditions are baroclinic, NWCBs can appear consecutively for more than a week. If the baroclinicity is reduced, a cloudband may still form but

and evaporation and decrease the sea level pressure. These actions, which are characteristic of monsoons, encourage NWCB formation. Weak winds then decrease evaporation and increase the sea level pressure the following year. This quasi-biennial feature was sometimes observed in the NWCB index (1985 to 1990 in Figure 4). Meehl [24] noted that a tropospheric biennial mechanism involving air-sea interactions was connected with the South Asian or Indian monsoon. This biennial mechanism is quasi-biennial, occurring intermittently or over several years (also apparent in Figure 4 as a noncontinuous biennial feature).

The SST anomalies in the NWCB formation region of 90–100°E, 10–15°S, in the Eastern Indian Ocean, are referred to as GEN. The correlation between NWCB activity and GEN was 0.61 in April—one of the highest NWCB activity months. The atmospheric dynamical processes might have caused this correlation, as the SST and OLR in a region next to the NWCB formation region were uncorrelated [11].

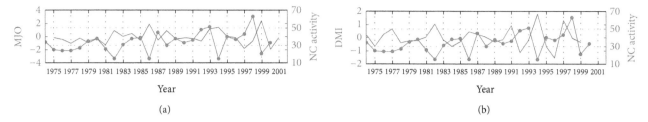

FIGURE 7: (a) The mean MJO anomaly and seasonal (April–October) NWCB frequency for 1975–1999 (MJO in W/m$^2$, based on OLR values) and (b) the mean dipole mode index anomaly and seasonal (April–October) NWCB frequency for 1975–1999. The dipole mode index represents the temperature anomaly difference (°C).

will usually dissipate after 24 hours. The frequency of NWCB days over Australia between April and October 1998 is shown in Figure 4. The synoptic conditions in April, which favoured cloudband activity, resulted in fourteen days of cloudband activity, followed by more cloudband activity less than a week later. Towards the end of the NWCB season, conditions were unfavourable for cloudband activity and cloudbands tended to dissipate after a day.

An inverse relationship between NWCB activity and the MJO is shown in Figure 7(a). The MJO index was created using OLR anomalies in the region of 90–100°E, 5°S–5°N. The MJO mainly occurs to the NW of the NWCB formation region, but also crosses into the NWCB formation region [14]. An increase in MJO activity was linked to a decrease in NWCB activity over Australia and more cloudiness (hence lower OLR) in the eastern Indian Ocean. The relationship between NWCB activity and the MJO was strongest in July (correlation of −0.48; see Table 2).

The NWCB formation region overlaps with the eastern part of the Indian Ocean dipole region. An inverse relationship was found between NWCB activity and the Indian Ocean dipole. The dipole mode index measures the difference in SST between the West and East Indian Ocean and is used to measure the Indian Ocean dipole. The Indian Ocean dipole event of 1994 coincided with one of the lowest recorded NWCB activity seasons. Low NWCB activity also coincided with a peak in warm SSTs in the Western Indian Ocean and cool SSTs in the Eastern Indian Ocean in 1982; this year also corresponded with the 1982-1983 El Niño event. The relationship between NWCB activity and the Indian Ocean dipole was strongest in July (correlation of −0.60; see Table 2).

*3.4. Australian Rainfall and Northwest Cloudbands.* NWCBs produce high rainfall amounts in Australia between April and October. In some inland regions where the annual rainfall is low (e.g., <200 mm/year), the contribution of >50 mm of rainfall from a cloudband can be significant. The interactions of NWCBs with other features, such as troughs, fronts, and convective activity, can also generate rainfall.

Two years of extreme cloudband activity, 1982 and 1998, were chosen for analysis. Cloudband activity was low in 1982, with only 15 NWCB days recorded. The cloudbands were mostly limited to the west and south coasts, with few NWCBS recorded in Central Australia (Figure 8(a)).

Cloudband-related rainfall was also mostly limited to coastal areas, especially the west coast (Figure 8(b)). Most of the inland regions received less than 30 mm of cloudband-related rainfall between April and October.

Cloudband activity was high in 1998, with 63 NWCB days recorded. The cloudband coverage was more widespread in 1998 (Figure 9(a)) than it was in 1982. The cloudband contribution to inland rainfall, especially over Western Australia, is shown in Figure 9(b). Some of the inland rainfall stations received 65% of their April–October rainfall from cloudband-related activity (see next section). Most of the rainfall stations in Western Australia received >100 mm of cloudband-related rainfall between April and October, which is a significant amount for the inland stations.

*3.5. Case Study: Southwest Australian Rainfall.* Winter rainfall in SW Australia has declined more than 25% since the 1950s [25]. The decline has affected farming and other industries in coastal and inland regions. Complex interactions between atmospheric and oceanic circulations affect rainfall in the SW, with cold fronts producing most of the winter rainfall. NWCBs and convective activity also generate winter rainfall, but to a lesser degree [26].

Wright [19] and Kuhnel [18] discussed the importance of NWCBs as rainfall generators. Cloudbands can cause heavy rainfall, especially when they interact with an extratropical front (e.g., the cloudband activity in NW Victoria discussed in [19]); however, this heavy rainfall is irregular. A study of the rainfall gauge data associated with cloudband activity revealed this irregularity. The SW rainfall stations were chosen from the high-quality data set that Lavery et al. [27] described. Of these stations, rainfall gauges that were spread evenly over the SW and were 100% complete for the study period (1950–1999) were selected. The high-quality rainfall stations for the SW Australian region are shown in Figure 10, with the shaded circles denoting the stations chosen for cloudband analysis. These nine rainfall stations, along with their latitudes and longitudes, are listed in Table 3.

When NWCBs cross SW Australia and interact with frontal systems, heavy, widespread rainfall can occur. Cloudbands, and their interactions with cold fronts, account for up to 25% of the SW winter rainfall [1]. The results from this study showed cloudband-related rainfall was sometimes as high as 65%, as seen with station 12052 (Menzies) in 1998 (Figure 10). The cloudband-related rainfall, however,

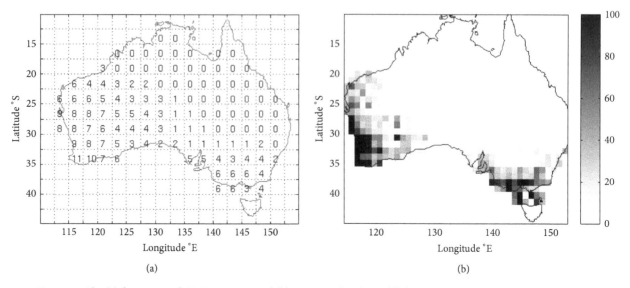

FIGURE 8: The (a) frequency of NWCB activity and (b) NWCB-related rainfall for 1982 (a low cloudband activity year).

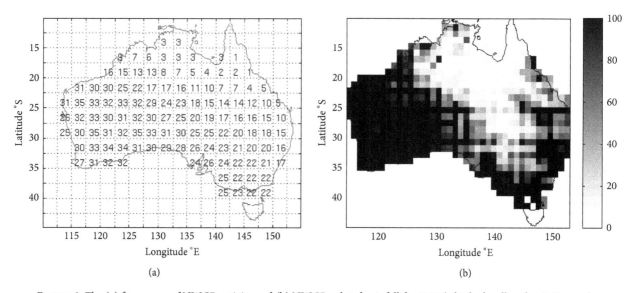

FIGURE 9: The (a) frequency of NWCB activity and (b) NWCB-related rainfall for 1998 (a high cloudband activity year).

TABLE 3: Details of the SW rainfall stations.

| Station number | Station name | Latitude (°S) | Longitude (°E) |
|---|---|---|---|
| 8147 | Yuna | 28.3261 | 114.9572 |
| 9519 | Cape Naturaliste | 33.5381 | 115.0183 |
| 9557 | Hopetoun P.O. | 33.9511 | 120.1247 |
| 9564 | King River | 34.9444 | 117.9228 |
| 9616 | Westbourne | 34.0928 | 116.6603 |
| 10092 | Merredin | 31.4769 | 118.2778 |
| 10592 | Lake Grace P.O. | 33.1017 | 118.4608 |
| 10795 | Avondale Research Station | 32.1189 | 116.8664 |
| 12052 | Menzies | 29.6933 | 121.0286 |

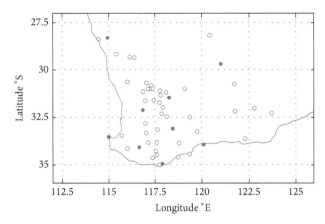

FIGURE 10: The locations of high-quality rainfall stations for the SW Australian region (courtesy of the Australian Bureau of Meteorology). The shaded circles denote the nine rainfall stations used in this study.

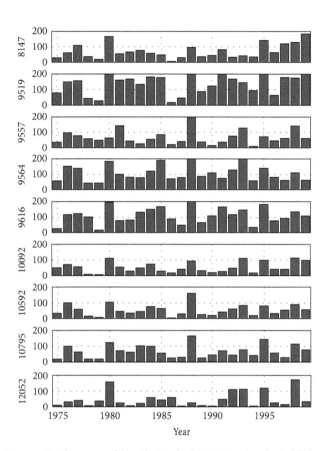

FIGURE 11: The seasonal (April–October) NWCB-related rainfall for 1975–1999 recorded at the nine SW rainfall stations.

was highly variable. For Menzies, for example, the percentage of cloudband-related rainfall was sometimes less than 5% of the total April–October rainfall.

Over SW Australia, fronts provide a form of uplift for NWCBs. The rainfall produced as a result of cold fronts is much less than that produced from the interaction of cold fronts with NWCBs. Cloudband-related rainfall can exceed 75 mm a day, whereas frontal activity may produce only 25 mm of rainfall a day. Kuhnel [18] and Wright [1] showed the rain-producing capacity of a cloudband decreased towards the south and east. In SW Australia, the interaction of cold fronts with NWCBs produced most of the cloudband-related rainfall. In the northern regions (e.g., the Gascoyne-Murchison region), cloudband-related rainfall may occur without these interactions.

The seasonal (April–October) NWCB-related rainfall data for 1975–1999 for the nine stations listed in Table 3 are shown in Figure 11. The coastal stations generally recorded higher rainfall than the inland stations. There was no corresponding upward trend in the rainfall compared with the upward trend in cloudband activity. Between 1975 and 1999, unusually heavy rainfall was recorded in 1980, 1988, and 1998, with 1988 and 1998 corresponding to La Niña years.

The NWCB-related rainfall contributions to the total April–October rainfall varied seasonally and spatially (Figure 12). Years that seem to have no cloudband contribution to rainfall (i.e., 0%) were due to missing seasonal data. Most of the rainfall stations that recorded high rainfall amounts were located near the coast (e.g., station 9557 in 1980). Cloudbands and their interactions with meteorological features contributed to 15–30% of the April–October rainfall. In an extreme year (e.g., 1980), up to 75% of the seasonal rainfall could be attributed to cloudbands and their interactions with meteorological features.

## 4. Conclusions

This study examined NWCB activity over Australia from 1975 to 1999 and an index representing cloudband activity over Australia between 1974 and 1999 was created and dissected to create regional indices, which showed the NWCB activity had a NW-SE bias. Atmospheric and oceanic influences on cloudband activity were examined and revealed that NWCB activity was related to the MJO, Indian Ocean dipole, El Niño southern oscillation, and SSTs in the NWCB formation region. These influences were not independent; thus NWCB activity over Australia was hard to predict.

The NWCB-related rainfall data were collated from the regional indices to show the cloudband-related rainfall for each 2.5° grid point across Australia. Most cloudband-related rainfall occurred in coastal regions, where the interaction of NWCBs with other features, such as frontal disturbances, increased the rainfall. Cloudband-related rainfall was also found to be spatially and temporally variable.

## Conflict of Interests

The authors declare that there is no conflict of interests regarding the publication of this paper.

## Acknowledgments

The Gascoyne-Murchison Strategy and the Climate Variability in Agriculture Program funded this research. The authors would like to thank the Climate Diagnostics Centre, Boulder,

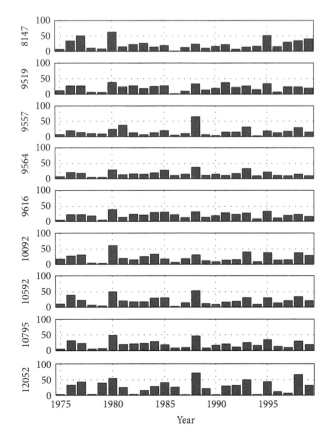

FIGURE 12: Percentages of NWCB-related rainfall contributions to the total April–October rainfall for 1975-1999 for the nine SW rainfall stations.

for kindly providing the Reynolds reconstructed SST and reanalysis data, and the Australian Bureau of Meteorology for their rainfall data sets and processed GMS-5 satellite images (courtesy of the Japan Meteorological Agency). They would also like to thank the following people for their advice and guidance with this research: Dr. Neville Nicholls and Dr. Wasyl Drosdowsky, Australian Bureau of Meteorology Research Centre; Dr. William Wright, NCC; John Cramb and Joseph Courtney, Australian Bureau of Meteorology; Dr Gary Meyers, CSIRO Marine Research; Dr. Ian Smith, CSIRO Division of Atmospheric Research; Professor Greg Ivey, UWA; and Alisa Krasnostein, UWA.

# References

[1] W. J. Wright, "Tropical-extratropical cloudbands and Australian rainfall: I. Climatology," *International Journal of Climatology*, vol. 17, no. 8, pp. 807–829, 1997.

[2] W. K. Downey, T. Tsuchiya, and A. J. Schreiner, "Some aspects of a northwestern Australian cloudband," *Australian Meteorological Magazine*, vol. 29, no. 3, pp. 99–113, 1981.

[3] N. C. G. Hart, C. J. C. Reason, and N. Fauchereau, "Cloud bands over southern Africa: seasonality, contribution to rainfall variability and modulation by the MJO," *Climate Dynamics*, vol. 41, no. 5-6, pp. 1199–1212, 2013.

[4] I. Kuhnel, "Tropical-extratropical cloudband climatology based on satellite data," *International Journal of Climatology*, vol. 9, no. 5, pp. 441–463, 1989.

[5] R. G. Tapp and S. L. Barrell, "The north-west Australian cloud band: climatology, characteristics and factors associated with development," *Journal of Climatology*, vol. 4, no. 4, pp. 411–424, 1984.

[6] B. Geerts and E. Linacre, "Northwest cloud bands," University of Wyoming, 1998, http://www-das.uwyo.edu/~geerts/cwx/notes/chap08/nw_cloud.html.

[7] J. S. Frederiksen and C. S. Frederiksen, "Twentieth century winter changes in Southern Hemisphere synoptic weather modes," *Advances in Meteorology*, vol. 2011, Article ID 353829, 16 pages, 2011.

[8] C. C. Ummenhofer, M. H. England, P. C. McIntosh et al., "What causes southeast Australia's worst droughts?" *Geophysical Research Letters*, vol. 36, no. 4, Article ID L04706, 2009.

[9] A. Hoell, M. Barlow, and R. Saini, "Intraseasonal and seasonal-to-interannual indian ocean convection and hemispheric teleconnections," *Journal of Climate*, vol. 26, no. 22, pp. 8850–8867, 2013.

[10] G. A. Meehl, "The annual cycle and interannual variability in the tropical Pacific and Indian Ocean regions," *Monthly Weather Review*, vol. 115, no. 1, pp. 27–50, 1987.

[11] P. J. Webster, V. O. Magaña, T. N. Palmer et al., "Monsoons: processes, predictability, and the prospects for prediction," *Journal of Geophysical Research C: Oceans*, vol. 103, no. 7, pp. 14451–14510, 1998.

[12] R. A. Madden and P. R. Julian, "Observations of the 40–50-day tropical oscillation—a review," *Monthly Weather Review*, vol. 122, no. 5, pp. 814–837, 1994.

[13] A. C. Subramanian, M. Jochum, A. J. Miller, R. Murtugudde, R. B. Neale, and D. E. Waliser, "The Madden-Julian oscillation in CCSM4," *Journal of Climate*, vol. 24, no. 24, pp. 6261–6282, 2011.

[14] P. J. Webster, "The Asian-Australian monsoon system: a prospectus for a research program. Draft recommendations from the St Michaels, Maryland, monsoon workshop July 28-30, 1998," U.S. Climate Variability and Predictability Research Program (CLIVAR) Working Group Report, 1998.

[15] N. H. Saji, B. N. Goswami, P. N. Vinayachandran, and T. Yamagata, "A dipole mode in the tropical Indian ocean," *Nature*, vol. 401, no. 6751, pp. 360–363, 1999.

[16] S. Sankar, M. R. R. Kumar, and C. Reason, "On the relative roles of El Nino and Indian Ocean Dipole events on the monsoon onset over Kerala," *Theoretical and Applied Climatology*, vol. 103, no. 3-4, pp. 359–374, 2011.

[17] I. D. Bell, *Extratropical cloud systems: satellite interpretation and three-dimensional airflows [Ph.D. thesis]*, Monash University, Melbourne, Australia, 1982.

[18] I. Kuhnel, "Tropical-extratropical cloudbands in the Australian region," *International Journal of Climatology*, vol. 10, no. 4, pp. 341–364, 1990.

[19] W. J. Wright, *Synoptic and climatological factors influencing winter rainfall variability in Victoria [Ph.D. thesis]*, The University of Melbourne, Melbourne, Australia, 1987.

[20] J. S. Godfrey, "The effect of the Indonesian throughflow on ocean circulation and heat exchange with the atmosphere: a review," *Journal of Geophysical Research C: Oceans*, vol. 101, no. 5, pp. 12217–12237, 1996.

[21] A. C. Hirst and J. S. Godfrey, "The response to a sudden change in Indonesian throughflow in a global ocean GCM," *Journal of Physical Oceanography*, vol. 24, no. 9, pp. 1895–1910, 1994.

[22] G. Meyers, "Variation of Indonesian throughflow and the El Niño-Southern Oscillation," *Journal of Geophysical Research C: Oceans*, vol. 101, no. 5, pp. 12255–12263, 1996.

[23] G. A. Meehl, "A coupled air-sea biennial mechanism in the tropical Indian and Pacific regions: role of the ocean," *Journal of Climate*, vol. 6, no. 1, pp. 31–41, 1993.

[24] G. A. Meehl, "Coupled land-ocean-atmosphere processes and South Asian monsoon variability," *Science*, vol. 266, no. 5183, pp. 263–267, 1994.

[25] N. Nicholls, L. Chambers, M. Haylock, C. Frederiksen, D. Jones, and W. Drosdowsky, *Climate Variability and Predictability for South-West Western Australia: Phase 1 Report to the Indian Ocean Climate Initiative*, Bureau of Meteorology Research Centre, Melbourne, Australia, 1999.

[26] J. Gentilli, *Australian Climate Patterns*, Thomas Nelson, Melbourne, Australia, 1972.

[27] B. Lavery, G. Joung, and N. Nicholls, "An extended high-quality historical rainfall dataset for Australia," *Australian Meteorological Magazine*, vol. 46, no. 1, pp. 27–38, 1997.

# 6

# Analysis and Comparison of Trends in Extreme Temperature Indices in Riyadh City, Kingdom of Saudi Arabia, 1985–2010

## Ali S. Alghamdi[1,2] and Todd W. Moore[2]

[1] *Department of Geography, King Saud University, P.O. BOX 2456, Riyadh 11451, Saudi Arabia*
[2] *Department of Geography and Environmental Planning, Towson University, 8000 York Road, Towson, MD 21252, USA*

Correspondence should be addressed to Ali S. Alghamdi; aorifi@ksu.edu.sa

Academic Editor: Ines Alvarez

This study employed the time series of thirteen extreme temperature indices over the period 1985–2010 to analyze and compare temporal trends at two weather stations in Riyadh city, Saudi Arabia. The trend analysis showed warming of the local air for the city. Significant increasing trends were found in annual average maximum and minimum temperatures, maximum of minimum temperature, warm nights, and warm days for an urban and a rural station. Significant decreasing trends were detected in the number of cool nights and cool days at both stations. Comparison of the trends suggests that, in general, the station closer to the city center warmed at a slower rate than the rural station. Significant differences were found in a lot of the extreme temperature indices, suggesting that urbanization and other factors may have had negative effects on the rate of warming at the urban station.

## 1. Introduction

A better understanding of trends in local extreme temperature has potential benefits to many practical problems. Increasing extremely high temperatures, for example, directly affect energy consumption [1]. In summer 2010, which was the warmest season in Saudi Arabia's record with temperatures reaching 52°C in Jeddah city, eight power plants throughout Saudi Arabia were forced to shut down, resulting in a loss of power in several cities [2], leaving people exposed and vulnerable to the extreme temperature. If observed trends indicate that such extreme temperatures are becoming more frequent, energy providers may be able to adapt their electricity supply to minimize the likelihood of widespread power outage. Evidence that localized high temperatures are becoming less frequent, on the other hand, along with assessment of possible causes of the cooling, can be used to develop and implement urban planning strategies to mitigate local- and larger-scale warming [3–5].

Multiple recent studies have been dedicated to gaining a better understanding of mean and extreme temperature

patterns and trends in Saudi Arabia (e.g., [2, 6–12]). One study [7] reported that annual mean air temperature in Saudi Arabia increased at an average rate of 0.60°C decade$^{-1}$ over the period 1978–2009 and another one [8] showed that the increasing trend was robust across the seasons, although the rate of change did vary seasonally. Annual maximum and minimum temperature also increased over the period 1978–2009 [8]. Other extreme temperature indices, such as extremely hot days (maximum temperature >90th percentile), warm spells (6 consecutive days when maximum temperature >90th percentile), and tropical nights (daily minimum >20°C) also had increasing trends between 1981 and 2010 whereas the number of hot nights (minimum temperature >90th percentile) and cool days/nights (maximum temperature <10th percentile/minimum temperature <10th percentile) had decreasing trends [9]. Another study [12] also reported trends in a suite of extreme temperature indices over the period 1979–2008, including increasing trends in warm spells, annual summer days (daily maximum temperature >35°C), tropical nights (daily minimum temperature >25°C), annual maximum and minimum values, annual mean values of maximum and minimum temperatures, diurnal temperature range, warm days (maximum temperature >90th percentile) and warm nights (minimum temperature >90th percentile), and decreasing trends in cool days (maximum temperature <10th percentile) and cool nights (minimum temperature <10th percentile).

Only a couple of studies have focused on local extreme temperature trends in Saudi Arabia. One study [6] focused on temperature variation over Dhahran, on the East Coast of Saudi Arabia, for the period 1970–2006 and reported that the number of hot days (maximum temperature $\geq 35°C$) and hot nights (minimum temperature $> 20°C$) increased by $0.828\,\mathrm{d\,yr^{-1}}$ and $0.552\,\mathrm{d\,yr^{-1}}$, respectively, while the frequency of cold days (daily maximum temperature $\leq 20°C$) and cold nights (daily minimum temperature $\leq 15°C$) decreased by $0.603\,\mathrm{d\,yr^{-1}}$ and $0.35\,\mathrm{d\,yr^{-1}}$, respectively. Another study [11] reported that the number of hot nights decreased by $0.525\,\mathrm{d\,yr^{-1}}$ and the number of hot days increased by $2.2\,\mathrm{d\,yr^{-1}}$ in Jeddah City, on the West Coast, over the period 1970–2006. In addition, [11] found that monthly and annual mean maximum temperatures increased more than minimum temperatures.

Temperature trend studies tend to focus on regional or larger scale patterns in mean and extreme temperature, but previous studies (e.g., [7, 13–15]) have shown that trends can vary over relatively small scales. In Saudi Arabia, [7] illustrated that mean temperature has not increased uniformly and [6, 11] reported disparate results in terms of the number of hot nights for Dhahran and Jeddah City. Variability in temperature trends at the micro-, local-, and mesoscales suggests that localized forcing may have a greater impact than regional and global forcing [14]. Urbanization (i.e., the urban heat island (UHI)), for example, has been shown to influence trends in time series of extreme temperatures recorded at urban stations relative to nearby rural stations [16, 17]. UHI is perhaps most well-known for its creation of warm urban "islands" in the surrounding "sea" of relatively cool rural areas [18]. Urbanization, however, does not have the same effect on temperature in all climates. In arid and semiarid climates, for example, urbanized cities often do not exhibit a strong UHI effect and some even are cooler than the surrounding areas (UHI sink) [19–23].

A previous study [24] found evidence of the UHI in Riyadh city when examining air temperature trends at four stations in Saudi Arabia and proposed that the observed increase in annual mean temperature in Riyadh city was likely due to the UHI effect. Another recent study [10] examined the effect of urbanization on air temperature trends in Saudi Arabia by associating population change and air temperature change and concluded that urbanization has had very little effect because the temperature significantly increased regardless of the population changes. These disparate findings warrant additional study to determine whether urbanization has substantially contributed to the warming observed in Saudi Arabia. There also remains the possibility to be explored that increased urbanization has lessened the larger-scale warming trend, as is the case in UHI sinks [19–23].

Riyadh city, the capital of the Kingdom of Saudi Arabia, has experienced significant population growth during the last decades. In 1952, the population was about 80 000 and, by 2006, it had increased to 4 600 000 [25]. The built environment of Riyadh covered less than $3.5\,\mathrm{km^2}$ before the 1950s, but today the built environment covers about $1785\,\mathrm{km^2}$ [25, 26]. Such population growth and urban expansion in Riyadh city may have been accompanied by changes in the local extreme temperature patterns. This study first analyzes temporal trends in thirteen extreme temperature indices over the period 1985–2010 and then compares the observed trends between two weather stations in Riyadh city to estimate the possible influence of urbanization on the trends.

## 2. Data and Methods

*2.1. Data Sources and Quality Control.* Two or more stations with relatively long time series, representing urban and rural areas, are needed to investigate the effect of urbanization on extreme temperature trends. Air temperature data for two weather stations in Riyadh city were obtained from the Saudi Presidency of Meteorology and Environment (SPME) as daily maximum and minimum temperature over the period 1985–2010 (Figure 1). The period of record for this study begins in 1985 because this is the year in which the King Khalid Airport station (New Station) began recording observations. These two stations are the only stations representative of Riyadh city that provide a sufficiently long time series suitable for temporal trend analysis and, furthermore, their time series' have been shown to be homogeneous (e.g., [9, 10, 27, 28]).

Quality control was applied to the daily data series to detect missing data and errors using RClimdex software [29] to obtain the highest level of accuracy for the analysis. The extreme temperature indices that were calculated and analyzed (Table 1) were drawn from a list of indices recommended by the Expert Team of Climate Change Detection and Index (ETCCDI) that have been used by numerous studies (e.g., [9, 12, 17, 30–32]).

*2.2. Trend Detection and Comparison.* Similar to [17, 30], the nonparametric Kendall-tau test was used to detect and assess the statistical significance of the linear trends and the least squares method, which is one of the most widely used methods of estimating the slope, or rate of change, of linear trends [33, 34], was used to estimate the slope of the linear trends. The Kendall-tau test is more powerful than the $t$-test when the data are skewed [35]. In addition to trends that are significant at the 99% and 95% level, those significant at the 90% level also are presented to minimize the likelihood of a Type II error (i.e., failing to reject the null hypothesis of no trend when one actually exists) while still maintaining confidence in the results. The coefficient of variation ($C_v$), defined as the ratio of the standard deviation to the mean, was calculated to describe the interannual variability of the extreme temperature indices.

The urbanization effect ($\Delta X_{\mathrm{ur}}$) analysis presented by [17] can be used to detect significant changes between two trends and was thus used to compare the trends of the two Riyadh city weather stations. Although both stations are located at

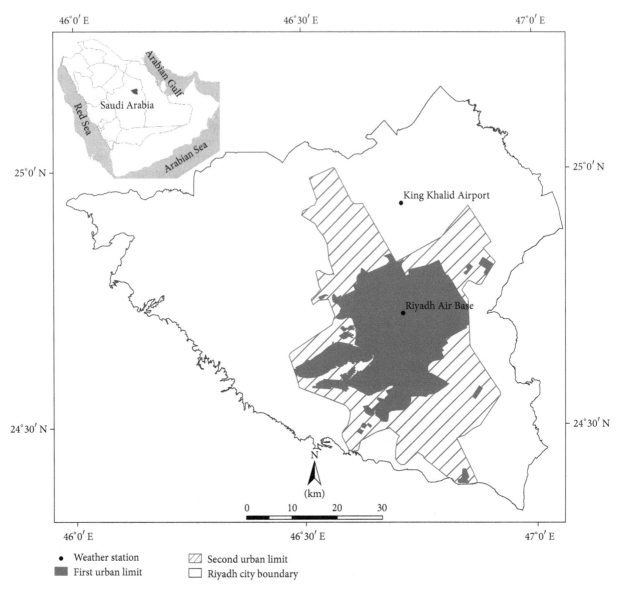

FIGURE 1: Maps showing the study area and the weather stations. Source: adapted from High Commission for the Development of Riyadh.

airports, they can be thought to represent relatively urban and rural locations. Riyadh Air Base (Old Station) (urban station) is located within the first urban limit at a distance of 150 m from the city edge and within the urban affected area whereas King Khalid Airport (New Station) (rural station) is approximately 20 km northeast of Riyadh Air Base (Old Station) beyond the second urban limit and is remote from urbanization [10, 24] (Figure 1); [24] used similar reasoning when using these two stations to study the UHI effect of Riyadh city.

$\Delta X_{ur}$ can be quantified with the following equation [16, 17, 36]:

$$\Delta X_{ur} = X_u - X_r, \tag{1}$$

where $\Delta X_{ur}$ is the urbanization effect, $X_u$ is the slope coefficient from the linear regression run on an extreme

temperature index at an urban station, and $X_r$ is the slope coefficient from the linear regression run on an extreme temperature index at a rural station. If $\Delta X_{ur} > 0$, the urbanization effect resulted in greater warming amid lesser warming or lesser cooling amid greater cooling, at the urban station; if $\Delta X_{ur} < 0$, the urbanization effect resulted in lesser warming amid greater warming or greater cooling amid lesser cooling, at the urban station; if $\Delta X_{ur} = 0$, there is no difference in the rates of change, and so the urbanization effect had no influence on the urban station's trend. Another way to estimate $\Delta X_{ur}$ is to create a time series of the difference in the annual mean values of the extreme temperature indices between Riyadh Air Base (Old Station) (the urban station) and King Khalid Airport (New Station) (the rural station) [36]. The same trend analysis methods previously discussed can then be applied to this time series—the Kendall-tau test can be used to assess the linear trend in the time series for statistical significance and the slope coefficient of the least squares linear regression approximates the difference ($\Delta X_{ur}$) derived from (1) [36].

In addition to detecting significant differences between two trends, (2) can be used to represent the level of the urbanization effects contributions to the overall trend of an extreme temperature index series ($E_u$) and can be expressed as a percentage as follows [16, 17, 36]:

$$E_u = \left| \frac{\Delta X_{ur}}{X_u} \right| \times 100. \tag{2}$$

The outcome of (2) will range from 0%, indicating no contribution, to 100%, indicating high contribution. Values of $E_u$ that exceed 100% reflect the influence of unknown local factors or data errors [17, 36].

# 3. Results

*3.1. Trends in Extreme Value Indices.* TXa increased significantly at both stations, at a rate of $0.45°C$ decade$^{-1}$ at the urban station and a rate of $0.69°C$ decade$^{-1}$ at the rural station (Figure 2(a)). TNa similarly increased significantly at urban and rural station, at rates of $0.68°C$ decade$^{-1}$ and $0.83°C$ decade$^{-1}$, respectively (Figure 2(b)). Comparison of TNa between the urban and the rural station illustrates well the existence of the nocturnal UHI in Riyadh city. Although TNa is greater at the urban station than at the rural station for any given year, the rate at which TNa has increased at the rural station is greater than the rate at which it increased at the urban station. DTR decreased over the period of record at both stations, although statistically significant trends were not detected at either station (Figure 2(c)). The decrease in DTR is caused by TNa increasing at a greater rate than TXa. Furthermore, because TNa is greater at the urban station and TXa is nearly the same at both stations, DTR is less at the urban station than at the rural one.

TXx increased significantly at the rural station at a rate of $0.49°C$ decade$^{-1}$, but no trend was detected at the urban station (Figure 3(a)). TNx increased significantly at the urban and rural station at rates of $0.96°C$ decade$^{-1}$ and $1.01°C$ decade$^{-1}$, respectively (Figure 3(b)). No trends were detected in TXn or TNn (Figures 3(c) and 3(d)). Based on these trends, or lack thereof, the observed increasing trends in TXa and TNa appear to be more the result of increases in maximums rather than minimums. TXx and TNx also displayed less interannual variability than TXn and TNn.

*3.2. Trends in Relative Indices.* Both urban and rural sites had significant increasing trends in TX90p, with rates of $2.20$ d decade$^{-1}$ and $3.85$ d decade$^{-1}$, respectively (Figure 4(a)). Both stations also experienced a notable leap in their TX90p in 1998 and recorded their highest TX90p in 1999 and the lowest one in 1992. TN90p also increased significantly at both the urban and the rural stations at rates of $3.27$ d decade$^{-1}$ and $4.68$ d decade$^{-1}$, respectively (Figure 4(b)). The rural station displayed the greatest interannual variability in TX90p whereas the urban station displayed the greatest interannual variability in TN90p.

TX10p significantly decreased by $-2.48$ d decade$^{-1}$ at the urban station and by $-4.14$ d decade$^{-1}$ at the rural station (Figure 4(c)). For both stations, the TX10p frequency of the last decade was only approximately half that of the previous decade. TN10p also significantly decreased at both sites by $-5.57$ d decade$^{-1}$ at the urban station and $-6.27$ d decade$^{-1}$ at the rural station (Figure 4(d)). As with TX10p, the frequency of TN10p during the last decade was substantially lower than that of the previous decade. The urban station had greater interannual variability than the rural station for both TX10p and TN10p.

*3.3. Trends in Duration Indices.* Neither WSDI nor CSDI had statistically significant trends (Figures 5(a) and 5(b)). Nevertheless, it is evident that WSDI is more frequent in the last decade—60% (53%) of the WSDI total occurred in the last decade at the urban (rural) station—and both sites had their maximum WSDI of 27 in 2010. It is also evident that CSDI was more frequent in the earlier portion of the period of record and is slightly less frequent in the latter portion—86% (65%) of the CSDI total occurred in the first decade at the urban (rural) station.

*3.4. Trend Comparisons.* $\Delta X_{ur}$ was less than zero for all indices except TN10p and TX10p, which represent the annual number of nights and days with daily minimum and maximum temperature <10th percentile, respectively, and significantly influenced the negative trends in the time series of TNa, TXx, TN90p, and TX90p (Table 2). $E_u$ ranged from a minimum of 22% (TNa) to a maximum of 167% (TXx) for the significant negative effects. $\Delta X_{ur}$ significantly influenced only one positive effect, TX10p, with an $E_u$ of 67%. Negative values of $\Delta X_{ur}$ along with the magnitude of the trends at both stations suggest that urbanization and other anthropogenic factors have lessened the rate of warming at the urban station within Riyadh city's first urban limit compared to the more rural station beyond the second urban limit. For example, increasing trends at the urban station in TXx and TX90p were $0.30°C$ decade$^{-1}$ and $1.65$ d decade$^{-1}$, respectively, less than the trends at the rural station and the decreasing trend at the urban station in TX10p was $1.66$ d decade$^{-1}$ greater than the trend at the rural station (Table 2).

The differential rate of warming between the urban and the rural station has changed the UHI intensity of Riyadh city. The difference between the urban and the rural stations' maximum temperature indices (i.e., TXa, TXx, and TX90p) suggests that the diurnal (i.e., daytime) UHI may have weakened through the late 1980s and became characteristic of a diurnal urban heat sink in the early 1990s (Figure 6). Significant $\Delta X_{ur}$ was found in the time series of TXx and TX90p, both of which illustrate this transition from a diurnal UHI to a diurnal urban heat sink. TX90p, however, again became characteristic of a diurnal UHI in the last few years of the period of record. Such variability is likely associated with local land cover changes near the weather stations over time. A significant $\Delta X_{ur}$ was not found with TXa, but the time series does provide visual evidence of the transition from a diurnal UHI to a diurnal urban heat sink. Nocturnal (i.e., nighttime) UHI intensity has not undergone the same

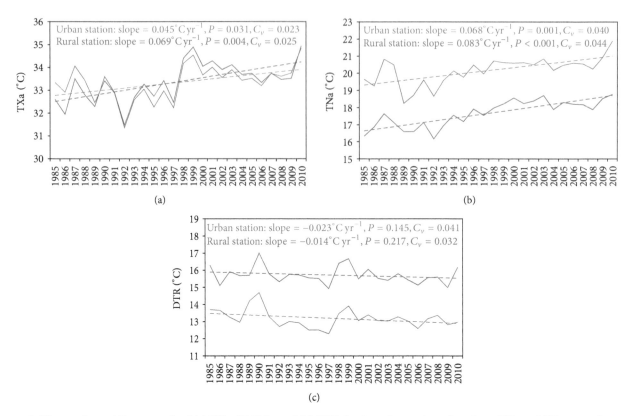

FIGURE 2: Time series and linear trends of (a) TXa, (b) TNa, and (c) DTR for the urban and rural stations. TXa and TNa: annual mean of monthly mean of daily maximum and minimum temperatures, respectively; DTR: diurnal temperature range.

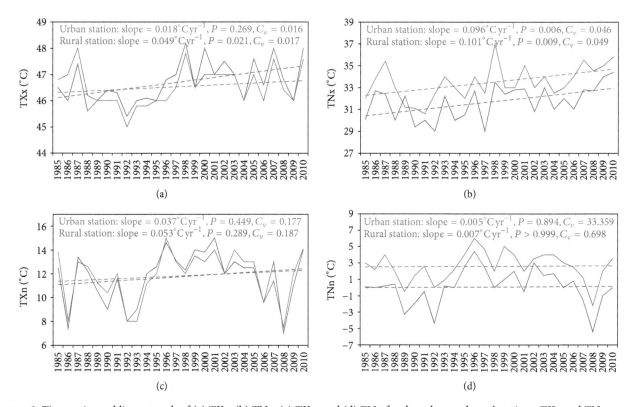

FIGURE 3: Time series and linear trends of (a) TXx, (b) TNx, (c) TXn, and (d) TNn for the urban and rural stations. TXx and TNx: annual mean of monthly maximum of daily maximum and minimum temperatures, respectively; TXn and TNn: annual mean of monthly minimum of daily maximum and minimum temperatures, respectively.

TABLE 1: Extreme temperature indices used in present study (adapted from [29]).

| Index | Descriptive name | Definition | Units |
|-------|------------------|------------|-------|
| | | Extreme value indices | |
| TXa | Mean TX | Annual mean of monthly mean TX | °C |
| TNa | Mean TN | Annual mean of monthly mean TN | °C |
| TXx | Max TX | Annual mean of monthly maximum of TX | °C |
| TNx | Max TN | Annual mean of monthly maximum of TN | °C |
| TXn | Min TX | Annual mean of monthly minimum of TX | °C |
| TNn | Min TN | Annual mean of monthly minimum of TN | °C |
| DTR | Diurnal temperature range | Annual mean of monthly mean difference between TX and TN | °C |
| | | Relative indices | |
| TN10p | Cool nights | Number of days when TN < 10th percentile | Days |
| TX10p | Cool days | Number of days when TX < 10th percentile | Days |
| TN90p | Warm nights | Number of days when TN > 90th percentile | Days |
| TX90p | Warm days | Number of days when TX > 90th percentile | Days |
| | | Range indices | |
| WSDI | Warm spell duration indicator | Annual count of days with at least 6 consecutive days when TX > 90th percentile | Days |
| CSDI | Cold spell duration indicator | Annual count of days with at least 6 consecutive days when TN < 10th percentile | Days |

T: temperature; Max.: maximum; N: minimum; TN: daily minimum temperature; TX: daily maximum temperature.

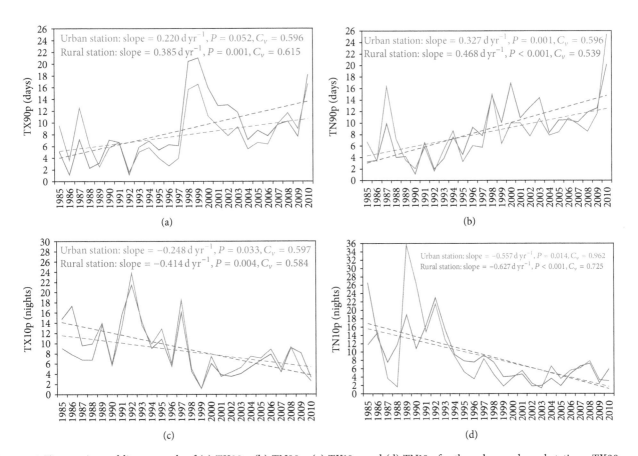

FIGURE 4: Time series and linear trends of (a) TX90p, (b) TN90p, (c) TX10p, and (d) TN10p for the urban and rural stations. TX90p and TN90p: number of days when daily maximum and minimum temperatures >90th percentile, respectively; TX10p and TN10p: number of days when daily maximum and minimum temperatures >10th percentile, respectively.

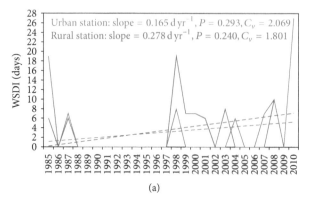 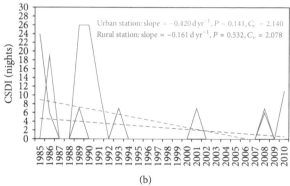

(a)                                                                                    (b)

FIGURE 5: Time series and linear trends of (a) WSDI and (b) CSDI for the urban and rural stations. WSDI: warm spell duration indicator; CSDI: cold spell duration indicator.

TABLE 2: Results of the trend, urbanization effect, and urbanization effect contribution analyses.

| Index | Urban trend ($X_u$) | Rural trend ($X_r$) | Urbanization effect ($\Delta X_{ur}$) | Urbanization contribution ($E_u$, %) |
|---|---|---|---|---|
| TXa ($^\circ$C yr$^{-1}$) | 0.045[b] | 0.069[c] | −0.023 | 51 |
| TNa ($^\circ$C yr$^{-1}$) | 0.068[c] | 0.083[c] | −0.015[a] | 22 |
| TXx ($^\circ$C yr$^{-1}$) | 0.018 | 0.049[b] | −0.030[b] | 167 |
| TNx ($^\circ$C yr$^{-1}$) | 0.096[c] | 0.101[c] | −0.005 | 5 |
| TXn ($^\circ$C yr$^{-1}$) | 0.037 | 0.053 | −0.017 | 46 |
| TNn ($^\circ$C yr$^{-1}$) | 0.005 | 0.007 | −0.002 | 40 |
| DTR ($^\circ$C yr$^{-1}$) | −0.023 | −0.014 | −0.009 | 39 |
| TN10p (d yr$^{-1}$) | −0.557[b] | −0.627[c] | 0.070 | 13 |
| TX10p (d yr$^{-1}$) | −0.248[b] | −0.414[c] | 0.166[a] | 67 |
| TN90p (d yr$^{-1}$) | 0.327[a] | 0.468[c] | −0.141[b] | 43 |
| TX90p (d yr$^{-1}$) | 0.220[a] | 0.385[c] | −0.165[b] | 75 |
| WSDI (d yr$^{-1}$) | 0.165 | 0.278 | −0.114 | 69 |
| CSDI (d yr$^{-1}$) | −0.420 | −0.161 | −0.259 | 62 |

[a]Statistically significant at the 90% level.
[b]Statistically significant at the 95% level.
[c]Statistically significant at the 99% level.

transition. A significant negative $\Delta X_{ur}$ was found with TNa, implying that urbanization has lessened the rate of warming at the urban station, but the decreasing nocturnal UHI intensity has remained positive throughout the period of record.

## 4. Summary and Discussion

This study analyzed and compared temporal trends in the time series of thirteen extreme temperature indices at two weather stations in Riyadh city. Statistically significant linear trends were detected in the time series of TXa, TNa, TNx, TN10p, TX10p, TN90p, and TX90p at both stations (and in TXx at King Khalid Airport) (New Station). These trends suggest that temperatures in the upper-end of the distribution are becoming more common and temperatures in the lower end of the distribution are becoming less common in Riyadh city. Such a shift in the temperature distribution is generally consistent with previous trend studies that reported increasing mean temperature in the region (e.g., [9, 10]).

The trends at Riyadh Air Base (Old Station) near the center of Riyadh city were significantly lower than the trends at King Khalid Airport (New Station) nearly 20 km to the northeast in several indices (Table 2). The trends in the TNa, TN90p, and TX90p indices, for example, were 0.15°C decade$^{-1}$, 1.41 d decade$^{-1}$, and 1.65 d decade$^{-1}$, respectively, lower at Riyadh Air Base (Old Station). These results support [10] conclusion that urbanization has not substantially contributed to the large-scale warming trend observed throughout Saudi Arabia and also suggest that urbanization may have lessened the rate of warming in urban areas such as Riyadh city compared to surrounding rural areas. This result is consistent with the results reported by [27] who found that small/rural towns warmed at a greater rate than large cities in the Arabian Peninsula and with other studies that have reported urban heat sinks in arid environments (e.g., [19–23]).

In arid environments, urbanization often is accompanied by an increase in the amount of vegetation cover compared to the surrounding areas, thereby increasing latent heat flux and

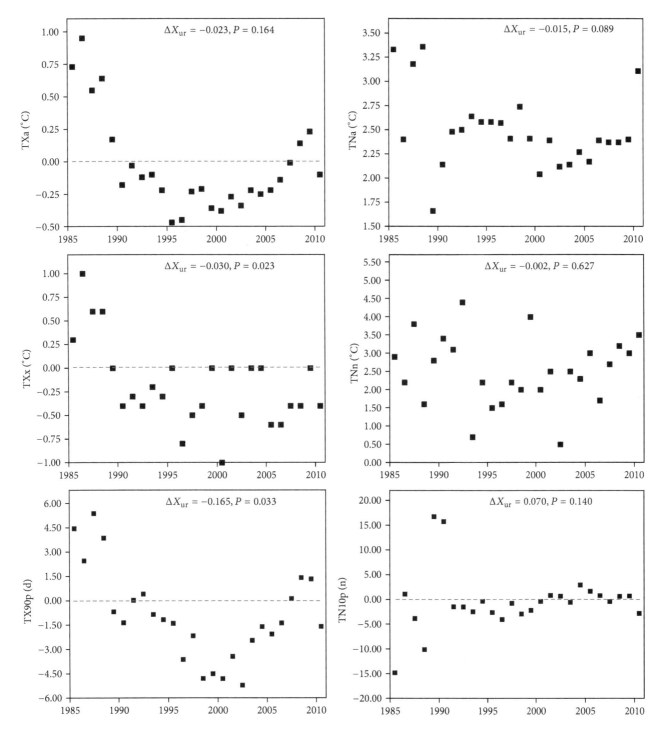

FIGURE 6: Plots of the difference between the urban station and the rural station for six extreme temperature indices. TXa, TXx, and TX90p are representative of diurnal maximum temperature and the diurnal UHI. TNa, TNn, and TN10p are representative of nocturnal minimum temperature and nocturnal UHI. Values greater than zero are characteristic of the UHI effect whereas values less than zero are characteristic of an urban heat sink. Results of the urbanization effect analysis are provided in the top-right corner of each plot.

shading and decreasing sensible heat flux [37]. This change in energy flux at the surface often lessens the warming trend observed in the surrounding area or even induces a cooling trend [3, 5, 19–23, 38–41]. The vegetation cover in Riyadh city has increased over the last several decades, for example, from 38.6 km² in 1979 to 58.3 km² in 1999 [42]. The increased

latent heat flux and shading and decreased sensible heat flux associated with this increase in vegetation cover may have played a role in lessening the rate of warming in the urban affected area of Riyadh city (i.e., near Riyadh Air Base) (Old Station) that was observed in this study. Furthermore, the cooling effect of vegetation is most pronounced during the daytime [43], which may explain why Riyadh city's diurnal UHI transitioned to an urban heat sink in the early 1990s. These results support the proposition by [5] that increased evapotranspiration in urban areas may be useful to UHI effect mitigation efforts. Additional research is needed, however, to identify which factors are responsible for the lower rate of warming near the urban station.

Lastly, it is important to note that the results of this study do not imply that all locations within Riyadh city have experienced the same rate of change in extreme temperature. Cities often are complex mixtures of various juxtaposed land covers that undoubtedly have different effects on temperature in the urban canopy layer. Additional study at the local scale is needed to gain a better understanding of UHI spatial variability and of the link between changes in the urban environment over time and temperature trends in the urban canopy layer. Such study requires site metadata that detail characteristics such as terrain aspect, land cover, and sky view factor [44]. Such analyses were beyond the scope of this study as the main purposes were to analyze and compare temporal trends in extreme temperature and not to assess the spatial variability of the UHI effect in Riyadh city. Understanding whether urbanization is mitigating or exacerbating larger-scale warming is important not only to human heat stress and comfort but also to various other sectors such as energy and water provision and urban planning, specifically regarding the incorporation of more vegetation cover and green infrastructure into arid cities to mitigate larger-scale warming.

## Conflict of Interests

The authors declare that there is no conflict of interests regarding the publication of this paper.

## References

[1] IPCC, "Climate change 2007: the scientific basis," in *Contribution of Working Group I to the Fourth Assessment Report of the Intergovernmental Panel on Climate Change*, S. Solomon, D. Qin, M. Manning et al., Eds., Cambridge University Press, New York, NY, USA, 2007.

[2] Met Office, *Climate: Observations, Projections and Impacts Saudi Arabia*, Met Office, Devon, UK, 2013.

[3] T. Susca, S. R. Gaffin, and G. R. Dell'Osso, "Positive effects of vegetation: urban heat island and green roofs," *Environmental Pollution*, vol. 159, no. 8-9, pp. 2119–2126, 2011.

[4] S. Peng, S. Piao, P. Ciais et al., "Surface urban heat island across 419 global big cities," *Environmental Science and Technology*, vol. 46, no. 2, pp. 696–703, 2012.

[5] G.-Y. Qiu, H.-Y. Li, Q.-T. Zhang, W. Chen, X.-J. Liang, and X.-Z. Li, "Effects of evapotranspiration on mitigation of urban temperature by vegetation and urban agriculture," *Journal of Integrative Agriculture*, vol. 12, no. 8, pp. 1307–1315, 2013.

[6] S. Rehman, "Temperature and rainfall variation over Dhahran, Saudi Arabia, (1970–2006)," *International Journal of Climatology*, vol. 30, no. 3, pp. 445–449, 2010.

[7] M. Almazroui, M. Nazrul Islam, H. Athar, P. D. Jones, and M. A. Rahman, "Recent climate change in the Arabian Peninsula: annual rainfall and temperature analysis of Saudi Arabia for 1978–2009," *International Journal of Climatology*, vol. 32, no. 6, pp. 953–966, 2012.

[8] M. Almazroui, M. N. Islam, P. D. Jones, H. Athar, and M. A. Rahman, "Recent climate change in the Arabian Peninsula: Seasonal rainfall and temperature climatology of Saudi Arabia for 1979–2009," *Atmospheric Research*, vol. 111, pp. 29–45, 2012.

[9] M. Almazroui, M. N. Islam, R. Dambul, and P. D. Jones, "Trends of temperature extremes in Saudi Arabia," *International Journal of Climatology*, vol. 34, no. 3, pp. 808–826, 2014.

[10] M. Almazroui, M. N. Islam, and P. D. Jones, "Urbanization effects on the air temperature rise in Saudi Arabia," *Climatic Change*, vol. 120, no. 1-2, pp. 109–122, 2013.

[11] S. Rehman and L. M. Al-Hadhrami, " Extreme temperature trends on the west coast of Saudi Arabia," *Atmospheric and Climate Sciences*, vol. 2, pp. 351–361, 2012.

[12] H. Athar, "Trends in observed extreme climate indices in Saudi Arabia during 1979–2008," *International Journal of Climatology*, vol. 34, no. 5, pp. 1561–1574, 2014.

[13] A. T. DeGaetano and R. J. Allen, "Trends in twentieth-century temperature extremes across the United States," *Journal of Climate*, vol. 15, no. 22, pp. 3188–3205, 2002.

[14] R. A. Pielke Sr., T. Stohlgren, L. Schell et al., "Problems in evaluating regional and local trends in temperature: an example from eastern Colorado, USA," *International Journal of Climatology*, vol. 22, no. 4, pp. 421–434, 2002.

[15] R. W. Dixon and T. W. Moore, "Trend detection in Texas temperature and precipitation," *Southwest Geography*, vol. 15, pp. 80–103, 2011.

[16] G. Y. Ren, Y. Q. Zhou, Z. Y. Chu et al., "Urbanization effects on observed surface air temperature trends in north China," *Journal of Climate*, vol. 21, no. 6, pp. 1333–1348, 2008.

[17] Y. Zhou and G. Ren, "Change in extreme temperature event frequency over mainland China, 1961–2008," *Climate Research*, vol. 50, no. 2-3, pp. 125–139, 2011.

[18] T. R. Oke, "The heat island of the urban boundary layer: characteristics, causes and effects," in *Wind Climate in Cities*, pp. 81–107, Kluwer Academic Publishers, Dordrecht, The Netherlands, 1995.

[19] A. Brazel, N. Selover, R. Vose, and G. Heisler, "The tale of two climates-Baltimore and Phoenix urban LTER sites," *Climate Research*, vol. 15, no. 2, pp. 123–135, 2000.

[20] M. A. Peña, "Relationships between remotely sensed surface parameters associated with the urban heat sink formation in Santiago, Chile," *International Journal of Remote Sensing*, vol. 29, no. 15, pp. 4385–4404, 2008.

[21] L. Bounoua, A. Safia, J. Masek, C. Peters-Lidard, and M. L. Imhoff, "Impact of urban growth on surface climate: a case study in Oran, Algeria," *Journal of Applied Meteorology and Climatology*, vol. 48, no. 2, pp. 217–231, 2009.

[22] M. L. Imhoff, P. Zhang, R. E. Wolfe, and L. Bounoua, "Remote sensing of the urban heat island effect across biomes in the continental USA," *Remote Sensing of Environment*, vol. 114, no. 3, pp. 504–513, 2010.

[23] M. Lazzarini, P. R. Marpu, and H. Ghedira, "Temperature-land cover interactions: the inversion of urban heat island phenomenon in desert city areas," *Remote Sensing of Environment*, vol. 130, pp. 136–152, 2013.

[24] F. M. Alkolibi, "Possible effects of global warming on agriculture and water resources in Saudi Arabia: impacts and responses," *Climatic Change*, vol. 54, no. 1-2, pp. 225–245, 2002.

[25] R. Municipality, "Riyadh development," 2014, http://www.alriyadh.gov.sa/en/alriyadh/Pages/development_of_population.aspx.

[26] M. A. Al-Saleh, "Variability and frequency of daily rainfall in Riyadh, Saudi Arabia," *The Geographical Bulletin*, vol. 39, no. 1, pp. 48–57, 1997.

[27] S. AlSarmi and R. Washington, "Recent observed climate change over the Arabian Peninsula," *Journal of Geophysical Research D: Atmospheres*, vol. 116, Article ID D11109, pp. 1984–2012, 2011.

[28] S. H. Alsarmi and R. Washington, "Changes in climate extremes in the Arabian Peninsula: Analysis of daily data," *International Journal of Climatology*, vol. 34, no. 5, pp. 1329–1345, 2014.

[29] X. Zhang and F. Yang, *RClimDex (1.0) User guide. Climate Research Branch Environment Canada, Downsview*, Ontario, Canada, 2004.

[30] G. Choi, D. Collins, G. Ren et al., "Changes in means and extreme events of temperature and precipitation in the Asia-Pacific Network region, 1955–2007," *International Journal of Climatology*, vol. 29, no. 13, pp. 1906–1925, 2009.

[31] A. M. G. Klein Tank, F. W. Zwiers, and X. Zhang, "Guidelines on analysis of extremes in a changing climate in support of informed decisions for adaptation," WMO-TD 1500, World Meteorological Organization, 2009.

[32] Q. You, S. Kang, E. Aguilar et al., "Changes in daily climate extremes in China and their connection to the large scale atmospheric circulation during 1961–2003," *Climate Dynamics*, vol. 36, no. 11-12, pp. 2399–2417, 2011.

[33] T. M. L. Wigley, "Appendix A: atatistical issues regarding trends," in *Temperature Trends in the Lower Atmosphere: Steps for Understanding and Reconciling Differences*, T. R. Karl, S. J. Hassel, C. D. Miller, and W. L. Murray, Eds., A report by the US Climate Change Science Program and the subcommittee on Global Change Research, Washington, DC, USA, 2006.

[34] Z. Wu, N. E. Huang, J. M. Wallace, B. V. Smoliak, and X. Chen, "On the time-varying trend in global-mean surface temperature," *Climate Dynamics*, vol. 37, no. 3, pp. 759–773, 2011.

[35] B. Önöz and M. Bayazit, "The power of statistical tests for trend detection," *Turkish Journal of Engineering & Environmental Sciences*, vol. 27, no. 4, pp. 247–251, 2003.

[36] L. Zhang, G. Y. Ren, Y. Y. Ren, A. Y. Zhang, Z. Y. Chu, and Y. Q. Zhou, "Effect of data homogenization on estimate of temperature trend: a case of Huairou station in Beijing Municipality," *Theoretical and Applied Climatology*, vol. 115, no. 3-4, pp. 365–373, 2014.

[37] M. E. Gregory, "Cooling urban heat islands with sustainable landscapes," in *The Ecological City: Preserving and Restoring Urban Biodiversity*, R. A. Rowntree, R. H. Platt, and P. C. Muick, Eds., pp. 151–171, University of Massachusetts Press, Amherst, Mass, USA, 1994.

[38] H. A. Nasrallah, A. J. Brazel, and R. C. Balling Jr., "Analysis of the Kuwait City urban heat island," *International Journal of Climatology*, vol. 10, no. 4, pp. 401–405, 1990.

[39] R. A. Spronken-Smith and T. R. Oke, "The thermal regime of urban parks in two cities with different summer climates," *International Journal of Remote Sensing*, vol. 19, no. 11, pp. 2085–2104, 1998.

[40] E. Jauregui, "Influence of a large urban park on temperature and convective precipitation in a tropical city," *Energy and Buildings*, vol. 15, no. 3-4, pp. 457–463, 1991.

[41] T. V. Ca, T. Asaeda, and E. M. Abu, "Reductions in air conditioning energy caused by a nearby park," *Energy and Buildings*, vol. 29, no. 1, pp. 83–92, 1998.

[42] N. A. Al Sahhaf, *The use of remote sensing and geographic information system technologies to detect, monitor, and model urban change in Riyadh, Saudi Arabia [Dissertation, thesis]*, Univers ity of California, Santa Barbara, Calif, USA, 2000.

[43] A. Buyantuyev and J. Wu, "Urban heat islands and landscape heterogeneity: linking spatiotemporal variations in surface temperatures to land-cover and socioeconomic patterns," *Landscape Ecology*, vol. 25, no. 1, pp. 17–33, 2010.

[44] I. D. Stewart and T. R. Oke, "Local climate zones for urban temperature studies," *Bulletin of the American Meteorological Society*, vol. 93, no. 12, pp. 1879–1900, 2012.

# 7

# Coastal Boundary Layer Characteristics of Wind, Turbulence, and Surface Roughness Parameter over the Thumba Equatorial Rocket Launching Station, India

**K. V. S. Namboodiri,**[1] **Dileep Puthillam Krishnan,**[1] **Rahul Karunakaran Nileshwar,**[1] **Koshy Mammen,**[1] **and Nadimpally Kiran kumar**[2]

[1] *Meteorology Facility, Thumba Equatorial Rocket Launching Station (TERLS), Vikram Sarabhai Space Centre (VSSC), Indian Space Research Organisation, Thiruvananthapuram 695 022, India*
[2] *Space Physics Laboratory, Vikram Sarabhai Space Centre (VSSC), Indian Space Research Organization, Thiruvananthapuram 695 022, India*

Correspondence should be addressed to K. V. S. Namboodiri; sambhukv@yahoo.com

Academic Editor: Elena Paoletti

The study discusses the features of wind, turbulence, and surface roughness parameter over the coastal boundary layer of the Peninsular Indian Station, Thumba Equatorial Rocket Launching Station (TERLS). Every 5 min measurements from an ultrasonic anemometer at 3.3 m agl from May 2007 to December 2012 are used for this work. Symmetries in mesoscale turbulence, stress off-wind angle computations, structure of scalar wind, resultant wind direction, momentum flux ($M$), Obukhov length ($L$), frictional velocity ($u_*$), w-component, turbulent heat flux ($H$), drag coefficient ($C_D$), turbulent intensities, standard deviation of wind directions ($\sigma_\theta$), wind steadiness factor-$\sigma_\theta$ relationship, bivariate normal distribution (BND) wind model, surface roughness parameter ($z_0$), $z_0$ and wind direction ($\theta$) relationship, and variation of $z_0$ with the Indian South West monsoon activity are discussed.

## 1. Introduction

The lowest layers of the earth's atmosphere based on wind variation with height are categorized into different layers such as (1) laminar sublayer, (2) Prandtl layer, and (3) Ekman spiral layer. (1) and (2) are together called surface boundary layer (SBL) or surface layer (SL) or the tower layer or constant flux layer. SBL studies can find applications in wind power meteorology, aviation meteorology, aerospace meteorology, structural loading, air pollution, flow modeling, biometeorology, and weather forecast models. In the SBL, more specifically in Prandtl layer, the eddy stress (due to turbulence) is an order of magnitude larger than the horizontal pressure gradient force. Several approximations are available regarding SBL [1–3] about the height, vertical stress, heat flux, wind direction, turbulent diffusion, and insignificance of Coriolis effect. In a diabatic (nonadiabatic) atmosphere,

the thermodynamic change of the state of the system is one in which there is transfer of heat across the boundaries of the system [4]. A general formula for the diabatic (nonadiabatic) wind profile in the SBL can be derived [3, 5] which provides the height above the earth's surface where the mean wind speed ($\overline{u}$) vanishes before the surface is called as the surface roughness parameter ($z_0$):

$$z_0 = \frac{z}{\exp\left((\overline{u}k/u_*) + \psi\left(z/L\right)\right)}, \tag{1}$$

where $z$ is the wind measuring level, $\psi$ is the universal function of height $z$ relative to the Monin-Obukhov (MO) similarity theory [6], $u_*$ is the frictional velocity, $k$ is the von Karman constant ($\approx 0.4$), and $L$ is the Obukhov length. To determine the surface roughness parameter, stability corrected method proposed by Paulson [7] is used. The stability corrected Businger [5] method uses constants 4.7 and

15 instead of 5 and 16 which are used in Paulson [7]. Businger method can be seen in Rao et al. [8].

If $(z/L) > 0$ (stable),

$$\psi\left(\frac{z}{L}\right) = -5\frac{z}{L} \tag{2}$$

and when $z/L < 0$ (unstable),

$$\psi\left(\frac{z}{L}\right) = 2\log\left[\frac{(1+x)}{2}\right] + \log\left[\frac{(1+x^2)}{2}\right] - 2\tan^{-1}(x) + \frac{\pi}{2}, \tag{3}$$

where $x \equiv [(1 - 16(z/L))]^{1/4}$. If $z/L = 0$, the $z_0$ value pertained to logarithmic wind profile in neutral case can be obtained as

$$z_0 = \frac{z}{\exp(\bar{u}k/u_*)}. \tag{4}$$

The relation of $z_0$ to various terrain types is extensively studied by Nappo [9], Smedman-Hogstrom and Hogstrom [10], Hicks et al. [11], Kondo and Yamazawa [12], Thompson [13], and Garratt [14] and occurrence of imaginary and small $z_0$ values was reported by Panofsky and Peterson [15] over a narrow peninsula surrounded by bays of varying width. The $z_0$ values may vary with the heights of wind measurements as profiles from tower will sense only local terrain irregularities of small horizontal size and also at the low level the roughness length is representative of the locally smooth surface crops [16, 17]. From a limited period data, Ramachandran et al. [18] have studied the variability of $z_0$ for Indian coastal station, Thumba, and reported that the $z_0$ values are low during the Indian South West monsoon (SW monsoon) season compared to other months. For the same station, Gupta et al. [19] have shown that $z_0$ is maximum when onshore flow is perpendicular to the coast. Namboodiri [20] has documented direct proportionality among $z_0$ and exponent in wind power law and diurnal variation of $z_0$ and its dependence with wind direction.

Surface roughness parameter values have application in oil sleek sedimentation and evaporation modeling. Studies on the influence of marine boundary layer on land boundary layer are very limited. Properties of marine boundary layer were studied [26, 27]. Lange et al. [28] suggested that MO similarity theory cannot be applied to the coastal marine boundary layer under stable and warm advections. Momentum and sensible heat exchange in an Arctic Fjord was carried out by Kilpelainen and Sjoblom [29] which has proved the application of MO similarity theory under "nonideal" conditions. Marine boundary layer changes are due to behaviors in wind speed, sea water surface tension, stress by young developing and decaying waves, changing wind direction, water depth, wave breaking, wave strength, fetch, and shoreline processes.

Sonic anemometer measures wind data at high frequency rate and eddy-correlation method facilitates estimation of turbulence parameters. From the high frequency (10 Hz.) $x$, $y$, and $z$ wind components, the total stress $\tau$, and thereby drag coefficient ($C_D$), and stress off-wind angle ($\alpha$) can be computed as

$$\tau = \rho u_*^2, \tag{5}$$

$$C_D = \left\{\frac{k}{[\ln(z/z_0) - \psi(z/l)]}\right\}^2, \tag{6}$$

$$\alpha = \tan^{-1}\left(\frac{\langle v'w'\rangle}{\langle u'w'\rangle}\right), \tag{7}$$

where $\rho$ is the air density, $u_* = [\langle u'w'\rangle^2 + \langle v'w'\rangle^2]^{1/4}$, and the vertical transport of horizontal momentum and $\langle u'w'\rangle$, $\langle v'w'\rangle$ are the components of the kinematic momentum flux in the along wind and cross wind directions, respectively. A positive (negative) $\alpha$ corresponds to the stress vector orientation to right (left) of the wind direction; that is, the stress vector is shifted clockwise (anticlockwise) from the wind direction. Using eddy-correlation method $L$, vertical momentum flux (negative $\tau$) and turbulent heat flux ($H$) are computed:

$$L = \frac{-u_*^3 T}{gk\langle w'T'\rangle},$$

$$H = \rho C_p \langle w'T'\rangle, \tag{8}$$

where $g$ is the acceleration due to gravity, $C_p$ is the specific heat of dry air at constant pressure, and $\langle w'T'\rangle$ is the kinematic temperature flux ($T'$ is the perturbations in sonic temperature, $T$).

This study presents the features of wind, turbulence, and $z_0$ over the peninsular coastal boundary layer over the Thumba Equatorial Rocket Launching Station, TERLS, ($8°32'$N/$76°52'$E, 6 m above msl), a place in Thiruvananthapuram, the capital city of the Kerala state of India.

## 2. Experimental Site and Data Base

Figure 1 shows the locations of TERLS in the west coast and Sriharikota in the east coast situated in the south Peninsular India. Sriharikota ($13°42'$N/$80°12'$E) is marked for the sake of an analysis mentioned about the local wind direction variability during the SW monsoon over both stations.

An Ultra Sonic Anemometer (USA1, M/s Metek, Germany) is installed in TERLS at a height of 3.3 m above the ground level. In the near vicinity, the soil is sandy with sparsely grown very short grass vegetation. Cashew plantations of less than 8 m in height are about 50 m far in the NNW to ENE sectors. Wide open sandy yard in the ENE to SSE sector is about 10000 m² in area. A height level of 3.3 m represents statistics of surface wind and turbulence for our applications of interest. The sonic anemometer is located at about 140 m from the Arabian seashore line (oriented 145°–325° azimuth). This provides a unique opportunity to capture coastal boundary layer wind modifications due to the marine SBL. Onshore winds arrive through very sparingly planted medium size coconut trees of less than

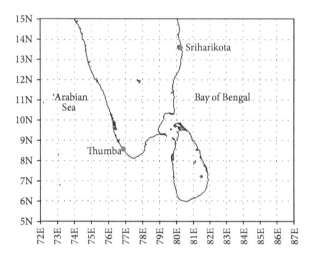

FIGURE 1: Peninsular India showing the locations of Thumba and Sriharikota.

8 m within a distance of 50 m to 140 m. Meteorology facility, TERLS, has the heaviest responsibility in supporting regular sounding rocket launches from the launch pad located on the seashore. Wind climatology, SBL wind variability due to mesosynoptic features, role of turbulent parameters in environmental assessments, namely, corrosion, evaporation, coastal sedimentation processes are highly significant for the engineering demands at TERLS and thereby this project is commissioned. Round the clock 5-minute observations of wind and turbulent parameters (based on MO similarity theory) are obtained by using 10 Hz data. The data from May 2007 to December 2012 (more than 5 years) is used for time section isopleth analyses to depict mean wind and turbulent parameters. Stress off-wind angle calculations are made to know the influence of marine boundary layer. The variability of $z_0$ in relation to the SW monsoon season has been investigated.

## 3. Data Validation through Symmetries and Distributions in Mesoscale Turbulence

Through dimensional analysis Brutsaert [30] identified that the normalized turbulence or statistics-variances, covariances, and gradients are universal functions of $z/L$. Also the relation between turbulent statistics and $z/L$ is to be empirically determined. Extensive data validation processes are made through universal symmetries and scatter distribution diagrams of mesoscale turbulence [3, 25, 31]. For the same station Ramachandran et al. [18] and Kunhikrishnan [32] have studied surface layer variables expressed in universal form by means of scaling parameters. Krishnan and Kunhikrishnan [22] have reported relations among normalized standard deviations of variables to $z/L$ within the tropic of cancer. Ramana et al. [24] have documented the same outside the tropic of cancer. In the present study $z/L$ values are categorized into stable-unstable cases for onshore and offshore winds. Kaimal and Finnigan [33] have discussed the significance of well-defined spectral properties in boundary

layer characterization studies. Prakash et al. [34] have shown distinct spectral features for the mesoscale circulations over this coastal station. Due to this finding, all the obtained data points have been taken in the analyses on the universal symmetry and distributions of turbulence. Onshore wind (from $145°$ to $325°$) and offshore wind (from $325°$ to $145°$) are separately considered while deriving the symmetries. +ve and −ve $L$ values correspond to stable and unstable SBL, respectively.

Figure 2 provides distribution of $u_*$ Vs $1/L$ for the months January and July represent winter ($D$, $J$) and SW monsoon ($J$, $J$, $A$, $S$), respectively, in India. +ve and −ve $u_*$ values adopted in discussions are reasoned as follows. Biltoft [35] reported that frictional velocity and $L$ cannot be properly defined if momentum flux is positive, even though these equations will produce what appears to be a mathematically correct result. As a convention, during the incidence of positive momentum flux, the $u_*$ values are reported as negative values [36]. Mathematically positive momentum flux has to generate a negative Reynolds stress term and hence negative $u_*$ and $C_D$ values. Convention adopted in $u_*$ during positive momentum flux is an embedded feature in the onboard computation of the instrument. One can expect either a positive or negative slope in the logarithmic wind profile with respect to wind speed increases or decreases, respectively, with height. Accordingly, more precise interpretation may be conventionally explained as +ve $u_*$/−ve $u_*$ for +vely/−vely slopped logarithmic wind profiles (+ve/−ve momentum flux corresponds to −vely/+vely slopped logarithmic wind profile by convention). $u_*$ peaks at $1/L \rightarrow 0$ ($\tau$ also peaks up due to (5)), for both onshore and offshore wind cases. $u_*$ becomes small for both negative and positive increases of $1/L$ values. Small $u_*$ values against high negative and positive $1/L$ values are due to dominant convection (highly unstable) and severely reduced mechanical turbulence (highly stable) due to temperature stratification, respectively. Medium values of both +ve and −ve values that fall in between $1/L \rightarrow 0$ and $<$ high $\pm 1/L$ values contribute to mechanical turbulence slightly damped by temperature stratification (weakly stable) and dominant mechanical (inertial) turbulence (weakly unstable). Peak values of $u_*$ when $1/L \rightarrow 0$ are due to neutral regime characterized by purely mechanical turbulence. Even with the momentum flux becomes positive (thereby negative $u_*$), the peaking of $u_*$ when $1/L \rightarrow 0$ is observed irrespective of onshore or offshore wind. A universal symmetry relation cannot be established between $u_*$ and $1/L$.

To see the contribution of $w$ component in $u_*$, $u_*$ values are plotted against standard deviation of $w$-components ($\sigma_w$) as in Figure 3. Steep linear slopes are observed during offshore and onshore events, which shows increase in $u_*$ values with increase in standard deviation of $w$ components. The near constancy in slopes with value about 1.3 provides a typical neutral $u_*$ value of the order of about 0.7 $ms^{-1}$ and the same may be noticed from Figure 2.

In contrast to the standard deviation of vertical velocity, those of the horizontal components over uniform terrain do not vary with height in the surface layer, not even in unstable air, but depend strongly on temperature stratification

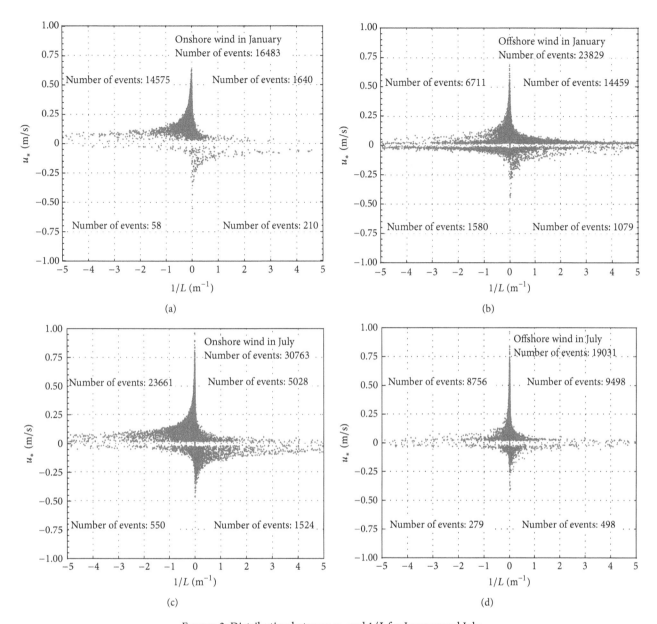

FIGURE 2: Distribution between $u_*$ and $1/L$ for January and July.

[3, 37, 38]. With this background knowledge, we have studied the dependence of variances of lateral ($p$), longitudinal ($q$), and vertical ($r$ or $w$) components normalized by $u_*$ and $z/L$. Studies among variances of wind components to $z/L$ are available mostly outside the tropics [39, 40]. In the present study, investigators could obtain power curves relations of the form $\sigma/u_* = a + b(z/L)^c$ among variances of $p$, $q$, and $r$ components normalized by $u_*$ and $z/L$ for both $z/L > 0$ and $z/L < 0$ similar to what Pahlow et al. [25] studied for $z/L > 0$. On the $x$-axis $\pm z/L$ values are depicted from 0 to 300. The power curve relations result in the dependence of ($p$, $q$, $r$) variances on $z/L$ by considering data points of more than 15000 samples individually for the stable and unstable regime of onshore and offshore cases (Figures 4(a)–4(c)). Towards the origin of the fitted curves among standard deviations of ($p$, $q$, and $r$) and $z/L$, magnitude of $\pm z/L$ values

decreases to extremely low values. Scattering of variances with respect to $z/L$ is at its minimum level from $-10$ to 10 and is characterized by maximum points of occurrence. Variances are not showing much $z/L$ dependence over this regime of weakly stable or weakly unstable conditions. Abrupt enhancement in $\sigma_p/u_*$, $\sigma_q/u_*$, and $\sigma_r/u_*$ can be seen from $z/L$ values $< -10$ and $>10$ in association with highly unstable or highly stable conditions, respectively. A 1/3 power relation for unstable regime and power as unity for stable regime exists between $\sigma_r/u_*$ and $z/L$. No relation among horizontal fluctuations (lateral and longitudinal) with $z/L$ due to the control of convective motions over horizontal velocity variances attributed to mixed layer scaling down to the surface was suggested by several investigators [21, 22, 24, 32]. Panofsky et al. [21] specifically mentioned only an approximate agreement of 1/3 power relation among

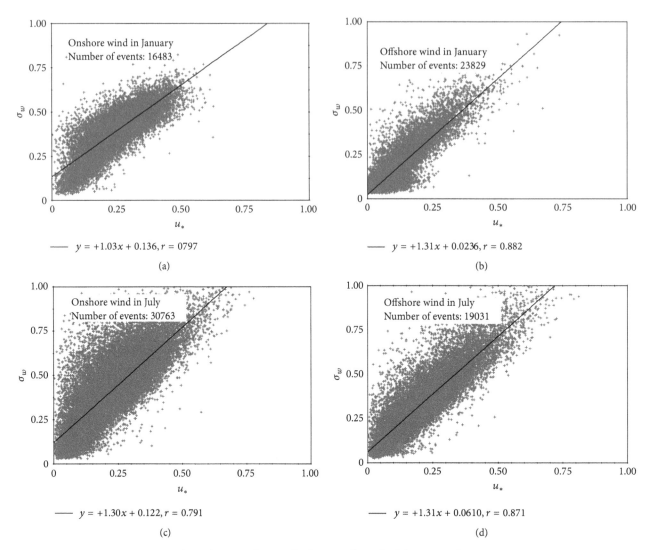

FIGURE 3: Relation between frictional velocity and standard deviation of $w$-component.

variances of wind components to $z/L < 0$. Table 1 provides various relations among normalized standard deviation of vertical component ($\sigma_r/u_*$ or $\sigma_w/u_*$) to $z/L$. The investigators could obtain power curve relations among ($p, q, r$) variances and $z/L$ with high correlation coefficient ($r$) values, for both stable and unstable flow regimes.

Stray points are observed in the negative portion of the standard deviation normalized by $u_*$. Interestingly these negative ordinate portions also have shown similar power relations. Approaching the neutral regime, negative $u_*$ values are drastically reduced due to positively slopped classical neutral logarithmic wind profile approximation. In this part of analysis near neutral approximations are obtained (through nonoccurrence of negative $u_*$) specifically within the region of $z/L$ from $-0.03$ to $0.02$. Rieder et al. [41] have mentioned $z/L$ between $-0.01$ and $0.01$ as the neutral regime and Ramana et al. [24] quantified it as $-0.1$ to $0.1$. Curve fittings from other investigators for the unstable and stable regimes are also incorporated in Figure 4.

Figure 5 represents the relation between $z/L$ and sonic temperature normalized by characteristic temperature ($T_*$). For comparison power curve fittings following Pahlow et al. [25] are depicted in the figure.

Figure 6 shows the scatter plot of wind directions against $z/L$ for offshore and onshore wind regimes separated by straight lines. +ve $z/L$ values are markedly lacking in the onshore wind regime, but both +ve and $-$ve $z/L$ contributions are seen in the offshore wind regime. This concludes that onshore SBL wind profiles are of mostly unstable nature, whereas offshore wind profiles can belong to either stable or unstable type.

As the sonic anemometer is very close to the shoreline, a detailed examination on stress off-wind angle ($\alpha$) (7) is carried out for $\alpha$ values, which vary from $-180°$ to $180°$. Similar studies can be seen elsewhere [27, 41–43]. These analyses can find application in estimating the direction of wind driven sea waves near to the coast over this region of the Arabian sea. Inverse relationship is seen between stress off-wind angle and wind speed (Figure 7).

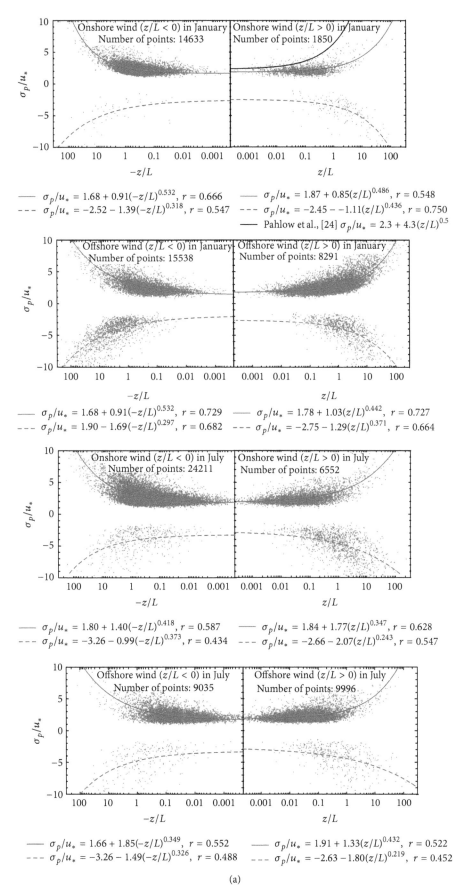

FIGURE 4: Continued.

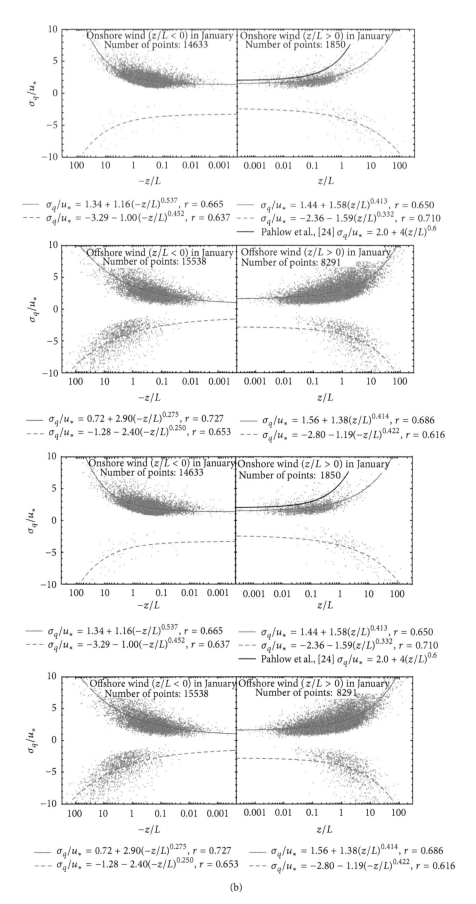

FIGURE 4: Continued.

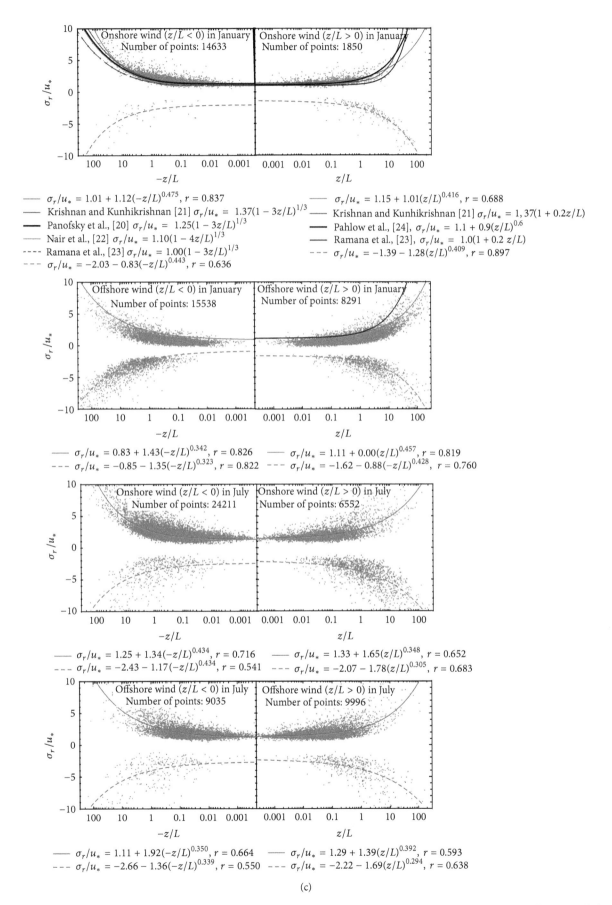

$\sigma_r/u_* = 1.01 + 1.12(-z/L)^{0.475}$, $r = 0.837$

Krishnan and Kunhikrishnan [21] $\sigma_r/u_* = 1.37(1 - 3z/L)^{1/3}$

Panofsky et al., [20] $\sigma_r/u_* = 1.25(1 - 3z/L)^{1/3}$

Nair et al., [22] $\sigma_r/u_* = 1.10(1 - 4z/L)^{1/3}$

Ramana et al., [23] $\sigma_r/u_* = 1.00(1 - 3z/L)^{1/3}$

$\sigma_r/u_* = -2.03 - 0.83(-z/L)^{0.443}$, $r = 0.636$

$\sigma_r/u_* = 1.15 + 1.01(z/L)^{0.416}$, $r = 0.688$

Krishnan and Kunhikrishnan [21] $\sigma_r/u_* = 1,37(1 + 0.2z/L)$

Pahlow et al., [24], $\sigma_r/u_* = 1.1 + 0.9(z/L)^{0.6}$

Ramana et al., [23], $\sigma_r/u_* = 1.0(1 + 0.2 z/L)$

$\sigma_r/u_* = -1.39 - 1.28(z/L)^{0.409}$, $r = 0.897$

$\sigma_r/u_* = 0.83 + 1.43(-z/L)^{0.342}$, $r = 0.826$

$\sigma_r/u_* = -0.85 - 1.35(-z/L)^{0.323}$, $r = 0.822$

$\sigma_r/u_* = 1.11 + 0.00(z/L)^{0.457}$, $r = 0.819$

$\sigma_r/u_* = -1.62 - 0.88(-z/L)^{0.428}$, $r = 0.760$

$\sigma_r/u_* = 1.25 + 1.34(-z/L)^{0.434}$, $r = 0.716$

$\sigma_r/u_* = -2.43 - 1.17(-z/L)^{0.434}$, $r = 0.541$

$\sigma_r/u_* = 1.33 + 1.65(z/L)^{0.348}$, $r = 0.652$

$\sigma_r/u_* = -2.07 - 1.78(z/L)^{0.305}$, $r = 0.683$

$\sigma_r/u_* = 1.11 + 1.92(-z/L)^{0.350}$, $r = 0.664$

$\sigma_r/u_* = -2.66 - 1.36(-z/L)^{0.339}$, $r = 0.550$

$\sigma_r/u_* = 1.29 + 1.39(z/L)^{0.392}$, $r = 0.593$

$\sigma_r/u_* = -2.22 - 1.69(z/L)^{0.294}$, $r = 0.638$

(c)

FIGURE 4: Normalized standard deviation of (a) lateral, (b) longitudinal, and (c) vertical velocity components as a function of $z/L$.

TABLE 1: Various relations among normalized standard deviations of vertical wind component to $z/L$.

| Unstable case ($z/L < 0$) | | Stable case ($z/L > 0$) | |
|---|---|---|---|
| Panofsky et al., 1977 [21] | $\dfrac{\sigma_w}{u_*} = 1.25 + \left(1 - 3\dfrac{z}{L}\right)^{1/3}$ | Krishnan and Kunhikrishnan, 2002 [22] | $\dfrac{\sigma_w}{u_*} = 1.37\left(1 + 0.2\dfrac{z}{L}\right)$ |
| Nair et al., 1990 [23] | $\dfrac{\sigma_w}{u_*} = 1.1 + \left(1 - 4\dfrac{z}{L}\right)^{1/3}$ | Ramana et al., 2004 [24] | $\dfrac{\sigma_w}{u_*} = 1.0\left(1 + 0.2\dfrac{z}{L}\right)$ |
| Krishnan and Kunhikrishnan, 2002 [22] | $\dfrac{\sigma_w}{u_*} = 1.37 + \left(1 - 3\dfrac{z}{L}\right)^{1/3}$ | Present data fit through power unity (for January) | $\dfrac{\sigma_r}{u_*} = 1.6\left(1 + 0.345\dfrac{z}{L}\right), r = 0.41$ |
| Ramana et al., 2004 [24] | $\dfrac{\sigma_w}{u_*} = 1.0 + \left(1 - 3\dfrac{z}{L}\right)^{1/3}$ | Power curve relation of the form $y = a + bx^c$ (for January) | $\dfrac{\sigma_r}{u_*} = 1.15 + 1.01\dfrac{z^{0.415}}{L}, r = 0.688$ |
| Present data fit through 1/3 power relation (for January) | $\dfrac{\sigma_r}{u_*} = 1.22 + \left(1 - 4.53\dfrac{z}{L}\right)^{1/3}, r = 0.845$ | Pahlow et al., 2001 [25] | $\dfrac{\sigma_w}{u_*} = 1.10 + 0.90\dfrac{z^{0.6}}{L}$ |
| Power curve relation of the form $y = a + bx^c$ (for January) | $\dfrac{\sigma_r}{u_*} = 1.01 + 1.12\left(-\dfrac{z}{L}\right)^{0.475}, r = 0.837$ | | |

Mathematically mean resultant wind direction (MRWD) can be expressed by

$$\text{MRWD} = \tan^{-1}\left\{\frac{\sum_{i=1}^{N}(u_i/N)}{\sum_{i=1}^{N}(v_i/N)}\right\}, \qquad (9)$$

where $u_i$ and $v_i$ are zonal and meridional wind components. $N$ is the number of observations. The structure of MRWD is shown in Figure 8. Due to the orientation of the coastline, clockwise wind directions which fall between 145° and 325° are considered as sea-breeze flow and from 325° to 145° are in the land-breeze regime. The changeover from sea to land breeze and land to sea breeze is well remarkable from October to May through January. The sea-breeze onset takes place around 1000 IST (IST = GMT + 5 hr 30 min), whereas the land-breeze onset is around 2000 IST during January to May and 1800 IST during October to December. In the SW monsoon months, between 1030 and 1930 IST, the sea breeze is in westerly direction and north-north-westerly during other timings. The north-north-westerly wind direction may be attributed to the interaction among the dominant synoptic scale monsoon flow and the mesoscale changes in the thermal environment due to land-sea temperature contrast.

Momentum flux ($M$) (Figure 9) shows higher negativity during sea-breeze flow, which means that the sea-breeze flow generated shearing force that transmits the $M$ downwards and the transfer mechanism was explained by Oke [44]. Due to the effect of forced mechanical convection generated by the surface roughness and mutual shearing between air flow layers moving at different velocities, turbulent eddies are continually moving up and down with respect to the measuring level. An eddy arriving at the measurement level developed above will, upon mixing, impart a net increase in velocity (and hence momentum) and thereby the anemometer would see this downdraft as an increase in wind speed during the course of a day as observed in mean scalar wind speed. The net effect of both updrafts and downdrafts is always to maintain a net flux of momentum downwards and results in negative values. Figure 10 represents the structure of stress off-wind angle ($\alpha$), which is low during onshore flow and high in offshore wind regime. $\alpha$ value provides the orientation

of wind direction vector with respect to shear vector and vice versa. The mutual orientation of wind and stress vector as coterminus/collinear corresponds to +ve/−ve $M$ values, respectively.

By adding mean resultant wind direction and the mean stress off-wind angle, the mean direction of stress is computed [43] in the meteorological coordinate system as depicted in Figure 11. For example, for a wind direction of 180° (southerly) and off-wind stress angle of 45°, the direction of the stress vector will be from 225° (southwesterly). The most dominant mean direction of the stress for this station is from about 242° during the daytime portion of SW monsoon season.

## 4. Mean Structure of Wind and Turbulent Parameters

In scalar wind speed (Figure 12), maxima are observed in the sea-breeze regime of the order of $\geq 2\,\text{m s}^{-1}$ and above and feeble wind speeds are observed during the land-breeze regime. The highest scalar speeds in the diurnal course correspond to the highest negativity in momentum flux values.

Figure 13 shows $u_*$ structure. During situations where turbulence created or modulated by wind shear near the ground is larger, the magnitude of the surface Reynolds stress is of larger and thereby higher $u_*$ values. Over the station, high $u_*$ is encountered during sea-breeze regime and highest values are observed of the order $0.35\,\text{m s}^{-1}$.

Large $u_*$ values during sea breeze compared to land breeze indirectly explain that wind shear generated by sea breeze is higher than that produced by land breeze.

Due to mechanical and thermal convection with respect to stability conditions, both +ve and −ve values are observed in the vertical wind component ($w$ component) (Figure 14). Higher −ve values are seen between 1200 IST and 1600 IST in the sea-breeze regime, except from May to July. Before 0900 IST and after 1700 IST +ve values are observed due to suppressed thermal activity. −ve values are attributed by the net downward flux of momentum created by shearing stress among layers. The portions of the greatest −ve

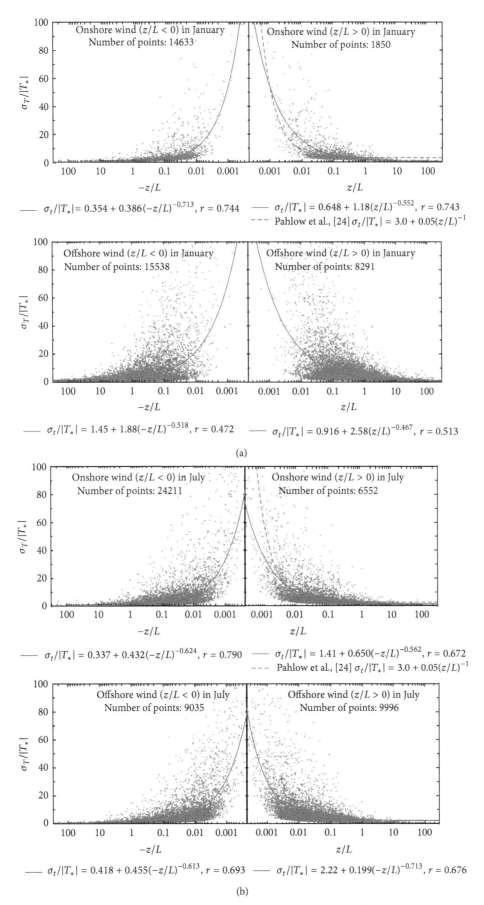

FIGURE 5: Normalized standard deviation of absolute value of sonic temperature as a function of $z/L$ for January and July.

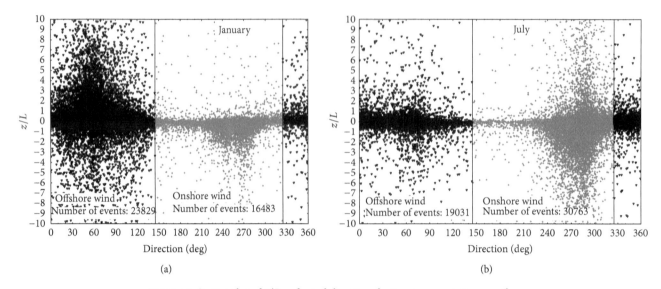

FIGURE 6: Scatter plot of $z/L$ and wind direction for two representative months.

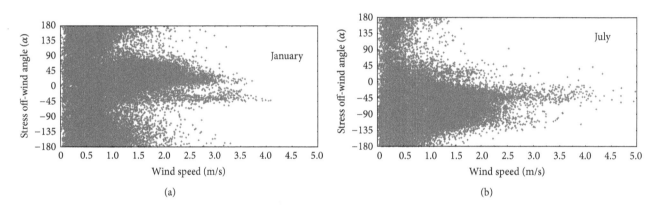

FIGURE 7: Scatter plot of stress off-wind angle ($\alpha$) and wind speed.

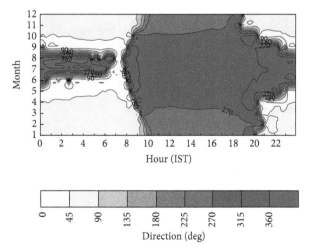

FIGURE 8: Structure mean resultant wind direction.

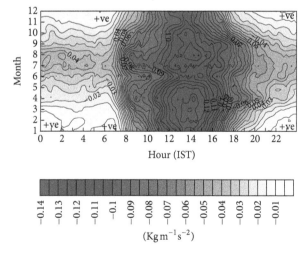

FIGURE 9: Mean structure of momentum flux ($M$).

values are associated with the greatest values in the net downward momentum flux. The opposite is true for +ve values.

Systematic occurrences of Obukhov length values (Figure 15) make a quantitative interpretation on mechanical turbulence and heat convection. From 0730 IST to

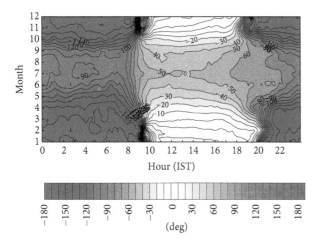

FIGURE 10: Mean structure of stress off-wind angle ($\alpha$).

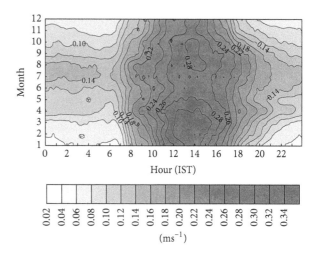

FIGURE 13: Mean structure of frictional velocity.

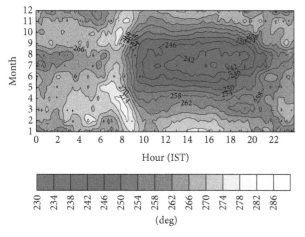

FIGURE 11: Mean direction of stress computed in the meteorological coordinate system.

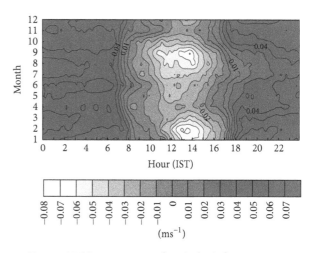

FIGURE 14: Mean structure of vertical wind component.

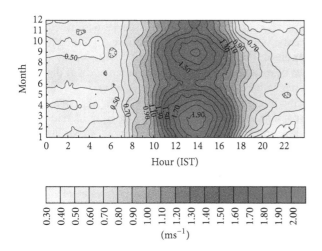

FIGURE 12: Mean scalar wind speed.

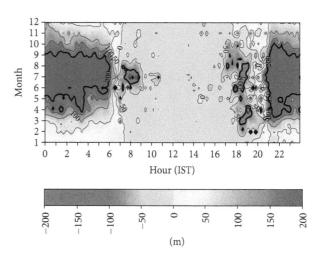

FIGURE 15: Mean structure of Obukhov Length ($L$).

TABLE 2: $L$ values under different stability characteristics.

| $L$ | Characteristics |
| --- | --- |
| >200 | Neutral |
| 50 to 100 | Weakly stable |
| 0 to 50 | Highly stable |
| 0 to −50 | Highly unstable |
| −50 to −100 | Weakly unstable |
| < −100 | Neutral |

1800/1900 IST it is observed that the $L$ values are negative and in the rest of the periods the values are positive. Strong/light negative $L$ values are the characteristics of highly/weakly unstable conditions and strong/light positive $L$ values represent highly/weakly stable conditions. A quantitative interpretation of $L$ may be inferred from Table 2. The changeover from negative to positive values and vice versa happens around 1930 IST and about 0730 IST, respectively. It is interesting to note the occurrence of neutral regimes for < −100 and >200 (corresponding to $z/L \approx 0.03$ and 0.02) values (presented with bold contour) is mostly during evening, night, and early hours of the day till about 0900 IST in SW monsoon. Overcast skies with high wind speeds are very common during these hours.

Turbulent sensible heat flux ($H$) values (Figure 16) in winter ($D, J, F$), summer ($M, A$), and premonsoon month May are higher than the SW monsoon ($J, J, A, S$) and NE monsoon ($O, N$) solely related to less cloudiness during former compared to later months. Another reason of $H$ increase around March and September is due to equinoctial solar activity. The highest values in turbulent heat flux during winter are of the order of above 280 W cm$^{-2}$. During the course of a day, the values increase steadily about an hour after the sunrise and decrease from about 1600 IST onwards. From the observations it is clear that heat flux values are totally controlled by cloudiness and solar activity.

Mean variation of $C_D$ is shown in Figure 17. The lowest values of less than 0.04 are observed during the sea-breeze regime between 1300 and 1600 IST. Though the wind speed and frictional velocity values are high during this period, increase in $C_D$ is not observed. The present result demonstrates the dominance of frictional velocity over the scalar speeds during daytime which results in low $C_D$ values (6) which may be inferred as changes in $C_D$ when the wind speed is rapid compared to $u_*$ [22].

Variations in turbulent intensities (longitudinal, i.e., collinear to the wind direction, transverse, i.e., perpendicular to the longitudinal and in the vertical) provide (Figures 18(a), 18(b), and 18(c)) excellent insight into SBL eddy formations. Turbulent intensities during the course of the day show similar features. This can be interpreted that an eddy which influences the measuring level will have equal contribution among the three perpendicular wind components. For an operational way to check turbulence, turbulent kinetic energy (TKE) helps to estimate the scale depth of eddy formation. Larger TKE indicates a higher intensity of turbulence. In the cascaded form of atmospheric energy spectra distribution,

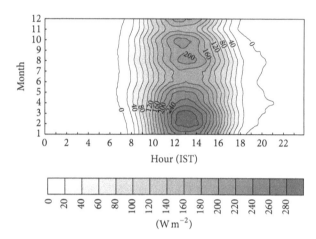

FIGURE 16: Structure of turbulent sensible heat flux values ($H$).

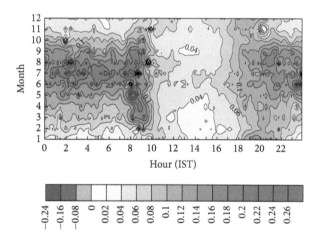

FIGURE 17: Structure of mean drag coefficient ($C_D$).

the TKE dissipation which takes place for internal energy of the system is through molecular viscosity. Mathematically

$$\text{TKE} = \frac{1}{2}\left(\sigma_u^2 + \sigma_v^2 + \sigma_w^2\right), \qquad (10)$$

where $\sigma_u$, $\sigma_v$, $\sigma_w$ are the standard deviations in wind components. Willis and Deardorff [45] have explained that, with a condition for standard deviation of scalar winds, $\sigma_s < 0.5$ times of mean scalar wind speed ($v$), an eddy of diameter $\lambda$ has negligible changes; then $P = \lambda/v$ (or $\lambda = Pv$) for the specific interval of time $P$ (also see Stull [46]). We assumed the passage of an eddy or part of the eddy of size ($\lambda$) past the sensor within a fixed time gap of $P = 5$ min with an average scalar speed of $v$ for cases $\sigma_s < 0.5v$. By inputting $v$, $\sigma_u, \sigma_v, \sigma_w$, and $P$, TKE and $\lambda$ can be estimated. TKE and eddy distribution in July and January are shown in Figures 19(a) and 19(b). Due to strong convection, high value of TKE/eddy size and eddy sizes are seen in July unlike air subsidence dominant winter month, January. Occurrences of TKE/eddy size of more than 2 correspond to eddy size above 600 m.

Diurnal variation of TKE and eddy size on a typical day is affected by mesoscale phenomena like sea/land breezes; thunderstorms are depicted in Figure 20. Abrupt increase

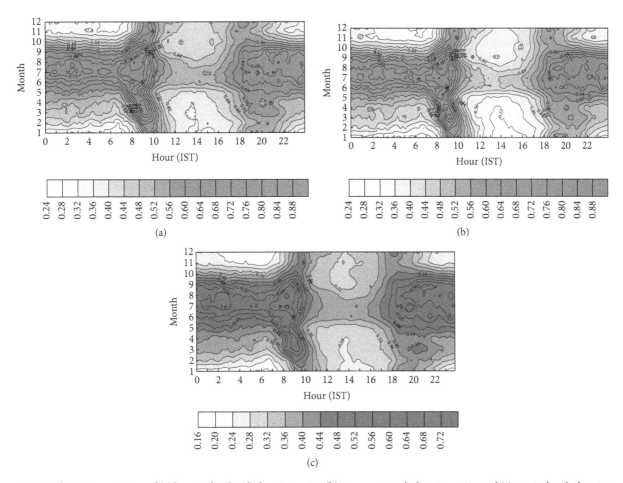

FIGURE 18: Similarity in variations of (a) longitudinal turbulent intensity, (b) transverse turbulent intensity, and (c) vertical turbulent intensity.

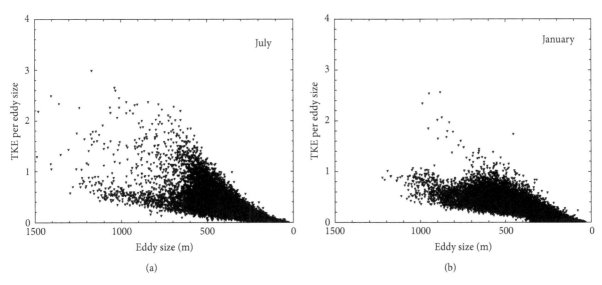

FIGURE 19: (a) Turbulent kinetic energy (TKE) and eddy size distribution in July and (b) January.

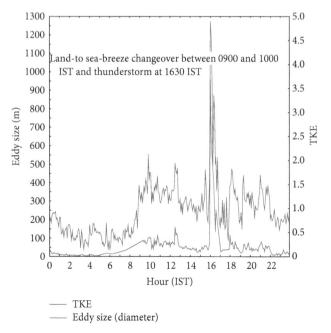

FIGURE 20: Diurnal variation of TKE and eddy size on May 09, 2010.

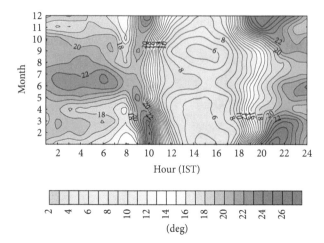

FIGURE 21: Structure of standard deviation of wind direction ($\sigma_\theta$).

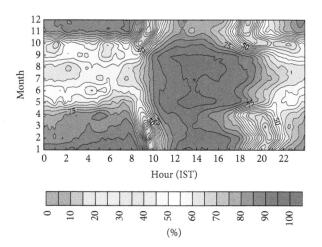

FIGURE 22: Structure of wind steadiness factor (WSF).

of TKE and thereby eddy size has been observed with changeover of land breeze to sea breeze (around 0900 IST) as well as a sudden shoot-up of TKE and eddy size during the passage of a thunderstorm (1630 IST).

## 5. Standard Deviation of Wind Direction and Wind Steadiness Factor

As wind direction is a circular function with a discontinuity of scale at 360°, special treatment is required to arrive the standard deviation of wind directions ($\sigma_\theta$) [47–50]. By the Ackermann [47] method, wind direction standard deviation is computed for an hour by using 12 five-minute interval wind data on speed and direction. In the same set of data wind steadiness factor (WSF) is also determined. WSF is a general quantitative meteorological technique to estimate the steadiness of wind direction [51]. Steadiness factor (expressed in %) is defined as the ratio of mean resultant wind speed (MRWS) to mean scalar wind speed (MSWS):

$$\text{MSWS} = \sum_{i=1}^{N} \frac{V_i}{N},$$

$$\text{MRWS} = \sqrt{\sum_{i=1}^{N} \left(\frac{u_i}{N}\right)^2 + \sum_{i=1}^{N} \left(\frac{v_i}{N}\right)^2}, \quad (11)$$

$$\text{WSF (\%)} = \frac{\text{MRWS}}{\text{MSWS}} \times 100,$$

where $V_i$, $u_i$, $v_i$, and $N$ are scalar wind speed, zonal component, meridional component, and number of observations, respectively. Mean structure of wind direction standard

deviation and WSF are shown in Figures 21 and 22, respectively. Inverse proportionality is detected among $\sigma_\theta$ and WSF as noticed by Namboodiri [20]. Standard deviation in wind direction ($\sigma_\theta$) provides a tool for atmospheric stability estimate for atmospheric dispersion modeling. To know the interaction between mesoscale onshore/offshore flow and the prevailing seasonal flow, diurnal $\sigma_\theta$ climatologies are compared between the west (Thumba) and east (Sriharikota) coast of the Peninsular India for July. In July both stations are under the influence of the same synoptic scale SW monsoonal flow. It is realized that whenever the mesoscale onshore/offshore flow and the synoptic scale seasonal flow are in opposite directions, the resultant $\sigma_\theta$ are high unlike low values for the same direction interactions (Figure 23). Half hourly variations of $\sigma_\theta$ and WSF are examined as a case study for an overhead quasistationary thunderstorm formation on May 09, 2010 from 1500 IST to 2100 IST (Figure 24). During the thunderstorm event, $\sigma_\theta$ increase and WSF decrease are observed.

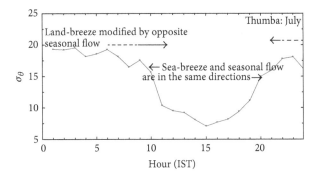

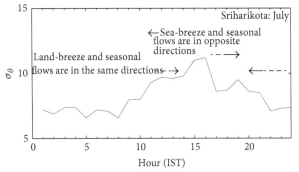

FIGURE 23: $\sigma_\theta$ variations due to mesoscale flow interaction with synoptic-scale flow.

## 6. Bivariate Normal Distribution (BND) Wind Model

When $u$ and $v$ are correlated, it is a necessary but not a sufficient condition for the marginal distribution $f(u)$ and $f(v)$ to be normal distribution and the joint distribution to be BND. Problems dealing with wind components require the integral of elliptical BND. An elaborate discussion on BND applicable to SBL wind components and the stepwise construction procedures can be seen in Namboodiri et al. [52]. The probability density function (pdf) for BND is

$$P(u, v)$$

$$= \frac{1}{2\pi\sigma_u\sigma_v\sqrt{\left(1 - \rho_{uv}^2\right)}}$$

$$\times \exp\left[-\frac{1}{2\left(1 - \rho_{uv}^2\right)}\right.$$

$$\times \left(\frac{(u - U)^2}{\sigma_u^2} - \frac{2\rho_{uv}(u - U)(v - V)}{\sigma_{uv}}\right.$$

$$\left.\left. + \frac{(v - V)^2}{\sigma_v^2}\right)\right] \qquad (12)$$

$$-\infty \leq u \leq \infty, \quad -\infty \leq v \leq \infty.$$

The pdf can be derived by setting

$$\left[\frac{(u - U)^2}{\sigma_u^2} - \frac{2\rho_{uv}(u - U)(v - V)}{\sigma_{uv}} + \frac{(v - V)^2}{\sigma_v^2}\right] = \lambda^2. \quad (13)$$

$U, V, \sigma_u, \sigma_v$, and $\rho_{uv}$ represent mean value of the component $u$, mean value of the component $v$, standard deviation of $u$, standard deviation of $v$, and correlation coefficient among $u$ and $v$, respectively.

Equation (13) is recognized as a family of ellipses depending on the values of $\lambda^2$. The density function has constant values on these ellipses. So the ellipses generated out of (13) are referred to as ellipses of equal probability. The percentile ellipses can be generated based on $\lambda$ value variations as

$$\lambda = \sqrt{2}\left\{\sqrt{-\ln\left(1 - \mathbf{P}\right)}\right\}, \qquad (14)$$

where $\mathbf{P}$ is the probability. For 90% probability ellipse $\lambda = 2.146$ and the same probability computations and thereby ellipse constructions are made in BND of wind components generated in the study. BND ellipses are available from Figures 25(a) and 25(b) for winter and SW monsoon months. Ellipse orientation is based on the prevailing wind direction of the season and thereby ellipses pertaining to same season are depicted in the same diagram. The larger measure of major axes in ellipses shows the dominance of zonal components. The ellipse quadrant having more area possesses more numbers of occurrences of events.

## 7. Structure of Surface Roughness Parameter ($z_0$)

$z_0$ values are computed by (1) by taking every 5 min values of different variables in the equation by considering unstable ($z/L < 0$) cases. Mean monthly structure of roughness parameter is in Figure 26, which shows the lowest and highest value regimes in sea breeze and land breeze, respectively. Mean structure of $z_0$ ranges from 0.02 m to 0.24 m with respect to wind direction. High values or imaginary values (not presented) mostly during land breeze and are associated with $z/L > 0$. $z_0$ values are computed for 5 min interval unstable case ($z/L < 0$) raw data by using (1). $z_0$ values observed during $z/L > 0$ have shown high or imaginary due to exponentiation in (1). Polar plot of 319656 numbers of $z_0$ values with respect to wind direction is shown in Figure 27 and most of the $z_0$ values are within 0.5 m, a representative value for sea coastline from the sonic anemometer measurements at 3.3 m level. For every 10° wind direction interval mean and standard deviation ($\sigma$) of $z_0$ are computed and the mean, mean + $\sigma$, mean + 2$\sigma$, and (mean + 3$\sigma$), plots are incorporated as thick lines in the polar diagram. A line shows 145°–325° is also made to demark sea-coastline orientation. Low and high $z_0$ values are clearly observed during onshore and offshore flows, respectively, with sharp decrease around 242° shown by an arrow. This interesting observation tells that minimization of $z_0$ values may not be the result of onshore flow perpendicular to the coast but due to orientation of the wind stress vector (242°). Buildup in $z_0$ starts as nearing to the coast at 325° orientation due to land frictional effect and maxima in $z_0$ happen around 35° azimuth. Onshore winds are arriving to the observation point through medium sized coconut plantations. Another low roughness regime encountered around 125° is attributed to the barren sandy land in the SE sector of the anemometer. High roughness

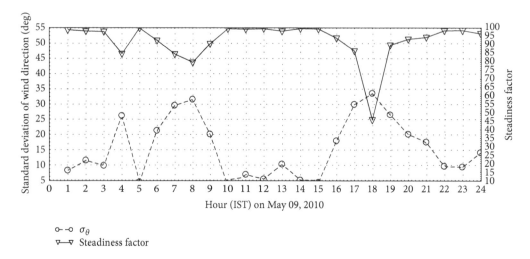

FIGURE 24: Half hourly variations of $\sigma_\theta$ and WSF during an overhead quasistationary thunderstorm formation on May 09, 2010 from 1500 to 2100 IST with severity around 1800 IST.

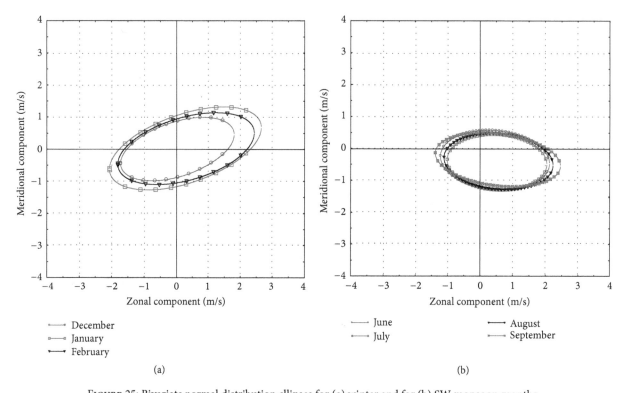

FIGURE 25: Bivariate normal distribution ellipses for (a) winter and for (b) SW monsoon months.

values are due to bushy and cashew vegetations between NNW and NE sector.

$z_0$ variations during different strengths of the SW monsoon activity over the region are studied. The strength of SW monsoon is called weak when the rainfall is less than half of the normal and vigorous if the monsoon rainfall is more than 4 times the normal under the conditions that the rainfall in at least two stations should be 8 cm (if the subdivision is along the west coast of India) and rainfall in the meteorological subdivision should be fairly widespread or widespread. The data on daily strength of SW monsoon and rainfall used in this study are made available from the daily weather report from Meteorological Centre, IMD, Thiruvananthapuram, Kerala. Figure 28 shows average $z_0$ values obtained during 10 weak and vigorous days from 2007 to 2012. Numbers ($N$) of $z_0$ observations are 2424 and 2404 for weak and vigorous monsoon, respectively. Average values of weak and vigorous monsoons are 0.056 m and 0.098 m, respectively, attributed to more northerliness of synoptic wind flow towards 325° coast azimuth.

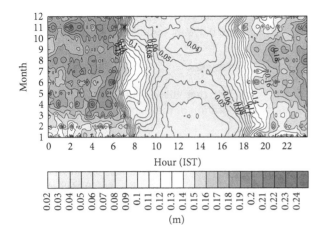

FIGURE 26: Mean monthly structure of roughness parameter ($z_0$) for unstable cases ($z/L < 0$).

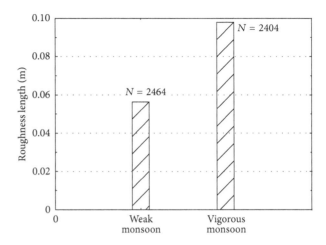

FIGURE 28: Average values of $z_0$ during 10 cases of weak and vigorous monsoon days from 2007 to 2012.

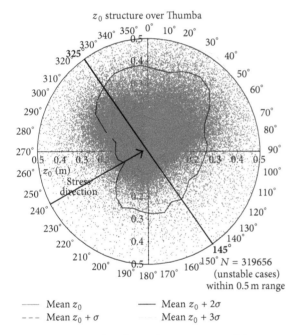

FIGURE 27: Polar plot of $z_0$ with respect to wind direction.

## 8. Conclusions

Detailed characterizations of lower SBL features and corresponding physical/meteorological mechanisms pertained are investigated and discussed over TERLS for potential applications and characterization of tropical SBL. The following are the salient results.

(i) $u_*$ peaks at $1/L \to 0$ and small $u_*$ values associated with high $-$ve $1/L$ values are due to heat convection and those with $+$ve $1/L$ values are due to severely reduced mechanical turbulence. Medium values of $u_*$ are due to mechanical turbulence caused by temperature stratification and inertia. Peak $u_*$ during $1/L \to 0$ is in the neutral regime.

(ii) $\sigma_w$ Vs $u_*$ distribution has shown typical neutral $u_*$ value.

(iii) Power curves relations of the form $\sigma/u_* = a + b(z/L)^c$ are fitted among normalized variances of $(p, q, r, T)$ and $z/L$. In the weakly unstable or stable regime of $z/L$ from $-10$ to $10$ with maximum $z/L$ points of occurrences, $z/L$ and variances are in poor dependence with constancy in variances. Beyond $z/L = \pm10$, variances enhance abruptly in the region of high stable or unstable regime. Neutral regime has been identified between $z/L$ from $-0.03$ to $0.02$.

(iv) Scatter of points for different wind directions against $z/L$ shows that onshore SBL wind profiles are of mostly unstable nature, whereas offshore wind profiles belong to either stable or unstable type.

(v) Significant decrease/increase in wind speed in stress off-wind angle ($\alpha$) is seen with respect to wind speed increase/decrease.

(vi) The structure of MRWD has shown wind direction climatology. Due to the dominant synoptic-scale SW monsoon system influence, sea-breeze flow has been seen throughout the day.

(vii) In the mutual orientation of wind and stress vectors coterminus/collinear orientation corresponds to $+$ve$/-$ve $M$ values, respectively.

(viii) The net effect of both updrafts and downdrafts in the SBL is always to maintain a net flux of momentum downwards and results in negative values.

(ix) The most dominant mean direction of the stress for this station is from about $242°$ during day time portion of the SW monsoon season.

(x) The highest scalar speeds in the diurnal course correspond to the highest negativity in momentum flux values.

(xi) Large $u_*$ values during sea breeze compared to land breeze indirectly explain that wind shear generated by sea breeze is higher than that produced by land breeze.

(xii) The portions of the greatest −ve $w$ components are associated with the greatest values in the net downward momentum flux.

(xiii) From 0730 IST to 1800/1900 IST it is observed that the $L$ values are negative, and in the rest of the periods the values are positive. Strong/light negative $L$ values are the characteristics of highly/weakly unstable conditions and strong/light positive $L$ values designate highly/weakly stable conditions.

(xiv) The occurrence of neutral regimes is mostly during evening, night, and early hours of the day till about 0900 IST in SW monsoon.

(xv) The highest values in turbulent heat flux encountered during winter are of the order of above 280 W m$^{-2}$.

(xvi) The lowest $C_D$ values of less than 0.04 are observed during the sea-breeze regime between 1300 IST and 1600 IST. Changes in $C_D$ with the wind speed are rapid compared to $u_*$.

(xvii) Variations in turbulent intensities show similar features, which can be interpreted that an eddy which influences the measuring level will have equal contribution among the three perpendicular wind components.

(xviii) In the SW monsoon featured by strong updraft currents (and thereby strong convection) greater values in both TKE per eddy size and eddy sizes are seen, whereas during air subsidence predominant winter month, January has contributed to lesser extent to TKE and in eddy size.

(xix) Abrupt increase of TKE and thereby eddy size has been observed with changeover of land breeze to sea breeze and during the passage of a thunderstorm.

(xx) Inverse proportionality is detected between $\sigma_\theta$ and WSF. Whenever the mesoscale onshore/offshore flow is opposite to the synoptic scale seasonal flow direction, the resultant $\sigma_\theta$ are higher unlike low values for same direction interactions. During thunderstorm event, $\sigma_\theta$ increase and WSF decrease are observed.

(xxi) BND ellipse orientation is based on the prevailing wind direction of the season and larger measure of major axes in ellipses shows the dominance of zonal components.

(xxii) Mean structure of $z_0$ ranges from 0.02 m to 0.24 m with respect to wind direction and most of the $z_0$ values are within 0.5 m. Low and high $z_0$ values are clearly observed during onshore and offshore flows, respectively, with sharp decrease around 242° due to orientation of the wind stress vector along this particular direction (242°). Build-up in $z_0$ starts as nearing to the coast at 325° orientation due to land frictional effect and maxima in $z_0$ happen around 35° azimuth.

(xxiii) Average $z_0$ values of weak and vigorous monsoons are 0.056 m and 0.098 m, respectively.

## Conflict of Interests

The authors declare that there is no conflict of interests regarding the publication of this paper.

## References

[1] J. C. Wyngaard, "On surface-layer turbulence," in *Workshop on Micrometeorology*, D. A. Haugen, Ed., pp. 101–149, American Meteorological Society, Boston, Mass, USA, 1973.

[2] G. A. McBean, Ed., *Planetary Boundary Layer: Technical Note No. 165*, World Meterological Organization, Geneva, Switzerland, 1979.

[3] H. A. Panofsky and J. A. Dutton, *Atmospheric Turbulence*, Academic Press, New York, NY, USA, 1984.

[4] R. E. Huschke, Ed., *Glossary of Meteorology*, American Meteorological Society, Boston, Mass, USA, 1959.

[5] J. A. Businger, "Turbulence transport in the atmospheric surface layer," in *Workshop on Micrometeorology*, D. A. Haugen, Ed., pp. 67–100, American Meteorological Society, Boston, Mass, USA, 1973.

[6] A. S. Monin and A. M. Obukhov, "The basic laws of turbulent mixing in the surface layer of the atmosphere," *Trudy Botanicheskogo Instituta Akademii Nauk SSSR*, vol. 24, no. 151, pp. 163–187, 1954.

[7] C. A. Paulson, "The mathematical representation of wind speed and temperature profiles in the unstable atmospheric surface layer," *Journal of Applied Meteorology*, vol. 9, no. 6, pp. 857–861, 1970.

[8] K. G. Rao, S. Raman, A. Prabhu, and R. Narasimha, "Turbulent heat flux variation over the Monsoon-Trough region during MONTBLEX-90," *Atmospheric Environment*, vol. 29, no. 16, pp. 2113–2129, 1995.

[9] C. J. Nappo, "Mesoscale flow over complex terrain during the Eastern Tennessee Trajectory Experiment (ETTEX)," *Journal of Applied Meteorology*, vol. 16, no. 11, pp. 1186–1196, 1966.

[10] A.-S. Smedman-Hogstrom and U. Hogstrom, "A practical method for determining wind frequency distributions for the lowest 200 m from routine meteorological data," *Journal of Applied Meteorology*, vol. 17, no. 7, pp. 942–954, 1978.

[11] B. B. Hicks, P. Hyson, and C. J. Moore, "A study of eddy fluxes over a forest," *Journal of Applied Meteorology*, vol. 4, pp. 58–66, 1975.

[12] J. Kondo and H. Yamazawa, "Aerodynamic roughness over an inhomogeneous ground surface," *Boundary-Layer Meteorology*, vol. 35, no. 4, pp. 331–348, 1986.

[13] R. S. Thompson, "Notes on the aerodynamic roughness length for complex terrain," *Journal of Applied Meteorology*, vol. 17, no. 9, pp. 1402–1403, 1978.

[14] J. R. Garratt, "Review of drag coefficients over oceans and continents," *Monthly Weather Review*, vol. 105, pp. 915–929, 1977.

[15] H. A. Panofsky and E. L. Peterson, "Wind profiles and change of surface roughness at Riso," *Quarterly Journal of the Royal Meteorological Society*, vol. 98, pp. 845–854, 1972.

[16] J. C. Kaimal, R. A. Eversole, D. H. Lenschow, B. B. Stankov, P. H. Kahn, and J. A. Businger, "Spectral characteristics of the convective boundary layer over uneven terrain," *Journal of the Atmospheric Sciences*, vol. 38, no. 5, pp. 1098–1114, 1982.

[17] A. Korrell, H. A. Panosky, and R. J. Rossi, "Wind profiles at the Boulder Tower," *Boundary-Layer Meteorology*, vol. 22, no. 3, pp. 295–312, 1982.

[18] R. Ramachandran, J. W. J. Prakash, K. S. Gupta, K. N. Nair, and P. K. Kunhikrishnan, "Variability of surface roughness and turbulence intensities at a coastal site in India," *Boundary-Layer Meteorology*, vol. 70, no. 4, pp. 385–400, 1994.

[19] S. K. Gupta, P. K. Kunhikrishnan, R. Ramachandran, J. W. J. Prakash, and K. N. Nair, "On the characteristic of coastal atmospheric boundary layer," Global Change Studies, Indian Space Research Organisation Scientific Report no. ISRO-GBP-SR-42-94, 1994.

[20] K. V. S. Namboodiri, *Studies on the vertical structure of horizontal wind variability in the surface boundary layer over Sriharikota [Ph.D. thesis]*, Cochin University of Science and Technology, Cochin, India, 2000.

[21] H. A. Panofsky, H. Tennekes, D. H. Lenschow, and J. C. Wyngaard, "The characteristics of turbulent velocity components in the surface layer under convective conditions," *Boundary-Layer Meteorology*, vol. 11, no. 3, pp. 355–361, 1977.

[22] P. Krishnan and P. K. Kunhikrishnan, "Some characteristics of atmospheric surface layer over a tropical inland region during southwest monsoon period," *Atmospheric Research*, vol. 62, no. 1-2, pp. 111–124, 2002.

[23] K. N. Nair, P. K. Kunhikrishnan, R. Ramachandran, K. S. Gupta, and J. W. J. Prakash, "Surface layer studies using tower based measurements at Thumba, India," in *Proceedings of the 5th International Symposium on Acoustic Remote Sensing of the Atmosphere and Oceans*, pp. 325–329, Tata McGraw-Hill, New Delhi, India, 1990.

[24] M. V. Ramana, P. Krishnan, and P. K. Kunhikrishnan, "Surface Boundary-Layer characteristics over a tropical inland station: seasonal features," *Boundary-Layer Meteorology*, vol. 111, no. 1, pp. 153–175, 2004.

[25] M. Pahlow, M. B. Parlange, and F. Porte-Agel, "On Monin-Obukhov similarity in the stable atmospheric boundary layer," *Boundary-Layer Meteorology*, vol. 99, no. 2, pp. 225–248, 2001.

[26] G. L. Geernaert, F. Hansen, M. Courtney, and T. Herbers, "Directional attributes of the ocean surface wind stress vector," *Journal of Geophysical Research*, vol. 98, no. 9, pp. 16571–16582, 1993.

[27] A. A. Grachev and C. W. Fairall, "Upward momentum transfer in the marine boundary layer," *Journal of Physical Oceanography*, vol. 31, no. 7, pp. 1698–1711, 2001.

[28] B. Lange, S. Larsen, J. Hojstrup, and R. Barthelmie, "The influence of thermal effects on the wind speed profile of the coastal marine boundary layer," *Boundary-Layer Meteorology*, vol. 112, no. 3, pp. 587–617, 2004.

[29] T. Kilpelainen and A. Sjoblom, "Momentum and sensible heat exchange in ice-free Arctic Fjord," *Boundary-Layer Meteorology*, vol. 134, no. 1, pp. 109–130, 2010.

[30] W. Brutsaert, "Aspects of bulk atmospheric boundary layer similarity under free-convective conditions," *Reviews of Geophysics*, vol. 37, no. 4, pp. 439–451, 1999.

[31] H. C. Friebel, A. Y. Benilov, J. L. Hanson, and D. T. Resio, "Long-term drag coefficient measurements in the coastal zone," USACE-FRP Report, US Army Corps of Engineers-NAP, 2009.

[32] P. K. Kunhikrishnan, *Studies of atmospheric boundary layer [Ph.D. thesis]*, Kerala University, Kerala, India, 1990.

[33] J. C. Kaimal and J. J. Finnigan, *Atmospheric Boundary Layer Flows, There Structure and Measurement*, Oxford University Press, New York, NY, USA, 1994.

[34] W. J. Prakash, R. Ramachandran, K. N. Nair, K. S. Gupta, and P. K. Kunhikrishnan, "On the spectral behaviour of atmospheric boundary-layer parameters at Thumba, India," *Quarterly Journal of the Royal Meteorological Society*, vol. 119, no. 509, pp. 187–197, 1993.

[35] C. Biltoft, "Momentum flux: gross, scalar, along wind or net?" Adiabat Meteorological Services Note 0303, Applied Technologies, 2003.

[36] K. V. S. Namboodiri, P. K. Dileep, K. Mammen et al., "Effects of annular solar eclipse of 15 January 2010 on meteorological parameters in the 0 to 65 km region over Thumba, India," *Meteorologische Zeitschrift*, vol. 20, no. 6, pp. 635–647, 2011.

[37] A. S. Monin and A. M. Yaglom, *Statistical Fluid Mechanics: Mechanics of Turbulence*, vol. 1, The MIT Press, Cambridge, Mass, USA, 1971.

[38] J. C. Wyngaard and O. R. Cote, "Cospectral similarity in the atmospheric surface layer," *Quarterly Journal of the Royal Meteorological Society*, vol. 98, no. 417, pp. 590–603, 1972.

[39] A. Smedman, "Observations of a multi-level turbulence structure in a very stable atmospheric boundary layer," *Boundary-Layer Meteorology*, vol. 44, no. 3, pp. 231–253, 1988.

[40] L. Mahrt, "Stratified atmospheric boundary layers and breakdown of models," *Theoretical and Computational Fluid Dynamics*, vol. 11, no. 3-4, pp. 263–279, 1998.

[41] K. F. Rieder, J. A. Smith, and R. A. Weller, "Observed directional characteristics of the wind, wind stress, and surface waves on the open ocean," *Journal of Geophysical Research*, vol. 99, no. 11, pp. 22589–22596, 1994.

[42] G. L. Geernaert, "Measurements of the angle between the wind vector and wind stress vector in the surface layer over the North Sea," *Journal of Geophysical Research: Oceans*, vol. 93, no. 7, pp. 8215–8220, 1988.

[43] P. O. G. Persson, B. Walter, and J. Hare, "Maritime differences between wind direction and stress: relationships to atmospheric fronts and implications," in *Proceedings of the 13th Conference on Interactions of the Sea and Atmosphere*, Portland, Me, USA, August 2004.

[44] T. R. Oke, *Boundary Layer Climates*, Halsted Press, New York, NY, USA, 1978.

[45] G. E. Willis and J. W. Deardorff, "On the use of Taylor's translation hypothesis diffusion in the mixed layer," *Quarterly Journal of the Royal Meteorological Society*, vol. 102, no. 434, pp. 817–822, 1976.

[46] R. B. Stull, *An Introduction to Boundary Layer Meteorology*, Kluwer Academic Publishers, Dordrecht, The Netherlands, 1994.

[47] G. R. Ackermann, "Means and standard deviations of horizontal wind components," *Journal of Climate and Applied Meteorology*, vol. 22, no. 5, pp. 959–961, 1983.

[48] K. A. Verrall and R. L. Williams, "A method of estimating the standard deviation of wind directions," *Journal of Applied Meteorology*, vol. 21, no. 12, pp. 1922–1925, 1982.

[49] R. J. Yamartino, "A comparison of several "single-pass" estimators of the standard deviation of wind direction," *Journal of Climate and Applied Meteorology*, vol. 23, no. 9, pp. 1362–1366, 1984.

[50] D. B. Turner, "Comparison of three methods for calculating the standard deviation of the wind direction," *Journal of Climate and Applied Meteorology*, vol. 25, no. 5, pp. 703–707, 1986.

[51] K. V. S. Namboodiri, P. K. Dileep, and K. Mammen, "Wind steadiness up to 35 km and its variability before the southwest

monsoon onset and the withdrawal," *Mausam*, vol. 63, no. 2, pp. 275–282, 2012.

[52] K. V. S. Namboodiri, G. V. Rama, and K. Mohan Kumar, "Distribution of horizontal wind components in the Surface Boundary Layer (SBL) over Sriharikota," *Mausam*, vol. 57, no. 2, pp. 301–306, 2006.

# 8

# On the Differences in the Intraseasonal Rainfall Variability between Western and Eastern Central Africa: Case of 10–25-Day Oscillations

**Alain Tchakoutio Sandjon,**[1,2] **Armand Nzeukou,**[1] **Clément Tchawoua,**[3] **and Tengeleng Siddi**[1,4]

[1] *Laboratory of Industrial Systems and Environmental Engineering, Fotso Victor Technology Institute, University of Dschang, Cameroon*
[2] *Laboratory for Environmental Modeling and Atmospheric Physics, Department of Physics, Faculty of Sciences,*
   *University of Yaoundé 1, P.O. Box 812, Yaoundé, Cameroon*
[3] *Laboratory of Mechanics, Department of Physics, Faculty of Science, University of Yaoundé 1, Cameroon*
[4] *Higher Institute of Sahel, University of Maroua, Cameroon*

Correspondence should be addressed to Alain Tchakoutio Sandjon; stchakoutio@yahoo.com

Academic Editor: Maxim Ogurtsov

In this paper, we analyze the space-time structures of the 10–25 day intraseasonal variability of rainfall over Central Africa (CA) using 1DD GPCP rainfall product for the period 1996–2009, with an emphasis on the comparison between the western Central Africa (WCA) and the eastern Central Africa (ECA) with different climate features. The results of Empirical Orthogonal Functions (EOFs) analysis have shown that the amount of variance explained by the leading EOFs is greater in ECA than WCA (40.6% and 48.1%, for WCA and ECA, resp.). For the two subregions, the power spectra of the principal components (PCs) peak around 15 days, indicating a biweekly signal. The lagged cross-correlations computed between WCA and ECA PCs time series showed that most of the WCA PCs lead ECA PCs time series with a time scale of 5–8 days. The variations of Intraseasonal Oscillations (ISO) activity are weak in WCA, when compared with ECA where the signal exhibits large annual and interannual variations. Globally, the correlation coefficients computed between ECA and WCA annual mean ISO power time series are weak, revealing that the processes driving the interannual modulation of ISO signal should be different in nature or magnitude in the two subregions.

## 1. Introduction

The monitoring and prediction of climate in the tropics remain a crucial problem in the scientific community. It is well-known that many regions in the tropics are vulnerable to climate change because their resources are highly rainfall dependent. Amongst these regions, Central Africa (CA) is particularly vulnerable because the majority of its population is rural and practice rain-fed agriculture [1]. A strong rainfall intensity can result in devastating floods, whereas a weak rainfall is usually associated with droughts, thus affecting living conditions and the economy of the densely populated region in many sectors such as agriculture, livestock, and energy [2, 3].

The CA extends from 15°S to 15°N and 0–50°E mainly over the land and part of Atlantic and Indian Oceans on its edges (Figure 1). The topography of the region is quite various, including highlands, mountains, and Plateaus. The western part (15°S–15°N; 0–30°E) is consisting of the zones of intense precipitation, especially over the Congo Basin [4]. Some of the highest rainfall totals are reported over Mountain Cameroon, at its western edge, where mean annual rainfall exceeds 10 meters. The Congo basin was proven to experience one of the world's most intense thunderstorms and highest frequency of lightning flashes [5, 6]. The western central Africa is almost covered by the Congo forest, which keep this region quite wet within the year. The Eastern part (15°S–15°N; 30–50°E) is characterized by widely diverse climates ranging

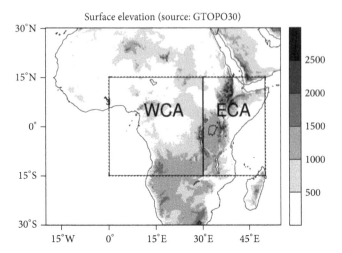

FIGURE 1: Surface elevation over the study area based on 30 min topographic data (m) from Digital Elevation Model (DEM) of the US Geological Survey. The study domain (15°S–15°N, 0–50°E) and the division line between WCA and ECA are shown on the plot.

TABLE 1: Comparison of some geographical features between ECA and WCA.

|  | Western Central Africa (west of Rift Valley) | Eastern Central Africa (east of Rift Valley) |
| --- | --- | --- |
| Average topography | Lower than 700 m | Greater than 1500 m |
| Vegetation | Almost covered by the Congo forest | From desert to small forests over relatively small areas |
| Mean annual rainfall | Greater than 1600 mm | Lower than 700 mm |
| Borders | Rift Valley-Atlantic Ocean | Rift Valley-Indian Ocean |

from desert to forest over relatively small areas. The complex topography of the region is an important contributing factor to climate because it leads to the orographically induced rainfall and affects local climates [7]. This region has suffered both excessive and deficient rainfall in recent years [8, 9]. In particular, the frequency of anomalously strong rainfall causing floods has increased. The difference between the east and west boundaries of CA is of approximately 3000 m in surface elevation and 3000 mm in annual mean rainfall. Then the two subregions (ECA and WCA), separated from each other by the Rift Valley, are very different in terms of topography, surface conditions, and precipitation (Table 1).

Rainfall variability in the tropics is complex, including different timescales ranging from few hours to a few months. Before the 1970s, the lack of high temporal resolution datasets leads many authors to the study of the rainfall fluctuations in the tropics in terms of annual cycle and interannual variability. The processes that induce rainfall variability within the season could not be investigated because it requires a high temporal resolution data. But since the beginning of the 1970s, the emergence of satellite data produced capability to

estimate regional or global rainfall fields at higher resolution, from the satellite measurements. Therefore, the notion of intraseasonal variability (ISV) arose in the scientific community and many authors used different rainfall or rainfall proxy datasets to study ISV in the tropics [10, 11].

Some of these authors have studied the ISV in precipitation for different geographical regions of Africa. For example, significant spectral peaks around 15- and 40-day period were found for Sahel precipitation by [12]. Analysis was made for Africa and signals over 10–25- and 25–60-day period were found in convection and precipitation in the western region [12–14]. [15] using outgoing long-wave radiation data developed climatology for tropical intraseasonal convective anomalies. [16] investigated the influence of the Madden-Julian oscillation on East African rainfall and showed that MJO phases leading to wet spells in the eastern (coastal) region are often those associated with overall suppressed deep convection in the Africa/Indian Ocean region. [17] have investigated the spatial and temporal structure of intraseasonal oscillations over Central Africa using 21 years (1980–2000) of pentad (5 day) rainfall and low level wind and they found three dominant modes for intraseasonal rainfall with the period of around 40–50 days. [18] used principal components analysis and Satellite-Derived Rainfall data to show that major cycles of intraseasonal rainfall events over Southern Africa are around 20 and 40 days.

Our understanding of the climate processes in CA is limited. Indeed, the literature on African climate variability is strongly biased towards West Africa (including the Sahel), East Africa, and Southern Africa. Thus, the CA region represents a notable gap in the study of the tropical climate system. Significantly, the structure of the intraseasonal oscillations (ISO) in Central Africa is only discussed by few authors [4, 17, 19, 20]. More recently, [19] used one-Degree Daily Global Precipitation Climatology Project (1 DD GPCP) higher resolution rainfall product to study the ISO patterns in CA, but they considered the whole area (15°S–15°N, 0–50°N) and did not take into account the contrast patterns between western and eastern parts of the region with different climate features.

Reference [4] also considered the whole area, and they showed that the East African modes (Ethiopian and Tanzanian modes) exhibit strong interannual variations, when compared with West African mode (Congo mode). Then the major issue of this study is the comparison of intraseasonal oscillations (ISO) patterns between western Central Africa (WCA) and eastern Central Africa (ECA). From the difference between physical features of ECA and WCA, one central question we may ask is *"what could be the difference in the ISO patterns between WCA and ECA?"* Some of the previous studies revealed two dominant frequency bands (10–25 and 25–60 days) at the intraseasonal timescale in tropical Africa [12–14], but for this study, we focused on the case of 10–25-day band. We use simple mathematical tools to assess the detailed spatial and temporal patterns of the leading modes of ISV of rainfall over the region, both throughout the year and from year to another.

In the next section, the data and method of analysis will be described. Section 3 will present the main results obtained and their analysis. Finally, Section 4 is devoted to a few discussions and conclusions.

## 2. Data and Methods

For this study, we used the 1DD GPCP rainfall data. The GPCP algorithm combines precipitation estimates from several sources, including infrared (IR) and passive microwave (PM) rain estimates and rain gauge observations [22, 23]. The IR data came mainly from the different Geostationary Meteorological Satellites but data from polar-orbiting satellites were also used to fill in the gaps at higher latitudes. The IR based estimates used the Geostationary Operational Environmental Satellite (GOES) precipitation index (GPI) described by [24]. The microwave data come mainly from the Special Sensor Microwave Imager (SSM/I) onboard the Defense Meteorological Satellite Program. The PM estimates were used to adjust the GPI estimate. Then, the multisatellite estimate was adjusted towards the large-scale gauge average for each grid box. The gauge-adjusted multisatellite estimates were then combined with gauge analysis using a weighted average, where the weights are the inverse error variances of the respective estimates. The current products include a monthly analysis at $2.5° \times 2.5°$ longitude-latitude grids, a 5-day (pentad) analysis at the same spatial resolution, and a daily product at a special resolution of one degree. The one-degree daily (1DD) product does not use PM rain estimates and gauge measurements directly [22]. SSM/I data were used within the framework of the threshold-matched precipitation index (TMPI) to delineate rain areas in the IR data. Gauge data were involved indirectly when the 1DD product was scaled so that monthly accumulations of 1DD matched the monthly GPCP product. The monthly and pentad analyses extend from 1979 to current, while the daily product is available starting from October 1996. The daily products are made available 2 to 3 months after the end of each month. The product used in this paper is the one-degree daily precipitation data available on the NOAA website http://www.esrl.noaa.gov/psd/data/gridded/data.gpcp.html. After some global and regional validations [25–27], the GPCP rainfall estimates are actually being used widely in place of gauge observations or to supplement gauge observations in the climate variability studies.

In the atmospheric sciences, climate is affected by the many phenomena with different spatial and temporal scales. To study the climate variability within a given scale, we generally use a digital filter to extract only the desired frequencies in a data time series. The Lanczos filtering [28, 29] used in this study is one of Fourier methods of filtering digital data. Its principal feature is the reduction of the amplitudes of Gibbs oscillation. Fourier coefficients for the smoothed response function are determined by multiplying the original weight function by a function that Lanczos called "sigma factors" (1):

$$\overline{w}_k = \frac{\sin 2\pi k f_c}{\pi k} \cdot \frac{\sin \pi k/n}{\pi k/n}, \tag{1}$$

where the first fraction is the original weight function and the second is Lanczos factor.

The wavelet analysis (WA) is a time series analysis method that has increasingly been applied in geophysics during the last three decades. It is becoming a common tool for analyzing temporal variations of power within a time series. The transformation in time series from the time space into the time-frequency space show that the WA is able to determine both the dominant timescales of variability and how they vary with time. WA has several attractive advantages over the traditional Fourier analysis, especially when dealing with time series with time-varying amplitudes. In contrast, to Fourier transform, that generate values of amplitudes and phases averaged over the entire time series for each frequency component or harmonic, wavelet transform provide a localized instantaneous estimate of the amplitude and phase for each spectral component of the series. This gives WA an advantage in the analysis of nonstationary data in which the amplitude and phase of the harmonic components may change rapidly in time or space. While Fourier transform of the nonstationary time series would smear out any detailed information on the changing features, the WA keeps track of the evolution of the signal characteristics throughout the time. Further details on wavelet analysis can be found in [30].

Empirical Orthogonal Functions (EOF) analysis is a powerful tool for identifying coherent patterns that explains the largest fraction of the total variance of a field and it has been used extensively in studies of intraseasonal variations in tropical convection [31–33]. The EOF analysis is used in this study to define space-time structures of the dominant modes of intraseasonal variability in our study area. In the EOF analysis, the time-dependent deviations from the long-term mean are decomposed into a sum of products of fixed spatial patterns $p_k$ and time-dependent amplitudes $\alpha_k$ (the principal components) as follows:

$$f(x_i, t) = \sum p_k(x_i) \alpha_k(t). \tag{2}$$

In (2), the principal components, $\alpha_k$, are uncorrelated to one another and described subsequently a maximum of variance in the original anomalies field, $f$. Under these conditions, EOFs, $p_k$ are eigenvectors of the covariance matrix of $f$.

## 3. Results and Analysis

*3.1. Spectral Analysis.* For every day, we averaged the data of all grid points in each of the two subregions (WCA and ECA) of the study area to have a mean value. The spectral analysis was then performed separately on the two daily area mean time series obtained. In Fourier decomposition, the importance of a given frequency in a time series can be quantified by the spectral variance, defined as the amplitude of the harmonic corresponding to the this frequency. The power spectra in the two subregions reveal the importance of the intraseasonal variability in Central African climate (Figure 2). However, one should note that the dominance is less pronounced in WCA when many significant peaks are observed at different frequencies in synoptic scale. The fraction of total variance occurring in the 35- to 80-day range is 0.33 for WCA and 0.48 for ECA. Wavelet analysis clearly revealed that the processes inducing rainfall variability

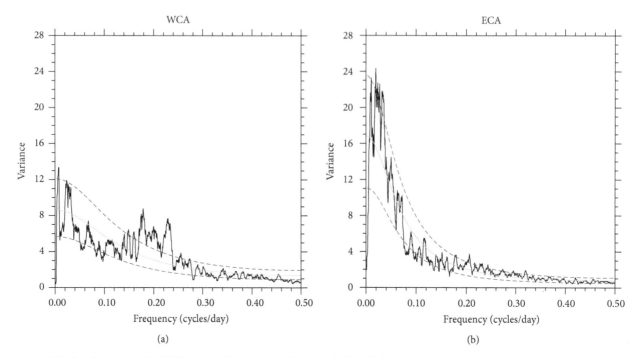

FIGURE 2: Calculated spectrum (solid black), red noise curve (green solid), and the curves indicating the upper (red dashed) and lower (blue dashed) bounds of 95% confidence level, for WCA (a) and ECA (b) subregions. The red noise spectrum is computed using the Markov processes [21].

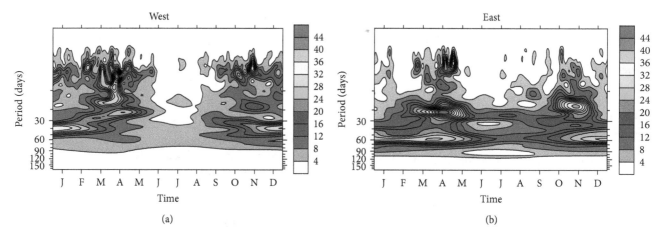

FIGURE 3: Annual mean wavelet power of the rainfall anomalies averaged over WCA (a) and ECA (b). A 120-day cutoff high pass filter was applied to long term anomalies in order to remove low frequencies variability such as interseasonal and interannual variability.

in the CA vary from synoptic (below 10 days) to longer timescales (Figures 3(a) and 3(b)). From the spectra, it is easy to identify the two bands of high wavelet powers in the ECA spectrum (10–25 days and 25–70 days), while in the WCA the relative importance of synoptic scale (below 10 days) does not allow to easily highlight the two intraseasonal periods bands. This shows that in WCA, the mechanisms that induce are much rainfall varied. At the synoptic scale, the spectral power peaks near 5 days in the two subregions. Nevertheless, the signal is predominantly contained within 10 and 70 days with two dominant timescales (10–25 days and 25–70 days), much highlighted in the ECA spectrum (Figure 3(b)). This

result is consistent to that found by other authors [14, 34] with daily OLR datasets over West Africa. The seasonality of wavelet power is clearly defined, showing the maximum power during the beginning and end of the year. In fact, this seasonality can be explained by the location of study area along the equator [35]. The peak of ITCZ precipitation belt shifts from north to south of the equator during November-December and returns to north during March-April.

3.2. *Intraseasonal Oscillations.* The spectral analysis performed above leads us to filter the rainfall anomalies either between 10 and 25 days to derive the space-time structures

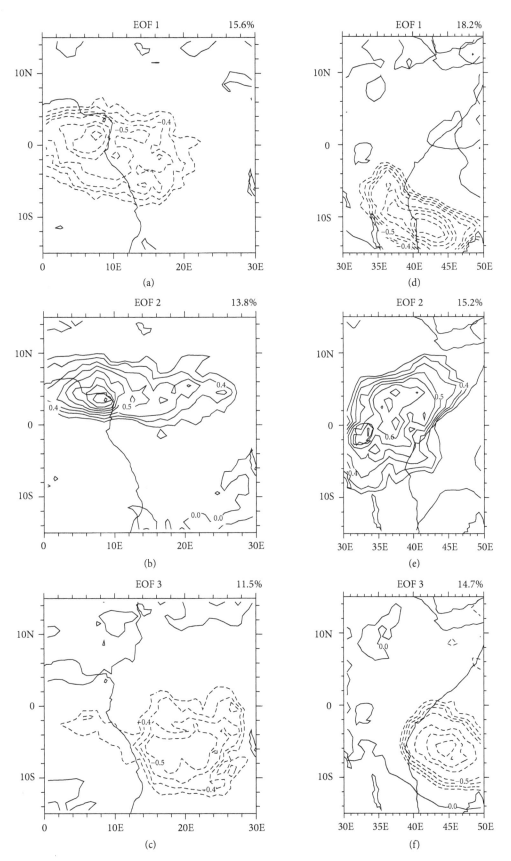

FIGURE 4: Correlation coefficient map (loadings) between rainfall anomalies field and PCs for WCA (a, b, c) and ECA (d, e, f). PCs were computed using the whole time series of daily rainfall for the period 1996–2009. Contour intervals are plotted every 0.05 and correlations in between −0.2 and 0.2 are omitted. The negative correlations are represented as dashed line and the positive correlation the solid lines.

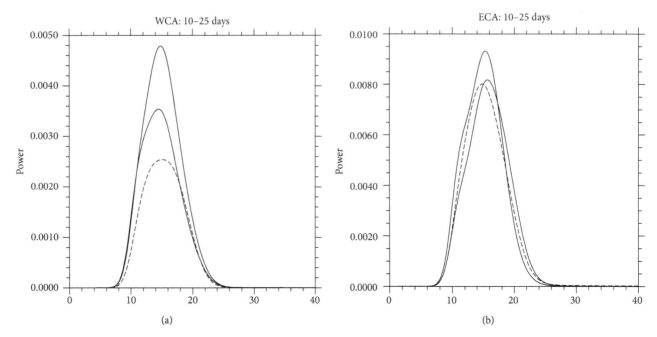

FIGURE 5: Power spectra of PC1 (Bold solid), PC2 (thin solid), and PC3 (thin dashed), respectively, over WCA (left top and left bottom) and ECA (right top and right bottom). The scale of the $x$-axis is in days for all plots and the power dimensionless because the anomalies field is normalized.

of the leading EOFs of 10–25-day intraseasonal variability of rainfall over CA. All these analyses are performed separately in ECA and WCA. The leading modes are retained according to the Scree test [36] and the North criteria [37], and spatial loadings associated with a given mode are obtained by computing the correlation coefficients between the corresponding PC's time series and the filtered data field.

The leading eigenvector and the next two are accounting for, respectively, 15.8%, 13.8%, and 11.3% of the variance for WCA and 18.2%, 15.2%, and 14.7% for ECA (Figure 4). For WCA, all three leading EOFs have positive loadings (Figures 4(a)–4(c)). The first two modes are centered over the Atlantic Ocean near the equator and extend to cover a part of the land and the Atlantic Ocean, while the third is centered over the land and extends over the Congo Forest. Over ECA, the spatial loadings of first and third EOFs are negative, while the second is positive (Figures 4(d)–4(f)). The amount of the total variance explained by the leadings EOFs is smaller in WCA (40.6%) than ECA (48.1%), suggesting that rainfall time series in WCA have more degree of freedom. These results are consistent with the spectral analysis of Section 3.1.

The power spectra of the three principal components (PCs) peak around 15 days both in WCA and ECA (Figure 5). The amplitude time series reveals a difference between ECA and WCA 10–25-day rainfall oscillations. WCA experiences intraseasonal oscillations (ISO) almost throughout the year and the contrast between the seasons is less pronounced, but there are relative bursts of signal amplitude during the October–April season (Figures 6(a)–6(c)). However, the seasonality of oscillations is well defined in ECA (Figures 6(d)–6(f)), as the oscillation amplitudes maximize during

October–April and there are weakened or no oscillation during the rest of the year.

It is well known that the slow eastward propagation at an average speed of $5 \text{ m·s}^{-1}$ [31] is one of the most fundamental features that distinguishes the MJO from other phenomena in the tropical atmosphere, especially, convectively coupled Kelvin waves, which propagate eastward at greater speeds of $15$–$17 \text{ m s}^{-1}$. This eastward propagation could potentially be the relationship between WCA and ECA modes of intraseasonal variability. In Figure 7, the lagged correlations are displayed between the WCA and ECA PCs time series. This figure clearly reveals that the observed PC's time series exhibit a characteristic lead/lag structure. The relatively high maximum positive correlation indicates that most of the WCA PC's lead ECA PCs time series with a time scale of 5–8 days. This result suggests a zonal propagation of EOF modes found above. Averaging filtered data between 15°N and 15°S, and plotting lagged correlation as a function of longitude, succinctly captures ISO propagation (Figure 8). Regression of the ISO indices [4], with filtered GPCP and 850 hPa zonal wind, are performed using the data over Central Africa (15°N–15°S; 0–50°E).

The eastward propagation of the ISO is associated with low-level convergence. The 850 hPa convergence anomalies lead the convection, showing enhanced easterlies (westerlies) ahead (behind) enhanced rainfall as the ISO propagates through the global tropics. The average duration of an intraseasonal event to move from 0 to 45°E is approximately 7 days.

*3.3. Interannual Variations in the ISO Amplitude.* In the previous section, it was shown that the ISO amplitude exhibits

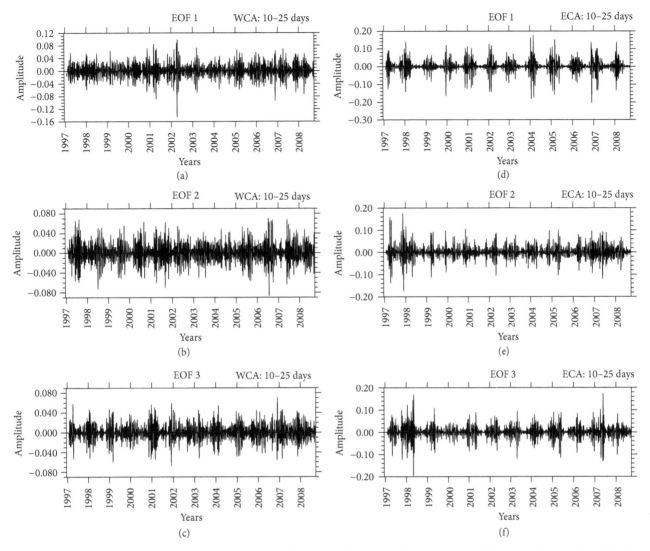

FIGURE 6: Principal components (scores) time series of the 10–25-day filtered rainfall anomalies for WCA ((a)–(c)) and ECA ((d)–(f)).

large interannual variations in ECA, when compared with WCA. Moreover, it was proved by many previous studies that an important characteristic of the MJO (ISO) is its irregularity from year to another [4, 38, 39]. Figure 9 shows the inter-annual variations in the ISO power over WCA (Figure 9(a)) and ECA (Figure 9(b)). For every year, the ISO strength is accessed by taking the mean squared amplitude of each PCs time series. The interannual variations of ISO strength are coherent for all the three PCs in each of the two subregions, as can be seen by the high values of the correlation coefficients (Table 2). But the correlation coefficients computed between the WCA and ECA ISO power time series are relatively low (Table 3), showing that the processes driving the interannual modulation of ISO signal are different in nature or magnitude in the two subregions. It can also be seen that the ISO strength experienced the highest fluctuations during the end of the 1990s and the beginning of the 2000s. For the two subregions, the variations of ISO activity during the end of the 1990s and early 2000s are similar. This period is marked

by above normal ISO activity in the year 1997, followed by the below normal ISO activity during 1997–2001. The difference clearly appears toward the end of the study period. For the two timescales, ECA is characterized by below normal ISO activity during the years 2005–2008, while during the same period, the WCA experienced above normal ISO activity in many years. Further investigations are required to identify the causation of the coherent interannual variations of ISO signal during this period.

## 4. Summary and Conclusions

ISV of rainfall over Central Africa was examined using 1DD GPCP rainfall product for the period 1996–2009, with an emphasis on the comparison between western Central Africa (WCA) and eastern Central Africa (ECA), with different climate features. The classical Fourier analysis revealed the importance of ISV in Central African climate, but the dom-inance of intraseasonal band is more highlighted in ECA,

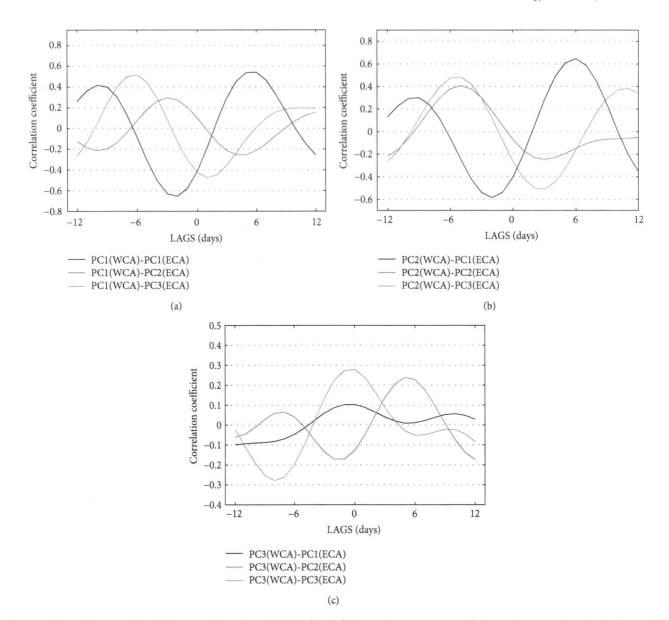

FIGURE 7: Lead-lag correlation between ECA and WCA 10–25-day PC's time series. Positive correlations at positive time lags indicate that convection over WCA leads that over the ECA.

TABLE 2: Correlation coefficients between WCA and ECA annual mean ISO power.

|         | WCA  | ECA  |
|---------|------|------|
| PC1-PC2 | 0.43 | 0.51 |
| PC1–PC3 | 0.61 | 0.12 |
| PC2-PC3 | 0.73 | 0.47 |

TABLE 3: Cross-correlation coefficients between WCA and ECA annual mean ISO power.

| WCA | ECA | | |
|-----|-----|-----|-----|
|     | PC1 | PC2 | PC3 |
| PC1 | −0.09 | 0.06 | −0.05 |
| PC2 | −0.12 | 0.15 | 0.13 |
| PC3 | 0.1 | −0.11 | −0.01 |

when compared with WCA. A wavelet analysis showed that intraseasonal precipitation variability in CA is dominated by two distinct timescales (10–25 days and 25–70 days) with significant spectral peaks centered, respectively, around periods of 15 days and 50 days, in consistency with the classical Fourier analysis. The seasonality of the ISO power is evident, showing the maximum power at the beginning and

end of the year. But this seasonality is well defined in ECA, unlike WCA where the signal persists almost throughout the year.

For 10–25-day band, the result of EOF analysis of the filtered data has shown that the leading eigenvectors vectors explain about 40.6% of total intraseasonal rainfall variability

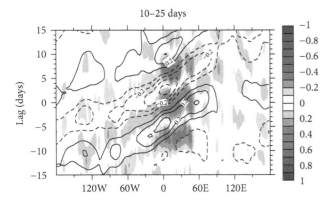

FIGURE 8: ISO composite based on the correlation between bandpass (10–25 days) filtered 850-hPa zonal wind (contours) and precipitation (colors) upon ISO index time series. The reference time series is Central Africa precipitation time series. The ISO index was computed using the average wavelet power.

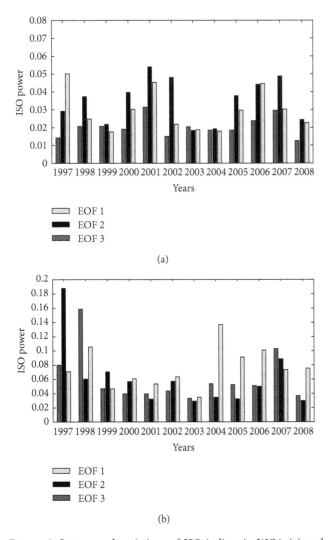

(a)

(b)

FIGURE 9: Interannual variations of ISO indices in WCA (a) and ECA (b), within the period 1997–2008, and for the three EOFs retained.

for WCA and 48.1% for ECA, suggesting that filtered rainfall time series in WCA have more degree of freedom than ECA. The spatial loadings are coherent, showing either positive or negative correlation coefficient between each eigenvector and rainfall. The power spectra of PCs peak around 15 days both in WCA and ECA.

The amplitude time series confirmed that rainfall variability in ECA is highly seasonal, revealing significant oscillations (wet/dry spells) with greater amplitudes during the beginning and end of the year (October–April) and weakened or no oscillation during the rest of the year. Once again, this seasonality is less pronounced in WCA PC's time series, but nevertheless the ISO amplitudes are relatively higher during October–April than other months.

An ISO strength index was built by averaging the wavelet power corresponding to each PC and for the 10–25-day timescales to form a daily indices time series. The plots of interannual variations of ISO strength clearly show that the signal over ECA exhibits larger interannual variations than WCA. Further investigations are needed to explain the causation of these patterns found.

The picture that emerges from these results is that the 10–25-day intraseasonal rainfall variability detected in 1DD GPCP rainfall shows some different patterns in ECA and WCA. The main difference between WCA and ECA modes is the variations of ISO strength. The variations in WCA are weak, compared to ECA where the signal exhibits large annual and interannual variations.

## Conflict of Interests

The authors declare no conflict of interests.

## Acknowledgments

Most of the scripts used in this study were produced from the NCL website http://www.ncl.ucar.edu/. The GPCP data was obtained from the NOAA website, http://www.esrl.noaa.gov/. All the administrator members of these websites are gratefully acknowledged for maintaining the updated data. The authors also thank the senior researchers of LEMAP and LISIE for helpful comments.

## References

[1] UNEP DEWA/PAN-AFRICAN START, "Vulnerability of water resources to environmental change in Africa," River Basin Approach, UNEP, 2004.

[2] T. James and P. Karl, "Economic losses and poverty effects of droughts and floods in Malawi," OER Africa, 2009.

[3] P. Dierk, M. Charles, and H. Stefan, "Diagnosing the droughts and floods in equatorial east Africa during boreal autumn 2005–08," Journal of Climate, vol. 23, no. 3, pp. 813–817, 2010.

[4] A. S. Tchakoutio, A. Nzeukou, and C. Tchawoua, "Intraseasonal atmospheric variability and its interannual modulation in Central Africa," Meteorology and Atmospheric Physics, vol. 117, no. 3-4, pp. 167–179, 2012.

[5] E. J. Zipser, D. J. Cecil, C. Liu, S. W. Nesbitt, and D. P. Yorty, "Where are the most intense thunderstorms on earth?" Bulletin

*of the American Meteorological Society,* vol. 87, no. 8, pp. 1057–1071, 2006.

[6] G. Hong and G. Heygster, "Intense tropical thunderstorms detected by the special sensor microwave imager/sounder," *IEEE Transactions on Geoscience and Remote Sensing,* vol. 46, no. 4, pp. 996–1005, 2008.

[7] M. Thomas, C. H. C. John, G. Alexander, and J. C. Nicolas, "Temporal precipitation variability versus altitude on a tropical high mountain: observations and mesoscale atmospheric modelling," *Quarterly Journal of the Royal Meteorological Society,* vol. 135, no. 643, pp. 1439–1455, 2009.

[8] A. M. Moore, J. P. Loschnigg, P. J. Webster, and R. R. Leben, "Coupled ocean-atmosphere dynamics in the Indian Ocean during 1997-98," *Nature,* vol. 401, no. 6751, pp. 356–360, 1999.

[9] C. Mutai, S. Hastenrath, and D. Polzin, "Diagnosing the 2005 drought in equatorial East Africa," *Journal of Climate,* vol. 20, no. 18, pp. 4628–4637, 2007.

[10] R. A. Madden and P. R. Julian, "Description of global-scale circulation cells in the tropics with a 40–50 day period," *Journal of the Atmospheric Sciences,* vol. 29, pp. 1109–1123, 1972.

[11] T. Yamagata and Y. Hayachi, "A simple diagnostic model for the 30–50 days oscillation in the tropics," *Journal of the Meteorological Society of Japan,* vol. 62, no. 5, pp. 709–717, 1984.

[12] B. Sultan and S. Janicot, "Intra-seasonal modulation of convection in the West African monsoon," *Geophysical Research Letters,* vol. 28, no. 3, pp. 523–526, 2001.

[13] B. Sultan and S. Janicot, "The West African monsoon dynamics—part II: the pre-onset and onset of the summer monsoon," *Journal of Climate,* vol. 16, pp. 3407–3427, 2003.

[14] S. Janicot and F. Mounier, "Evidence of two independent modes of convection at intraseasonal timescale in the West African summer monsoon," *Geophysical Research Letters,* vol. 31, no. 16, Article ID L16116, 2004.

[15] R. W. Higgins, D. E. Waliser, J.-K. E. Schemm, C. Jones, and L. M. V. Carvalho, "Climatology of tropical intraseasonal convective anomalies: 1979–2002," *Journal of Climate,* vol. 17, no. 3, pp. 523–539, 2004.

[16] P. Camberlin and P. Benjamin, "Influence of the Madden-Julian oscillation on East African rainfall—I: intraseasonal variability and regional dependency," *Quarterly Journal of the Royal Meteorological Society,* vol. 132, no. 621, pp. 2521–2539, 2006.

[17] L. Tazalika and M. R. Jury, "Intra-seasonal rainfall oscillations over central Africa: space-time character and evolution," *Theoretical and Applied Climatology,* vol. 94, no. 1-2, pp. 67–80, 2008.

[18] M. R. Jury and C. Hector, "Intraseasonal variability of satellite-derived rainfall and vegetation over Southern Africa," *Earth Interactions,* vol. 14, no. 3, pp. 1–26, 2010.

[19] A. S. Tchakoutio, A. Nzeukou, C. Tchawoua, F. M. Kamga, and D. Vondou, "A comparative analysis of intraseasonal variability in OLR and 1DD GPCP data over central Africa," *Theoretical and Applied Climatology,* vol. 116, no. 1-2, pp. 37–49, 2014.

[20] A. S. Tchakoutio, A. Nzeukou, C. Tchawoua, B. Sonfack, and T. Siddi, "Comparing the patterns of 20–70 days intraseasonal oscillations over Central Africa during the last three decades," *Theoretical and Applied Climatology,* 2013.

[21] D. L. Gilman, F. J. Fuglister, and J. M. Mitchell, "On the power spectrum of 'red noise'," *Journal of the Atmospheric Sciences,* vol. 20, no. 2, pp. 182–184, 1963.

[22] G. J. Huffman, M. Morrissey, D. Bolvin et al., "Global precipitation at one degree daily resolution from multisatellite observations," *Journal of Hydrometeorology,* vol. 2, pp. 36–50, 2001.

[23] R. F. Adler, "The version-2 Global Precipitation Climatology Project (GPCP) monthly precipitation analysis (1979-present)," *Journal of Hydrometeorology,* vol. 4, pp. 1147–1167, 2003.

[24] P. A. Arkin and B. N. Meisner, "The relationship between large-scale convective rainfall and cold cloud over the Western Hemisphere during 1982–84," *Monthly Weather Review,* vol. 115, no. 1, pp. 51–74, 1987.

[25] J. McCollum, E. Nelkin, D. Klotter et al., "Validation of TRMM and other rainfall estimates with a high-density gauge dataset for West Africa—part I: validation of GPCC rainfall product and pre-TRMM satellite and blended products," *Journal of Applied Meteorology,* vol. 42, no. 10, pp. 1337–1354, 2003.

[26] T. Dinku, P. Ceccato, E. Grover-Kopec, M. Lemma, S. J. Connor, and C. F. Ropelewski, "Validation of satellite rainfall products over East Africa's complex topography," *International Journal of Remote Sensing,* vol. 28, no. 7, pp. 1503–1526, 2007.

[27] F. Ruiz, T. Dinku, S. J. Connor, and P. Ceccato, "Validation and intercomparison of satellite rainfall estimates over Colombia," *Journal of Applied Meteorology and Climatology,* vol. 49, no. 5, pp. 1004–1014, 2010.

[28] C. Lanczos, *Applied Analysis,* Prentice-Hall, 1956.

[29] C. E. Duchon, "Lanczos filtering in one and two dimensions," *Journal of Applied Meteorology,* vol. 18, no. 8, pp. 1016–1022, 1979.

[30] C. Torrence and G. P. Compo, "A practical guide to wavelet analysis," *Bulletin of the American Meteorological Society,* vol. 79, no. 1, pp. 61–78, 1998.

[31] K. M. Weickmann and T. R. Knuston, "30–60 day atmospheric oscillations: composite life cycle of convection and circulation anomalies," *Monthly Weather Review,* vol. 115, no. 7, pp. 1407–1436, 1987.

[32] H. H. Harry and Z. Chindong, "Propagating and standing components of the intraseasonal oscillation in tropical convection," *Journal of the Atmospheric Sciences,* vol. 54, pp. 741–751, 1996.

[33] A. J. Matthews, "Intraseasonal variability over tropical Africa during northern summer," *Journal of Climate,* vol. 17, no. 12, pp. 2427–2440, 2004.

[34] E. D. Maloney and S. Jeffrey, "Intraseasonal variability of the West African monsoon and Atlantic ITCZ," *Journal of Climate,* vol. 21, no. 12, pp. 2898–2918, 2008.

[35] H. Rui and B. Wang, "Development characteristics and dynamic structure of tropical intraseasonal convection anomalies," *Journal of the Atmospheric Sciences,* vol. 47, no. 3, pp. 357–379, 1990.

[36] R. B. Cattell, "The scree test for the number of factors," *Multivariate Behavioral Research,* vol. 1, no. 2, pp. 245–276, 1966.

[37] G. R. North, T. L. Bell, R. F. Cahalan, and F. J. Moeng, "Sampling errors in the estimation of empirical orthogonal functions," *Monthly Weather Review,* vol. 110, no. 7, pp. 699–6706, 1982.

[38] H. H. Hendon, C. Zhang, and J. D. Glick, "Interannual variation of the Madden-Julian oscillation during austral summer," *Journal of Climate,* vol. 12, no. 8, pp. 2538–2550, 1999.

[39] J. M. Slingo, D. P. Rowell, K. R. Sperber, and F. Nortley, "On the predictability of the interannual behaviour of the Madden-Julian oscillation and its relationship with El Niño," *Quarterly Journal of the Royal Meteorological Society,* vol. 125, no. 554, pp. 583–609, 1999.

# 9

# Climatic Variation at Thumba Equatorial Rocket Launching Station, India

**K. V. S. Namboodiri, P. K. Dileep, and Koshy Mammen**

*Meteorology Facility, Thumba Equatorial Rocket Launching Station (TERLS), Vikram Sarabhai Space Centre (VSSC), Indian Space Research Organization, Thiruvananthapuram 695 022, India*

Correspondence should be addressed to K. V. S. Namboodiri; sambhukv@yahoo.com

Academic Editors: A. V. Eliseev and E. Paoletti

Long-term (45 years) diversified surface meteorological records from Thumba Equatorial Rocket Launching Station (TERLS), India, were collected and analysed to study the long-term changes in the overall climatology, climatology pertained to a particular observational time, mean daily climatology in temperature, inter-annual variability in temperature, interannual variability in surface pressure, and rainfall for the main Indian seasons—South West and North East monsoons and inter-annual mean monthly anomaly structure in temperature. Results on various analyses show strong and vivid features contributed by climate change for this South Peninsular Indian Arabian Sea Coastal Station, and this paper may be a first time venture which discusses climate change imparted perturbations in several meteorological parameters in different time domains, like a specific time, daily, monthly, and interannually over a station. Being a coastal rocket launching station, climatic change information is crucial for long-term planning of its facilities as well as for various rocket range operational demands.

## 1. Introduction

Climate change refers to a statistically strong and significant variation in either the average state of the climate or in its variability, persisting for an extended period, typically in decades or longer. Short-period oscillations are statistically insignificant in the scenario of long-term climate change context. Climate change may be due to natural internal processes on earth (atmospheric, seismic, volcanic, and oceanic), external forces (variation in solar activity, rotation and revolution of earth), and more recently anthropogenic activities. It is now well established that anthropogenic activities cause intensification of the greenhouse effect and thereby contribute to the climate change. The paper by Oreskes [1] highlighted the listings in the ISI database on climate change from several abstracts published in refereed scientific journals between 1993 and 2003.

In India, the first time document on the concern of greenhouse effect and possible mitigation measures was published in the Science Reporter, CSIR, by Ayyar [2]. Several papers were published in a special session in Current Science (volume 90, 2006) focussing the concerns over the socioeconomic and scientific aspects attributed to climate change on Indian scenario. To develop high-resolution climate change simulations over the Indian region, Providing Regional Climates for Impact Studies (PRECIS) modelling system was applied [3], and it was observed that the warming is monotonously widespread over India with substantial spatial differences in the projected rainfall changes.

Different data sets have been used in order to examine the monotonic change in the level of global temperature by various investigators [4–7]. From the diversified methodologies adopted, it is observed that the global surface temperature has risen by 0.74°C with an error of ±0.18°C when estimated by a linear trend over the last 100 years (1906–2005) and 0.07°C ± 0.02°C per decade. An executive summary is made available by Solomon et al. [8] on climate change in the fourth assessment report of the Intergovernmental Panel on Climate Change (IPCC). Climate change features are local or regional highly influenced by agricultural practices, urbanisation, heat island effect, changing irrigation, aerosol modifications, desertification, deforestation, and decay processes of organic materials [9–14]. Long-term changes in maximum temperature, minimum temperature, and their range have been analyzed over two stations in Cyprus by Price et al. [15] who established a reduction in range and thereby a possibility in increased cloud cover and tropospheric aerosols. Recently seasonal variability in mean sea level pressure extremes over the 11 zones of India and adjoining sea area has been studied by Singh et al. [16].

In the present paper, features on surface climate change over Thumba Equatorial Rocket Launching Station (TERLS), a coastal station near the South tip of Peninsular India, are critically examined. TERLS was established as an international rocket launching station dedicated to the United Nations for conducting meteorological and sounding rocket experiments on the geomagnetic equator. Meteorology group, TERLS has one of the heaviest responsibilities in the range to cater (i) weather forecasting and data needs of sounding rocket launch operations (in planning, pad operations and launching), (ii) climatological briefing to working crews including scientists to plan experiments, (iii) Organise various met. Observations, (iv) processing, analyses, storage and retrieval of various met. data (Class I surface observatory, autoweather stations, sonic anemometers, 50 m micro meteorological flux tower facility, meteorological balloon launches, rocket chaff derived wind data) and provide the data for users demand. For the planning of a scientific mission, the variability of climate is very crucial and thereby the importance of the present study on the climate change perspective. Importance of weather factors affecting rocket operations can be found elsewhere in detail [17, 18]. Steady increase in temperature anomaly at different levels in the troposphere over TERLS was investigated and documented in the Space research in India [19]. Joseph and Simon [20] have investigated weakening trend of the South West monsoon current through Peninsular India. Dash et al. [21] reported the decreasing frequency of monsoon depressions over the Indian region, and Krishnakumar et al. [22] have analysed significant decrease in South West monsoon rainfall while increase in post monsoon season over the whole Kerala State (also a meteorological subdivision), India. From the voluminous data bank, the present study analyses daily, monthly, seasonal, and interannual variability on different meteorological parameters based on both eye reading instruments and non-instrumental practices in cloud and visibility observations as per World Meteorological Organization (WMO) practices. Till the inception of long-term uninterrupted data collection through instrument-based cloud and visibility measurements, data collection from such non-instrumental means solely vested on conventional methods in surface meteorological observations adopted worldwide is to be accepted for a reasonable climatology extraction. The present work examines the imprints made by the climatic change on several meteorological parameters in different time domains, like a specific time, daily, monthly, and interannually unlike the climate change associated variations on a single meteorological parameter like temperature or rainfall alone have dealt in most of the climatic change studies.

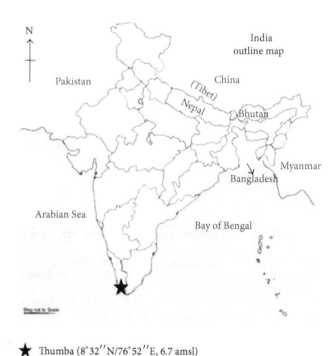

★ Thumba (8°32″N/76°52″E, 6.7 amsl)

FIGURE 1: Location shows the class-I surface meteorological observatory at Thumba.

## 2. Data and Methodology

A class I surface meteorological observatory was established at TERLS (8°32″N/76° 52″E, 6.7 amsl) by India Meteorological Department (IMD) approximately 70 m away from the coastline of the Arabian Sea coast (Figure 1). TERLS is situated hardly 8 km West North West of Thiruvananthapuram, the capital city of the Kerala state of India. Various meteorological data from 0830 IST (Indian Standard Time, IST = GMT + 0530 hr) observations for a period of 45 years (1964–2008), except for pressure where it is hourly data (1973–2008), are used in order to investigate changes encountered in meteorological parameters due to climate change. The data used are obtained from the following measurements at 0830 IST (1964–2008) namely, (i) maximum temperature (past 24 hr reported at 0830 IST), (ii) minimum temperature (past 24 hr reported at 0830 IST), (iii) past 24 hr rainfall (reported at 0830 IST), (iv) dry-bulb temperature at 0830 IST, (v) wet-bulb temperature at 0830 IST, (vi) relative humidity at 0830 IST, (vii) soil temperature at 0830 IST, (viii) non-instrumental eye observations on clouds and visibility at 0830 IST, (ix) wind speed and direction at 0830 IST, and (x) hourly surface pressure (1973–2008). Table 1 provides details on various instrumentation, exposure, maintenance, data span and the lowest and the highest in maximum absolute error values among various monthly statistics from 1964 to 2008.

The above mentioned 45 years of data have partitioned into two climatological slots of 30 years each, namely, 1964 to 1993 and from 1979 to 2008, respectively, called as past data for the former and present data for the later. Maximum absolute error values in the total data set (1962–2008) are the

TABLE 1: Details on instrumentation, data used, and the observed variation in maximum absolute error in monthly mean used in the discussion of climatology at 0830 IST.

| S. no | Parameter | Data used | Instrument | Height (agl) (cm) | Exposure | Lowest and highest maximum absolute error | Make | Maintenance and calibration protocol |
|---|---|---|---|---|---|---|---|---|
| 1 | Maximum temperature (°C) | 1964–2008 | Maximum thermometer (Mercury-in-glass thermometer) | 122 | Stevenson screen | 0.087 to 0.132 (°C) | IMD | |
| 2 | Minimum temperature (°C) | 1964–2008 | Minimum thermometer (spirit-in-glass) | 122 | Stevenson screen | 0.061 to 0.145 (°C) | IMD | |
| 3 | Dry-bulb temperature (°C) | 1964–2008 | Drybulb thermometer (mercury-in-glass thermometer) | 122 | Stevenson screen | 0.088 to 0.125 (°C) | IMD | Daily, weekly, monthly, half yearly, annual, and periodic |
| 4 | Rainfall (mm) | | Nonrecording rain gauge of collector area 20 square cm | 0 | Ground | | | |
| 4 | Wet-bulb temperature (°C) | 1964–2008 | Wet-bulb thermometer | 122 | Stevenson screen | 0.06 to 0.146 (°C) | IMD | |
| 5 | Relative humidity (%) | 1964–2008 | Drybulb thermometer, wetbulb thermometer, and psychrometer table. | 122 | Stevenson screen | 0.512 to 0.916 (%) | IMD | |
| 6 | Soil temperature (°C) | 1964–2008 | Soil thermometer in metallic tube | −5 | | 0.201 to 0.261 (°C) | IMD | |
| 7 | Pressure (mb) | 1974–2008 | Aneroid barograph (diaphragm type) | 6.70 m amsl | | | IMD | |
| 8 | Wind speed | 1964–2008 | Anemometer | 300 | | 0.043 to 0.24 (m/s) | IMD/Dynalab | |
| 9 | Wind direction | 1964–2008 | Wind vane | 300 | | Within 1 degree | IMD/Dynalab | |

summation of errors [23] of past and present data sets. Individual error is computed as the ratio of standard deviation to the number of observations of a particular data set. Trenberth [24] has established the climate shift as a prominent upward trend in temperature after 1976. In order to include this observation influence, half portion (15 years) period from 1979 to 1993 is made common in the latter half of past data and in the former part of present data. The difference obtained among present data and past data has been represented as Δ in the discussions throughout. To see genuineness in findings attributed to climate change, Δ values are compared with absolute errors so as to see that Δ do not fall within absolute errors of corresponding meteorological parameters.

## 3. Results and Discussion

*3.1. Δ Climatology of the Past 24 hr Observations.* Maximum temperature, minimum temperature, and 24 hr accumulated rainfall are the only 3 meteorological parameters registered at 0830 IST having effect of the past 24 hr variability behaviour. Climatology for various parameters has been derived for past data and present data and presented in Tables 2, 3, 4, and 5 in which months that constitute the different seasons are mentioned. Derived Δ values are also incorporated in the above tables. Comparison between past data and present data maximum temperature climatology shows marked increase in present data climatology in relation to the past data counterpart, and thereby, positive values alone are observed in Δ. Δ values vary from +0.01°C in the North East monsoon (NE monsoon) month November to +0.43°C in the pre-South West monsoon (SW monsoon) month May. The NE monsoon season comprises October and November, whereas the four months from June to September designate SW monsoon season, which are the main two Indian seasons. In minimum temperature, all Δ values are positive with maximum value in the winter month February with a value +0.53°C. The observations among maximum and minimum temperature climatology derived in past data and present data along with Δ values can infer that the minimum daily temperatures have generally increased at a larger rate than the maximum daily temperatures. Karl et al. [25] studied the possible cause of increased urbanisation effects and Rebetez, and Beniston [26] justified the result of increased cloud cover in the large increase in minimum temperature and small increase in maximum temperatures.

From each individual daily maximum and minimum temperature, mean daily temperatures ((maximum temperature + minimum temperature)/2) are computed and further derive the past data and present data climatologies of mean temperatures. Only positive Δ values are observed throughout the year, which clearly indicate that the temperature over this tropical coastal station is increasing. Temperature climatologies have shown the tropical monsoonal characteristic of two maxima and two minima in monthly variation.

The diurnal temperature range is calculated from individual maximum and minimum temperature value and calculated past data and present data temperature range climatologies. Range climatologies are featured by annual variation. The prominent feature of negative Δ values in range is seen, which shows that the difference between maximum temperature and minimum temperature in a day is decreasing. Also, the larger rate of increase in minimum temperatures than maximum temperatures results in these negative Δ values in the diurnal temperature range.

The rainfall Δ shows decrease in rainfall in present data for the summer months April and May. Well-marked negative Δ values of about 21 mm are observed in the SW monsoon months (June–August). An increase in rainfall activity is detected in the SW monsoon month September and in the NE monsoon months October and November (debted to monsoonal climate) with maximum value of +36.50 mm in October. The station weather is dominated by convective mesoscale systems during the NE monsoon. Probably this investigation poses the doubt about whether such meso-scale weather phenomena like thunderstorm formations are on increasing trend during NE Monsoon and thereby an increase in Δ.

*3.2. Climatology of a Particular Observation at 0830 IST.* Analyses are carried out on different parameters collected at 0830 IST. Δ in dry bulb temperature, wet bulb temperature, and dew point temperature obtained from the two climatologies show only positive values. It ranges from +0.60°C in July to +1.01°C in January for dry bulb temperature. Variations in Δ among the SW monsoon months are less in general for all the above three variables, even though a maximum value in Δ is observed in dry bulb temperature. The less variation of wet bulb and dew point temperatures may be as a result of moisture laden atmosphere during the period which imparts corresponding variations in the moisture parameters like wet bulb temperature and dew point temperature.

Station relative humidity values at 0830 IST show more than 70% throughout the year with maximum in SW monsoon of the order 89%. Low negative values in Δ relative humidity are detected as the air becomes drier in present data compared to past. As the air is getting dried, both dew point and the wet bulb temperatures are supposed to show negative Δ values, but the feature is not detected. This may be inferred as the dry bulb temperature is the key parameter in elevating Δ values of both dew point and wet bulb temperatures, and the meagre relative humidity decrease in present data than past data is insignificant to impart Δ changes in dew point and wet bulb temperatures. Climatologies of RH also have shown two maxima and two minima, whereas the secondary minima around the NE monsoon is shallow in character.

Significant positive Δ values are observed among the two climatologies in soil temperature with a low value of +0.99°C to a high of +2.35°C in July and January, respectively. Perhaps the Δ increase observed in the winter months January and February may contribute reduction in radiational cooling which leads to a reduction in fog formation probability over the station. Soil temperature climatologies also have shown tropical monsoonal type with two maxima and minima.

Wind speed (both scalar and resultant) is decreased in present climatologies. Mathematically resultant wind speed shall be determined as the square root of the sum of the squares of zonal and meridional wind components. The encountered Δ values that rise to a maximum value of

TABLE 2: Climatology and Δ of maximum temperature, minimum temperature, mean daily temperature, mean range in temperature, and monthly rainfall.

| Month | Maximum temperature °C | | | Minimum temperature °C | | | Mean daily temperature °C | | | Mean range in temperature °C | | | Monthly rainfall mm | | |
|---|---|---|---|---|---|---|---|---|---|---|---|---|---|---|---|
| | 1964–1993 | 1979–2008 | Δ | 1964–1993 | 1979–2008 | Δ | 1964–1993 | 1979–2008 | Δ | 1964–1993 | 1979–2008 | Δ | 1964–1993 | 1979–2008 | Δ |
| JAN | 32.59 | 32.76 | +0.17 | 21.30 | 21.58 | +0.28 | 26.94 | 27.16 | +0.22 | 11.29 | 11.17 | −0.12 | 11.80 | 11.10 | −0.70 |
| FEB | 33.12 | 33.32 | +0.20 | 21.89 | 22.42 | +0.53 | 27.51 | 27.87 | +0.36 | 11.24 | 10.91 | −0.33 | 16.20 | 16.40 | +0.20 |
| MAR | 33.84 | 34.07 | +0.23 | 23.75 | 24.13 | +0.38 | 28.78 | 29.09 | +0.31 | 10.06 | 9.92 | −0.14 | 33.20 | 34.80 | +1.60 |
| APR | 34.25 | 34.41 | +0.16 | 25.24 | 25.51 | +0.27 | 29.76 | 29.95 | +0.19 | 9.05 | 8.88 | −0.17 | 104.10 | 99.40 | −4.70 |
| MAY | 33.03 | 33.46 | +0.43 | 25.14 | 25.50 | +0.36 | 29.12 | 29.48 | +0.36 | 7.97 | 7.96 | −0.01 | 179.20 | 165.80 | −13.40 |
| JUN | 30.36 | 30.71 | +0.35 | 23.99 | 24.36 | +0.37 | 27.20 | 27.53 | +0.33 | 6.43 | 6.35 | −0.08 | 292.70 | 272.20 | −20.50 |
| JUL | 29.38 | 29.48 | +0.10 | 23.36 | 23.79 | +0.43 | 26.38 | 26.63 | +0.25 | 6.03 | 5.68 | −0.35 | 203.00 | 181.30 | −21.70 |
| AUG | 29.27 | 29.60 | +0.33 | 23.39 | 23.80 | +0.41 | 26.35 | 26.70 | +0.35 | 5.92 | 5.79 | −0.13 | 141.50 | 120.90 | −20.60 |
| SEP | 30.34 | 30.70 | +0.36 | 23.68 | 24.00 | +0.32 | 27.04 | 27.34 | +0.30 | 6.71 | 6.67 | −0.04 | 158.20 | 171.00 | +12.80 |
| OCT | 31.19 | 31.53 | +0.34 | 23.74 | 24.04 | +0.30 | 27.50 | 27.76 | +0.26 | 7.53 | 7.44 | −0.09 | 204.90 | 241.40 | +36.50 |
| NOV | 31.76 | 31.77 | +0.01 | 23.21 | 23.57 | +0.36 | 27.48 | 27.67 | +0.19 | 8.53 | 8.20 | −0.33 | 166.80 | 170.70 | +3.90 |
| DEC | 32.30 | 32.56 | +0.26 | 22.44 | 22.44 | +0.00 | 27.35 | 27.50 | +0.15 | 9.83 | 10.11 | +0.28 | 66.60 | 53.70 | −12.90 |

DEC, JAN, and FEB: winter; MAR, APR, and MAY: summer; JUN, JUL, AUG, and SEP: SW monsoon; OCT, NOV: NE monsoon.

TABLE 3: Climatology and Δ of dry-bulb temperature, wet bulb temperature, dew point temperature, relative humidity, and soil temperature.

| Month | Dry-bulb temperature (0830 IST) °C | | | Wet-bulb temperature (0830 IST) °C | | | Dew point temperature (0830 IST) °C | | | Relative humidity (%) (0830 IST) | | | Soil temperature (0830 IST) °C | | |
|---|---|---|---|---|---|---|---|---|---|---|---|---|---|---|---|
| | 1964–1993 | 1979–2008 | Δ | 1964–1993 | 1979–2008 | Δ | 1964–1993 | 1979–2008 | Δ | 1964–1993 | 1979–2008 | Δ | 1964–1993 | 1979–2008 | Δ |
| JAN | 26.06 | 27.07 | +1.01 | 22.33 | 22.90 | +0.57 | 20.35 | 20.69 | +0.34 | 71.36 | 68.85 | −2.51 | 29.03 | 31.38 | +2.35 |
| FEB | 26.77 | 27.71 | +0.94 | 23.04 | 23.55 | +0.51 | 21.12 | 21.43 | +0.31 | 71.76 | 69.41 | −2.35 | 30.41 | 32.54 | +2.13 |
| MAR | 28.31 | 29.20 | +0.89 | 24.58 | 24.94 | +0.36 | 22.81 | 22.94 | +0.13 | 72.55 | 69.53 | −3.02 | 32.80 | 34.25 | +1.45 |
| APR | 29.29 | 30.00 | +0.71 | 26.09 | 26.30 | +0.21 | 24.71 | 24.72 | +0.01 | 76.56 | 73.61 | −2.95 | 33.73 | 35.11 | +1.38 |
| MAY | 28.70 | 29.43 | +0.73 | 26.12 | 26.43 | +0.31 | 25.02 | 25.17 | +0.15 | 80.94 | 78.33 | −2.61 | 32.46 | 34.16 | +1.70 |
| JUN | 26.61 | 27.33 | +0.72 | 25.20 | 25.57 | +0.37 | 24.55 | 24.81 | +0.26 | 88.51 | 86.43 | −2.08 | 29.55 | 30.99 | +1.44 |
| JUL | 25.76 | 26.36 | +0.60 | 24.55 | 24.88 | +0.33 | 23.95 | 24.21 | +0.26 | 89.80 | 88.03 | −1.77 | 29.01 | 30.00 | +0.99 |
| AUG | 25.90 | 26.60 | +0.70 | 24.60 | 24.96 | +0.36 | 23.97 | 24.22 | +0.25 | 89.40 | 86.98 | −2.42 | 29.60 | 30.96 | +1.36 |
| SEP | 26.62 | 27.37 | +0.75 | 24.84 | 25.21 | +0.37 | 24.02 | 24.25 | +0.23 | 85.77 | 83.45 | −2.32 | 31.10 | 32.23 | +1.13 |
| OCT | 26.81 | 27.59 | +0.78 | 24.90 | 25.33 | +0.43 | 24.02 | 24.32 | +0.30 | 85.01 | 82.61 | −2.40 | 31.40 | 32.45 | +1.05 |
| NOV | 26.90 | 27.63 | +0.73 | 24.40 | 24.86 | +0.46 | 23.27 | 23.58 | +0.31 | 80.84 | 79.35 | −1.49 | 30.99 | 32.02 | +1.03 |
| DEC | 26.70 | 27.59 | +0.89 | 23.35 | 23.72 | +0.37 | 21.61 | 21.74 | +0.13 | 74.48 | 71.39 | −3.09 | 30.52 | 32.09 | +1.57 |

DEC, JAN, and FEB: winter; MAR, APR, and MAY: summer; JUN, JUL, AUG, and SEP: SW monsoon; OCT, NOV: NE monsoon.

TABLE 4: Climatology and Δ of visibility code, low cloud, high cloud, and total cloud.

| Month | Visibility code (0830 IST) | | | Low cloud (0830 IST) | | | Middle cloud (0830 IST) | | | High cloud (0830 IST) | | | Total cloud (0830 IST) | | |
|---|---|---|---|---|---|---|---|---|---|---|---|---|---|---|---|
| | 1964–1993 | 1979–2008 | Δ | 1964–1993 | 1979–2008 | Δ | 1964–1993 | 1979–2008 | Δ | 1964–1993 | 1979–2008 | Δ | 1964–1993 | 1979–2008 | Δ |
| JAN | 96.79 | 96.98 | +0.19 | 0.55 | 0.47 | −0.08 | 0.81 | 0.78 | −0.03 | 1.18 | 1.35 | +0.17 | 2.51 | 2.53 | +0.02 |
| FEB | 96.75 | 96.91 | +0.16 | 0.48 | 0.47 | −0.01 | 0.66 | 0.58 | −0.08 | 1.08 | 1.16 | +0.08 | 2.21 | 2.17 | −0.04 |
| MAR | 96.61 | 96.88 | +0.27 | 0.62 | 0.71 | +0.09 | 0.54 | 0.37 | −0.17 | 1.06 | 1.21 | +0.15 | 2.20 | 2.24 | +0.04 |
| APR | 96.77 | 96.86 | +0.09 | 1.25 | 1.57 | +0.32 | 1.00 | 0.87 | −0.13 | 1.72 | 1.79 | +0.07 | 3.91 | 3.94 | +0.03 |
| MAY | 96.70 | 96.68 | −0.02 | 2.02 | 2.08 | +0.06 | 1.45 | 1.43 | −0.02 | 1.56 | 1.64 | +0.08 | 4.87 | 4.68 | −0.19 |
| JUN | 96.26 | 96.43 | +0.17 | 2.86 | 2.84 | −0.02 | 2.31 | 2.37 | +0.06 | 1.16 | 1.49 | +0.33 | 5.96 | 5.90 | −0.06 |
| JUL | 96.10 | 96.37 | +0.27 | 2.74 | 2.74 | +0.00 | 2.50 | 2.82 | +0.32 | 1.26 | 1.67 | +0.41 | 5.96 | 6.06 | +0.10 |
| AUG | 96.41 | 96.58 | +0.17 | 2.27 | 2.38 | +0.11 | 2.31 | 2.28 | −0.03 | 1.31 | 1.86 | +0.55 | 5.63 | 5.68 | +0.05 |
| SEP | 96.48 | 96.63 | +0.15 | 2.16 | 2.15 | −0.01 | 1.63 | 1.62 | −0.01 | 1.34 | 1.62 | +0.28 | 4.93 | 4.90 | −0.03 |
| OCT | 96.76 | 96.70 | −0.06 | 1.75 | 1.96 | +0.21 | 2.07 | 2.14 | +0.07 | 1.29 | 1.73 | +0.44 | 4.92 | 5.06 | +0.14 |
| NOV | 96.68 | 96.65 | −0.03 | 1.44 | 1.55 | +0.11 | 1.86 | 1.94 | +0.08 | 1.26 | 1.72 | +0.46 | 4.40 | 4.65 | +0.25 |
| DEC | 96.81 | 96.79 | −0.02 | 1.01 | 0.91 | −0.10 | 1.40 | 1.15 | −0.25 | 1.20 | 1.40 | +0.20 | 3.56 | 3.25 | −0.31 |

DEC, JAN, and FEB: winter; MAR, APR, and MAY: summer; JUN, JUL, AUG, and SEP: SW monsoon; OCT, NOV: NE monsoon.

TABLE 5: Climatology and Δ of wind speed (scalar), wind speed (resultant), and wind steadiness factor.

| Month | Wind speed (scalar) (0830 IST) ms$^{-1}$ | | | Wind speed (resultant) (0830 IST) ms$^{-1}$ | | | Wind steadiness factor (0830 IST) (%) | | |
| --- | --- | --- | --- | --- | --- | --- | --- | --- | --- |
| | 1964–1993 | 1979–2008 | Δ | 1979–2008 | Δ | | 1964–1993 | 1979–2008 | Δ |
| JAN | 1.04 | 0.75 | −0.29 | 0.71 | 0.33 | −0.38 | 68.89 | 44.48 | −24.41 |
| FEB | 1.07 | 0.81 | −0.26 | 0.73 | 0.38 | −0.35 | 67.95 | 47.36 | −20.59 |
| MAR | 1.21 | 1.00 | −0.21 | 0.83 | 0.56 | −0.27 | 68.49 | 56.15 | −12.34 |
| APR | 1.29 | 1.00 | −0.29 | 0.92 | 0.63 | −0.29 | 71.46 | 63.15 | −8.31 |
| MAY | 1.45 | 1.04 | −0.41 | 1.14 | 0.75 | −0.39 | 78.40 | 72.44 | −5.96 |
| JUN | 1.49 | 1.09 | −0.40 | 1.03 | 0.8 | −0.23 | 68.97 | 73.65 | +4.68 |
| JUL | 1.68 | 1.35 | −0.33 | 1.15 | 0.94 | −0.21 | 68.08 | 69.60 | +1.52 |
| AUG | 1.61 | 1.31 | −0.30 | 1.36 | 1.05 | −0.31 | 84.68 | 79.99 | −4.69 |
| SEP | 1.57 | 1.10 | −0.47 | 1.17 | 0.81 | −0.36 | 74.57 | 74.06 | −0.51 |
| OCT | 1.10 | 0.82 | −0.28 | 0.64 | 0.40 | −0.24 | 58.69 | 49.26 | −9.43 |
| NOV | 0.97 | 0.74 | −0.23 | 0.49 | 0.23 | −0.26 | 50.29 | 31.17 | −19.12 |
| DEC | 1.01 | 0.75 | −0.26 | 0.67 | 0.37 | −0.30 | 66.44 | 48.88 | −17.56 |

DEC, JAN, and FEB: winter; MAR, APR, and MAY: summer; JUN, JUL, AUG, and SEP: SW monsoon; OCT, NOV: NE monsoon.

the order 0.5 m/s are due to decrease in wind speeds in the present data climatology in comparison with past data climatology. Wind steadiness factor values at 0830 IST are computed as the ratio of mean resultant wind speed to mean scalar wind (average value of scalar wind speeds) expressed in %, and the two slot climatologies are made. Wind steadiness factor is a general meteorological technique in order to derive the quantitative estimate of steadiness of wind direction [27]. The feature of losing controllability of blowing winds from a specific direction (computed from wind steadiness) as the demand of a particular season might be justified as a result of heat island effect of the very fast developing urbanisation hardly 8 km East South East of the observatory and the increasing vehicle populations and thereby drastic modifications in at 0830 IST observations which belong to land breeze prevalence over the station in the non-SW monsoonal months (up to a maximum of 20% decrease). Throughout the day, strong seabreeze is in prevalence in the SW monsoon months. Even in the SW monsoon which is dominantly associated with unidirectional wind flow, slight deteriorations in the steadiness of the wind directions have observed. Both the above observations brought out the phenomenal changes in the steadiness of offshore winds (land breeze) at 0830 IST during the non-SW monsoonal months at greater extend and onshore wind (sea breeze) at 0830 IST during the SW monsoon steadiness in smaller grades. The feature of very prominent decrease in direction steadiness in preset data to the past is clearly detected. As negative Δ dominance exists in the non-SW monsoon months where change over of land-sea breezes are so pronounced for the station, this observation may be further studied in order to check any possible modifications in the meso-scale meteorological phenomena like sea/land breeze, thunderstorms due to associated thermal and moisture environment changes.

Visibility and cloud climatologies are also studied. As these observational procedures are non-instrumental, the data is too subjective and quality observation is solely vested

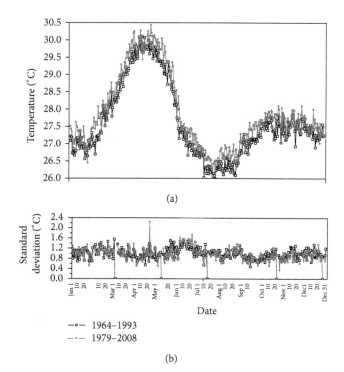

FIGURE 2: (a) Mean daily temperature climatologies in 1964–1993, 1979–2008, and their differences, along with (b) standard deviation of corresponding climatologies.

on the experience of the weather observer. The observation shows, better visibility codes in the present data climatology. A visibility code of 96 and above signifies that a well-defined object (like a mobile phone tower) at least 1 km far can be distinctly differentiable. Among various cloud climatology, dominant positive Δ values are existed in high clouds through out the year which may be pointed out as one of the reason for subdued albedo effect which enhances greenhouse effect, and

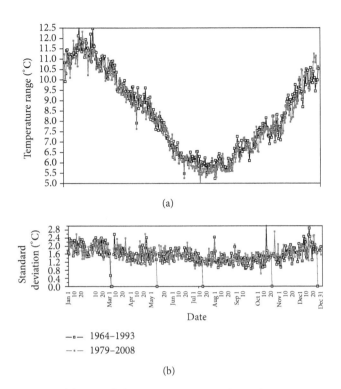

(a)

(b)

FIGURE 3: (a) Mean daily temperature range climatologies in 1964–1993, 1979–2008, and their differences, along with (b) standard deviation of corresponding climatologies.

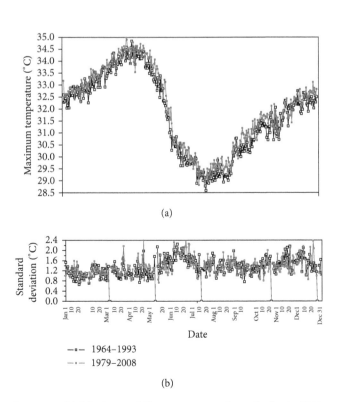

(a)

(b)

FIGURE 4: (a) Maximum daily temperature climatologies in 1964–1993, 1979–2008, and their differences, along with (b) standard deviation of corresponding climatologies.

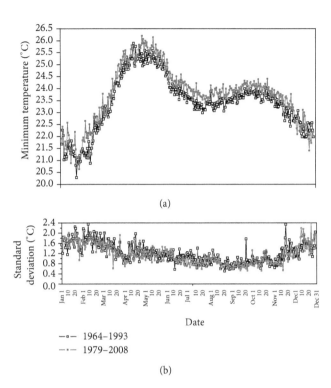

(a)

(b)

FIGURE 5: (a) Minimum daily temperature climatologies in 1964–1993, 1979–2008, and their differences, along with (b) standard deviation of corresponding climatologies.

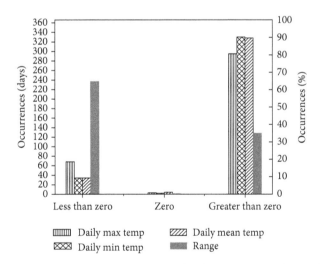

FIGURE 6: Difference statistics for 365 days between present data and past data on daily maximum, minimum, mean temperatures, and range in temperature within a day.

hereby surface temperature escalation. This may be the result of number of high cloud types (Ci and Cc) are less compared to the varieties of low and middle cloud types, which enables easy estimation for the observer while making counts on high clouds. The increase in high cloud climatology values in present data may be the result of increase in air traffic resulted contrails over the region which results an increase in cirrus cloud formations due to tropical dynamics unlike

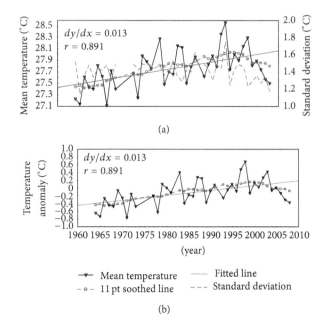

(a)

(b)

FIGURE 7: (a) Yearly mean temperature along with standard deviation and (b) anomaly.

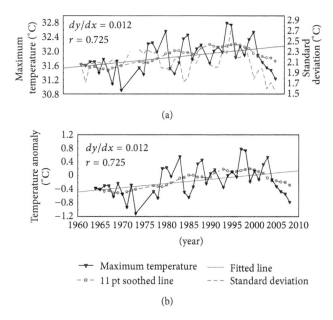

(a)

(b)

FIGURE 8: (a) Yearly mean maximum temperature along with standard deviation and (b) anomaly.

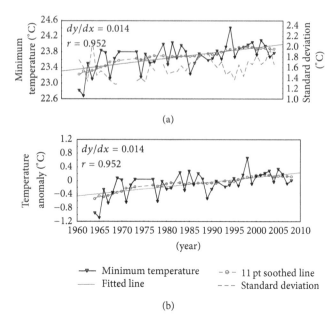

(a)

(b)

FIGURE 9: (a) Yearly mean minimum temperature along with standard deviation and (b) anomaly.

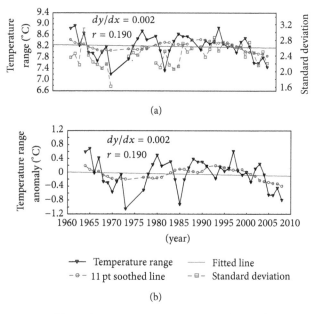

(a)

(b)

FIGURE 10: (a) Yearly mean temperature range along with standard deviation and (b) anomaly.

microphysical processes in mid-latitudes as in Minnis et al. [28], Zerefos et al. [29]. Eleftheratos et al. [30] have studied the variability of cirrus clouds over the tropics governed by the activity of El Nino Southern Oscillation, the Quasi Biennial Oscillation and solar activity. Possible radiation budget alterations, modifications at the time of occurrence of minimum and maximum temperatures are to be critically examined as a part of future investigations. Prominent two maxima (secondary maximum is very shallower compared to

the primary) and two minima were seen in middle, high and total cloud climatology.

*3.3. Mean Daily Climatology in Temperature.* Mean of daily maximum, minimum, mean and range in temperature between daily maximum and minimum temperatures are computed and depicted for the 365 days in a year (Figures 2, 3, 4 and 5) along with their standard deviations. The increase in mean daily values in present data temperatures are quite obvious and standard deviation values are fallen within

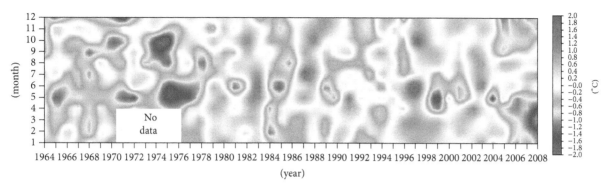

FIGURE 11: Maximum temperature anomaly.

2.7°C for all the four mean daily climatologies. A difference statistics for 365 days between present data to past data is made and is shown in Figure 6. Out of 365 days, more than 80% mean daily maximum, minimum and mean temperatures are distributed as greater than zero occurrences in the difference statistics. An elevated frequency of occurrence of more than 65% under less than zero class is a concrete feature which implies that the range among maximum and minimum temperature is alarmingly diminishing.

*3.4. Interannual Variability.* Inter annual variability has critically examined for annual mean of minimum, maximum and mean temperatures, annual rainfall in the SW and NE monsoon over the station and for the Kerala State and average annual surface pressure in the SW and NE monsoons for the station. Each inter annual variability data has undergone an 11 year moving average with end point extrapolation till the raw data availability to bring out long term trend and the trend line also has shown in each inter annual variability figures similar to Krishnakumar et al. [22] for the interannual variability of rainfall over Kerala. Interpolated values are obtained for each year from the trend and such a data set obtained, has been subjected to statistical significant test. interpolated data has not taken for a particular year, if the raw data point is missing for the corresponding year. This method has enabled no reduction in the degrees of freedom (*df*). To confirm the statistical significance on various 11 year moving averaged data and also to assess the strength of sample correlation coefficient (*r*), linear regression *t*-test were carried out and the results are presented (Table 6). Generally strong and significant linear trend variations are observed. Even though linear trends of maximum and minimum temperatures have shown strong and significant statistical trends, the derived variable-temperature range has resulted a weak and insignificant test statistic. This may be statistically inferred as a variable derived out of two variables which favours strong and significant test statistic separately need not show the same test static of strong and significant.

*3.4.1. Interannual Variability in Temperature.* To assess the interannual variability with temperature, annual mean minimum, mean maximum and mean of mean temperatures are found out for each year from 1964 to 2008. Also to

ascertain deviation from the normal, each yearly mean value is deducted from the mean of the entire values available for present data and thereby extracted the anomaly in temperature. Mean and standard deviation of daily mean, maximum, minimum and daily temperature range are computed for a year and depicted in Figures 7, 8, 9 and 10. The fitted linear trend line slope (*dy/dx*) as well as correlation coefficient (*r*) values are annotated in graphs. It is seen that from both mean daily temperature as well as from the maximum temperature an abrupt prominent shift or a discernible upward trend is occurred after 1976 may be like the "climate shift" as stated by Trenberth [24]. The 45 year rate in mean daily, maximum and minimum are obtained as +0.585°C, +0.540°C and +0.630°C respectively or with decadal rates of (+0.13 ± 0.01)°C, (+0.12 ± 0.02)°C and (+0.14 ± 0.0)°C accordingly. In the obtained value for the global mean surface temperature as a whole (+0.07 ± 0.02)°C, surface temperatures over land have risen at about double the ocean rate after 1979 (more than (+0.27 ± 0.06)°C per decade versus (+0.13 ± 0.04)°C per decade) as per the Trenberth et al. [31]. From 1950 to 2004, the global land trend in diurnal temperature range was extracted as negative (−0.07°C per decade) although tropical areas are still underrepresented [31]. Kumar et al. [32] showed trends of increase in diurnal temperature range, significant increase in annual maximum temperature, and practically trendless annual minimum temperature over India in the last century. Our inter annual variability results coincide mostly with former studied data period which supports the decrease of diurnal temperature range, and the same is very vivid from Figure 6 discussed earlier. It infers that for this tropical coastal station the obtained decadal rate of (+0.130 ± 0.01)°C, which is in agreement with global ocean rate, may be due to the station being coastal in origin. The error value reduction by ±0.03°C for TERLS can be due to more stabilized single station data. Also rate of increase in trend for minimum temperature is high in comparison to maximum temperature. A decreasing trend in maximum temperature is noticed after 2005 with decreasing standard deviation values. This observation might be correct as a result of shifting the observatory since 1964 to hardly 30 m towards north in 2005 were smooth and free sea-breeze flow which enables abrupt reach of maximum temperature in the diurnal course around the time of sea breeze onset. The feature of attaining the maximum temperature of the day around the time of onset of

TABLE 6: Linear trends of 11 year moving averaged data on various inter-annual variability. Level of significance was choosen at 0.05% to read out critical $t$ values ($t_c$) corresponding to the degree of freedom ($df$). If $r > 0.5$ strong, $r < 0.5$ weak, $t > t_c$ significant, and $t < t_c$ not significant.

| No. | Interannual variability in | $r$ | Slope $(dy/dx) \pm$ error | Calculated $t$ | $df$ | Comments | Data missing |
|---|---|---|---|---|---|---|---|
| 1 | Mean temperature | 0.891 | $0.013 \pm 0.001$ | 12.08154 | 38 | Strong and significant. (Figures 7(a) and 7(b)) | 1974, 1975, 1976, 1982, and 1992 |
| 2 | Mean maximum temperature | 0.725 | $0.0121 \pm 0.002$ | 6.50392 | 38 | Strong and significant. (Figures 8(a) and 8(b)) | 1974, 1975, 1976, 1982, and 1992 |
| 3 | Mean minimum temperature | 0.952 | $0.014 \pm 0.000$ | 19.29677 | 38 | Strong and significant. (Figures 9(a) and 9(b)) | 1974, 1975, 1976, 1982, and 1992 |
| 4 | Mean temperature range | $-0.190$ | $-0.002 \pm 0.002$ | $-1.19078$ | 38 | Weak and not significant (Figure 10(a)). | 1974, 1975, 1976, 1982, and 1992 |
| 5 | South West monsoon rainfall (Thumba) | $-0.477$ | $-2.145 \pm 0.616$ | $-3.483$ | 41 | Weak and significant (Figure 14) | 1964, 1998 |
| 6 | North East monsoon rainfall (Thumba) | 0.848 | $3.928 \pm 0.393$ | 10.000 | 39 | Strong and significant (Figure 15) | 1964, 1977, 1992, and 1998 |
| 7 | South West monsoon rainfall (Kerala) | $-0.721$ | $-6.149 \pm 0.934$ | $-6.587$ | 40 | Strong and significant (Figure 16) | 1964, 2007, and 2008 |
| 8 | North East monsoon rainfall (Kerala) | 0.737 | $3.430 \pm 0.491$ | 6.986 | 41 | Strong and significant (Figure 17) | 1964, 2008 |
| 9 | South West monsoon surface pressure | 0.445 | $0.006 \pm 0.002$ | 2.899 | 34 | Weak and significant (Figure 18) | 1964 to 1972 |
| 10 | North East monsoon surface pressure | $-0.616$ | $0.010 \pm 0.002$ | $-4.555$ | 34 | Strong and significant (Figure 19) | 1964 to 1972 |

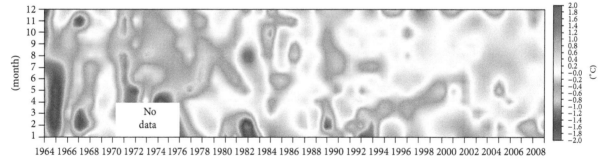

FIGURE 12: Minimum temperature anomaly.

sea breeze is known [33]. As the station is of coastal in origin, the occurrence of daily maximum temperature is very much linked to the setting and strong establishment of sea breeze front towards the land areas.

Mean monthly anomaly structures in minimum, maximum, and mean temperatures are presented in time section contour diagrams (Figures 11, 12 and 13). All contour depictions show well-marked temperature increase during the course of years and are so pronounced after 1976.

*3.4.2. Interannual Variability in SW and NE Monsoon Rainfall and Surface Pressure.* Interannual variability in rainfall and surface pressure particularly for SW monsoon and NE monsoon are studied. A decrease in trend in SW monsoon

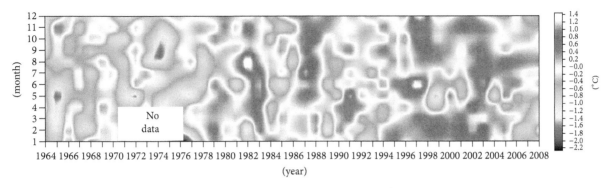

FIGURE 13: Mean temperature anomaly.

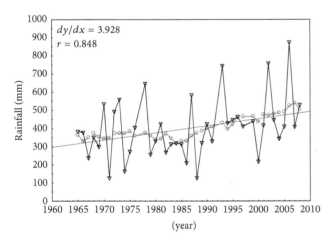

FIGURE 14: South West monsoon rainfall over Thumba.

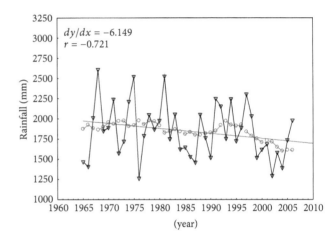

FIGURE 16: South West monsoon rainfall over Kerala.

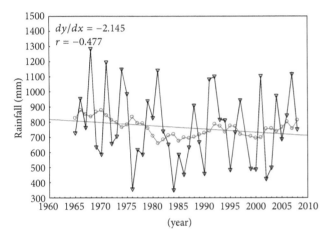

FIGURE 15: North East monsoon rainfall over Thumba.

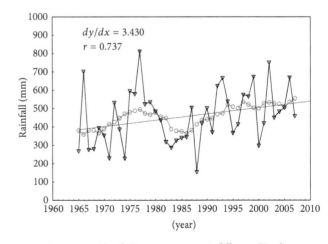

FIGURE 17: North East monsoon rainfall over Kerala.

and an increase in trend in NE monsoons are the observed features over the station with decadal variability of $(-21.5 \pm 6.2)$ mm and $(+39.3 \pm 3.9)$ mm, respectively (Figures 14 and 15). Analysis by considering initial rainfall values (at the beginning year on the fitted straight line) of South West monsoon, North East monsoon, corresponding slope of them as 817.822 mm, 296.535 mm, −21.5 mm/decade and

39.3 mm/decade, respectively, shows the likelihood exceed of North East monsoon rainfall over South West monsoon over this region after the year 2050 if the current trend continues. Similar analyses are carried out on the slot 1965 to 2007 for the Kerala meteorological subdivision from the longest instrumental rainfall series of the Indian regions (1813–2006) ([34]; http://www.tropmet.res.in/). The actual rainfall amount over Kerala during 2007 is obtained from the

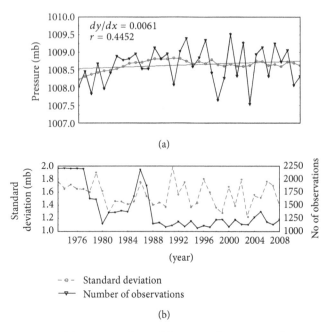

$$dy/dx = 0.0061$$
$$r = 0.4452$$

(a)

- ⊘ - Standard deviation
- ▽ - Number of observations

(b)

FIGURE 18: (a) Interannual variability in surface pressure for South West monsoon season along with (b) standard deviation and number of observations.

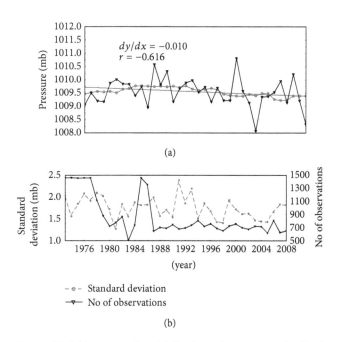

$$dy/dx = -0.010$$
$$r = -0.616$$

(a)

- ⊘ - Standard deviation
- ▽ - No of observations

(b)

FIGURE 19: (a) Interannual variability in surface pressure for North East monsoon season along with (b) standard deviation and number of observations.

quarterly journal the MAUSAM (July 2008 and October 2008 issues). Decadal variations show a decreasing trend in SW monsoon by (−61.5 ± 9.3) mm and increasing trend in NE monsoon by (+34.0 ± 4.9) mm (Figures 16 and 17).

Interannual variability in surface pressure was derived and depicted for SW monsoon (Figure 18) and NE monsoon (Figure 19). For this, every hourly data available since 1973

was subjected to mean monthly value computation. Number of observations available ranges from 1100 to 2245 in SW monsoon and 500 to 1495 in NE monsoon, respectively, and provided along with the standard deviation values in the above figures. Since atmospheric pressure is a highly stabilized property over a region for a particular season, the decadal trend variability observed is (+0.06 ± 0.02) mb and (−0.10±0.02) mb for SW monsoon and NE monsoon, respectively. The positive (increase) and negative (decrease) interannual trends observed in SW monsoon and NE monsoon, respectively, might have contributed possible implications in the development of associated synoptic scale meteorological systems (position and orientation of troughs/ridges and low pressures) and thereby corresponding decrease and increase in SW monsoon and NE monsoon rainfall trends.

## 4. Conclusion

Long-term surface meteorological records from TERLS were collected and analysed to study the long-term changes in overall climatology, climatology pertained to a particular time of observation, mean daily climatology in temperature, interannual variability in temperature, interannual variability in SW and NE monsoon rainfall, and mean monthly anomaly structure in temperature. Results on various analyses show strong and vivid features contributed by climate change. The salient features observed contributed by climate change are as follows:

(1) All the temperatures show increase in their behaviour with slight decrease in RH between the present climatology and past climatology.

(2) Steadiness of wind decreases especially in non-SW monsoon months, which gives an indirect signature of possible modifications in meso-scale systems.

(3) High cloud climatology increases and thereby probable modifications in radiation budget, time of occurrence of maximum and minimum temperatures.

(4) Daily statistics for 365 days show more than 80% exceedance in present climatology in comparison to past in minimum, maximum, and mean temperature.

(5) Interannual variability analysis in mean temperature for this tropical coastal station shows +1.3°C/100 years which is same as that of global ocean rate.

(6) Observed interannual rainfall variability in SW monsoon and North East monsoon shows decrease in trend by −21.45 mm/decade and increase in trend by +39.3 mm/decade, respectively. Prognosis on this observation points out the likelihood exceedance of NE monsoon rainfall over the South West monsoon after the year 2050 for the station. But it is prejudicial to infer that the rainfall increase in NE monsoon can compensate the forecasted decrease in SW monsoon, since NE monsoon is mostly of heavy rainfall in nature as a result of meso-scale convective systems. Such rainfall runoff is very fast than the slow water infiltration feature in association with SW monsoon

rainfall, which is essential for irrigation and agricultural needs.

(7) Decadal variability in surface pressure has shown +0.06 mb and −0.10 mb for SW monsoon and NE monsoon, respectively.

The imprints due to climate change on various meteorological parameters over this coastal station proved that the region also faces vulnerable situation due to the phenomena, and therefore, necessary engineering methods/socioeconomic steps are to be undertaken.

## Acknowledgments

The availability of this unique data for a continuous period of 45 years is the outcome of concerted and whole hearted efforts put in by the staffs of TERLS/India Meteorological Department. It is a pleasure to acknowledge with esteemed gratitude the team that worked since 1963 at the TERLS, Thumba, whose efforts have yielded the voluminous unique data bank which is available for investigations. The authors are extremely grateful to Dr. K Krishnamoorthy, Director, Space Physics Laboratory, VSSC, and the reviewers for their valuable comments that helped to improve the paper to the final form.

## References

[1] N. Oreskes, "The scientific consensus on climatic change," *Science*, vol. 306, p. 1686, 2004.

[2] H. Ayyar, Effect of Pollution on Weather. Science Reporter. Council of Scientific and Industrial Research (CSIR), 1973.

[3] K. Rupa Kumar, A. K. Sahai, K. Krishna Kumar et al., "High-resolution climate change scenarios for India for the 21st century," *Current Science*, vol. 90, no. 3, pp. 334–345, 2006.

[4] P. Brohan, J. J. Kennedy, I. Harris, S. F. B. Tett, and P. D. Jones, "Uncertainty estimates in regional and global observed temperature changes: a new data set from 1850," *Journal of Geophysical Research D*, vol. 111, no. 12, Article ID D12106, 2006.

[5] T. M. Smith and R. W. Reynolds, "A global merged land-air-sea surface temperature reconstruction based on historical observations (1880–1997)," *Journal of Climate*, vol. 18, no. 12, pp. 2021–2036, 2005.

[6] K. M. Lugina, P. Y. Groisman, K. Y. Vinnikov, V. V. Koknaeva, and N. A. Speranskaya, "Monthly surface air temperature time series area-averaged over the 30-degree latitudinal belts of the globe, 1881-2004 In Trends: a compendium of data on global change," Carbon Dioxide Information Analysis Centre, Oak Ridge National Laboratory, US Department of Energy, Oak Ridge, Tenn, USA, 2005, http://cdiac.esd.ornl.gov/trends/temp/lugina/lugina.html.

[7] J. Hansen, R. Ruedy, M. Sato et al., "A closer look at United States and global surface temperature change," *Journal of Geophysical Research D*, vol. 106, no. 20, pp. 23947–23963, 2001.

[8] S. Solomon, D. Quin, M. Manning et al., "Contribution of working group I to the Fourth Assessment Report of the IPCC," Tech. Rep. IPCC-AR4, 2007.

[9] R. C. Balling Jr. and S. W. Brazel, "Recent changes in Phoenix, Arizona summertime diurnal precipitation patterns," *Theoretical and Applied Climatology*, vol. 38, no. 1, pp. 50–54, 1987.

[10] R. Bornstein and Q. Lin, "Urban heat islands and summertime convective thunderstorms in Atlanta: three case studies," *Atmospheric Environment*, vol. 34, no. 3, pp. 507–516, 2000.

[11] P. G. Dixon and T. L. Mote, "Patterns and cause of Atlanda's urban heat island-initiated precipitation," *Journal of Applied Meteorology*, vol. 42, pp. 1273–1284, 2003.

[12] F. Fujibe, "Long-term surface wind changes in the Tokyo metropolitan area in the afternoon of sunny days in the warm season," *Journal of the Meteorological Society of Japan*, vol. 81, no. 1, pp. 141–149, 2003.

[13] T. Inoue and F. Kimura, "Urban effects on low-level clouds around the Tokyo metropolitan area on clear summer days," *Geophysical Research Letters*, vol. 31, no. 5, Article ID L05103, 2004.

[14] J. M. Shepherd, H. Pierce, and A. J. Negri, "Rainfall modification by major urban areas: observations from spaceborne rain radar on the TRMM satellite," *Journal of Applied Meteorology*, vol. 41, no. 7, pp. 689–701, 2002.

[15] C. Price, S. Michaelides, S. Pashiardis, and P. Alpert, "Long term changes in diurnal temperature range in Cyprus," *Atmospheric Research*, vol. 51, no. 2, pp. 85–98, 1999.

[16] H. N. Singh, S. D. Patil, S. D. Bansod, and N. Singh, "Seasonal variability in mean sea level pressure extremes over the Indian region," *Atmospheric Research*, vol. 101, no. 1-2, pp. 102–111, 2011.

[17] J. Kingwell, J. Shimizu, K. Narita, H. Kawabata, and I. Shimizu, "Weather factors affecting rocket operations: a review and case history," *American Meteorological Society*, vol. 72, pp. 778–793, 1991.

[18] R. E. Turner and C. K. Hill, *Terrestrial Environment (Climatic) Guidlines for Use in Aerospace Vehicle Development, 1982 Revision*, Technical Memor. 82473, National Aeronautics and Space Administration, 1982.

[19] "Space Research in India," in *Proceedings of the 38th COSPAR Meeting*, pp. 199–204, Bremen, Germeny, 2010.

[20] P. V. Joseph and A. Simon, "Weakening trend of the southwest monsoon current through peninsular India from 1950 to the present," *Current Science*, vol. 89, no. 4, pp. 687–694, 2005.

[21] S. K. Dash, J. R. Kumar, and M. S. Shekhar, "On the decreasing frequency of monsoon depressions over the Indian region," *Current Science*, vol. 86, no. 10, pp. 1404–1411, 2004.

[22] K. N. Krishnakumar, G. S. L. H. V. Prasada Rao, and C. S. Gopakumar, "Rainfall trends in twentieth century over Kerala, India," *Atmospheric Environment*, vol. 43, no. 11, pp. 1940–1944, 2009.

[23] National Council of Educational Research and Training (NCERT), India, Text book of physics—part I for class 11, 2006.

[24] K. E. Trenberth, "Recent observed interdecadal climate changes in the Northern Hemisphere," *Bulletin of the American Meteorological Society*, vol. 71, no. 7, pp. 988–993, 1990.

[25] T. K. Karl, H. F. Diaz, and G. Kukla, "Urbanization: its detection and effect in the United States climate record," *Journal of Climate*, vol. 1, pp. 1099–1123, 1988.

[26] M. Rebetez and M. Beniston, "Changes in sunshine duration are correlated with changes in daily temperature range this century: an analysis of Swiss climatological data," *Geophysical Research Letters*, vol. 25, no. 19, pp. 3611–3613, 1998.

[27] K. V. S. Namboodiri, P. K. Dileep, and K. Mammen, "Wind steadiness up to 35 km and its variability before the South West Monsoon onset and the withdrawal," *Mausam*, vol. 63, pp. 275–282, 2012.

[28] P. Minnis, A. J. Kirk, P. Rabindra, and P. Dung, "Contrails, cirrus trends, and climate," *Journal of Climate*, vol. 17, pp. 1671–1685, 2004.

[29] C. S. Zerefos, K. Eleftheratos, D. S. Balis, P. Zanis, G. Tselioudis, and C. Meleti, "Evidence of impact of aviation on cirrus cloud formation," *Atmospheric Chemistry and Physics*, vol. 3, no. 5, pp. 1633–1644, 2003.

[30] K. Eleftheratos, C. S. Zerefos, C. Varotsos, and I. Kapsomenakis, "Interannual variability of cirrus clouds in the tropics in el niño southern oscillation (ENSO) regions based on international satellite cloud climatology project (ISCCP) satellite data," *International Journal of Remote Sensing*, vol. 32, no. 21, pp. 6395–6405, 2011.

[31] K. E. Trenberth, P. D. Jones, and P. Ambenje, "Observations: surface and climate change," in *Climate Change 2007: The Physical Science Basis. Contribution of working group I to the Fourth Assessment Report of the Intergovernmental Panel on Climate Change*, S. Solomon, D. Qin, and M. Manning, Eds., Cambridge University Press, Cambridge, UK, 2007.

[32] K. R. Kumar, K. K. Kumar, and G. B. Pant, "Diurnal asymmetry of surface temperature trends over India," *Geophysical Research Letters*, vol. 21, no. 8, pp. 677–680, 1994.

[33] T. R. Sivaramakrishnan and P. S. Prakash Rao, "Sea-breeze features over Sriharikota, India," *Meteorological Magazine*, vol. 118, no. 1400, pp. 64–67, 1989.

[34] N. A. Sontakke, N. Singh, and H. N. Singh, "Instrumental period rainfall series of the Indian region (AD 1813–2005): revised reconstruction, update and analysis," *Holocene*, vol. 18, no. 7, pp. 1055–1066, 2008.

# 10

# Six Temperature Proxies of Scots Pine from the Interior of Northern Fennoscandia Combined in Three Frequency Ranges

**Markus Lindholm,**[1] **Maxim G. Ogurtsov,**[2] **Risto Jalkanen,**[1]
**Björn E. Gunnarson,**[3] **and Tarmo Aalto**[1]

[1] Metla, Rovaniemi Research Unit, P.O. Box 16, 96301 Rovaniemi, Finland
[2] A.F. Ioffe Physico-Technical Institute, St. Petersburg 194 021, Russia
[3] Bolin Centre for Climate Research, Department of Physical Geography and Quaternary Geology, Stockholm University, 106 91 Stockholm, Sweden

Correspondence should be addressed to Markus Lindholm; markus.lindholm@metla.fi

Academic Editor: Silvio Gualdi

Six chronologies based on the growth of Scots pine from the inland of northern Fennoscandia were built to separately enhance low, medium, and higher frequencies in growth variability in 1000–2002. Several periodicities of growth were found in common in these data. Five of the low-frequency series have a significant oscillatory mode at 200–250 years of cycle length. Most series also have strong multidecadal scale variability and significant peaks at 33, 67, or 83–125 years. Reconstruction models for mean July and June–August as well as three longer period temperatures were built and compared using stringent verification statistics. We describe main differences in model performance ($R^2 = 0.53$–$0.62$) between individual proxies as well as their various averages depending on provenance and proxy type, length of target period, and frequency range. A separate medium-frequency chronology (a proxy for June–August temperatures) is presented, which is closely similar in amplitude and duration to the last two cycles of the Atlantic multidecadal oscillation (AMO). The good synchrony between these two series is only hampered by a 10-year difference in timing. Recognizing a strong medium-frequency component in Fennoscandian climate proxies helps to explain part of the uncertainties in their 20th century trends.

## 1. Introduction

Several recent studies have discussed the potential of high-resolution proxies based on the growth of Scots pine from northern Fennoscandia for reconstruction of summer temperatures in particular at the low-frequency scale of variability [1–5]. The main concern has generally been the interesting temperature difference between medieval times, Little Ice Age, and the modern period viewing recent and projected warming within the context of natural variability. Less attention is usually paid to the strong multidecadal component of temperature variability in the observational as well as proxy records, which may seriously hamper the identification of an amplified warming signal in the last century in the Arctic and surrounding regions [6]. Some internal climate controls may have influenced regional climate simultaneously with the carbon dioxide induced warming, and Fennoscandian temperature proxies may have recorded both types of potentially coinciding, interacting, or even diverging signals in the decadal-to-centennial scales of variability. Multidecadal variability in Fennoscandian summertime climate may well be related to the AMO (sea surface temperatures (SST)), which has a period of about 40–80 years, suggested to arise from predictable internal variability of the ocean-atmosphere system [7, 8]. In addition Arctic air temperature and pressure have been shown to display strong multidecadal variability on similar time scales [9].

The goal here is to study periodicity and trends in growth variability of six recently published and updated millennia length proxies of Scots pine from the northern timberline. If the six series show consistent and synchronous interannual-to-decadal, decadal-to-multidecadal, as well as centennial and longer types of variability, three versions of each series will be built in order to highlight the three frequency bands separately in these data as well as in various combinations to regional averages (not mixing frequency classes). Growth signals are considered more or less frequency dependent if they all show similar and coherent behavior in the suggested frequency ranges and may thus be usefully combined to regional high-frequency (h-f), medium-frequency (m-f), and low-frequency (l-f) chronologies. Current knowledge is relatively limited regarding spectral details of the l-f and m-f trends and periodicities in these proxies (except for some older versions of individual proxies for relatively narrow frequency bands [10, 11]).

All the six series are known predominantly as summer temperature proxies [1, 4, 5, 12, 13]. This study will provide a synopsis of their rather complex potential as predictors of high-summer (July), standard summer season (June–August), and even some longer warm period temperatures in the three broad frequency ranges. The results will serve as practical guidelines for selecting proper proxy types and choice of indexing methods with respect to required response period as well as frequency band. Such knowledge is crucial in various multiproxy applications (see, e.g., [4]) where the amplitude, duration, and timing of changes are important. In particular, our goal is to reconstruct temperature variability at the multidecadal scale, which would be useful in interregional comparisons of meaningful periodicity in land surface temperatures in the subarctic region as well as between land and sea surface temperatures of larger fields in the search for periodic patterns (both intrinsic and external to the climate system). In order to gain insight into this topic we will here compare the well-known cycles of annual AMO with m-f periodicity in these data. Since Gray et al. [14] successfully reconstructed the AMO using a large network of tree-growth chronologies (including one early version of our data series), it will be interesting to test an AMO model now with these six longer and updated series.

In calibration model performance will be evaluated in each case based on known stringent verification tests. The aim is to provide feasible models in three scales from h-f to l-f with increasing proportion of low frequency variance (where h-f necessarily overlaps in m-f and m-f overlaps in l-f). The linear relationship between tree growth at the three frequency ranges and temperature over the whole range of target variations (unfiltered) is analyzed. It would be reasonable to expect model fit generally to increase if meaningful lower frequencies are added to the pool of predictors (from h-f to l-f). However the situation is generally more complicated as targets are relatively short series as compared to the proxies, and the longest trends in tree growth can only have a partial match in temperature. In an about 100-year temperature record (usual in calibrations in this region), the detection of verifiable trends is restricted to a maximum of about 50 years (interpolation). However, if a good partial fit of a longer trend is found, some extrapolation is usually reasonable. The results also aim to contribute to the largely missing debate on the uncertainties and even disagreement

of regional reconstructions or proxies as compared to the considerable attention over the discrepancies in hemispheric reconstructions [15].

## 2. Materials and Methods

The main interest here is in the inland region, which is occupied by the Fennoscandian Shield and forms a relatively homogeneous peneplane at the modest altitudes of about 200 to 500 m a.s.l. with sharp climatic and geobotanical boundaries in the east and west [16–20]. Climate in this interior region is arguably more homogeneous and continental without the more marine coastal regions. General location for all six data sets is the northern timberline, between the Swedish Scandes and the Khibiny Low Mountains region.

The data include the recently updated and bias-corrected ring width (SWR) and maximum density (SXD) from Sweden [5, 21, 22], ring width (FRW), height increment (FHI), and maximum density (FXD) from Finland [4, 23], as well as ring width (RRW) from the Kola peninsula, Russia [4, 24]. The ring width (SRW and FRW) data sets are large, including data from 650 and 536 trees, respectively. The others are smaller, consisting of samples from 167 (FHI) to 78 trees (FXD). Height data (FHI) are shifted by one year for making comparisons possible, because height growth reflects conditions in the previous year [25, 26].

Four long, monthly climate records represent the regional temperatures over their common period from 1908 to 2002 [17]. Tornedalen (Sweden) composite record is available as continuous between 1816 and 2002 [27], Karasjok (Norway) record since 1876, Karesuando (Sweden) record since 1890, and Sodankylä (Finland) record since 1908. Tornedalen and Karesuando are the closest to the western (Swedish RW and XD) sampling sites, Karasjok and Sodankylä to the central (Finnish RW, HI, and XD) sites, and Sodankylä is the closest to the eastern (Russian RW) site. Karasjok is the northernmost and Sodankylä and Tornedalen the southernmost climate records. Additional verification is obtained from the Bottenviken compilation series [28], which is based on the data from six stations (Abisko, Karesuando, Kvikkjokk, Jokkmokk, Haparanda, and Piteå, mainly from somewhat more western and southern locations than the ones used in calibrations here) in northern Sweden. Moreover, annual AMO anomalies [29] are used in multidecadal comparisons.

Regional curve standardization (RCS; see [5, 30–32]), 180-year and 30-year splines (see [33–35]) were used in indexing. They are well-known methods in dendroclimatology and frequently applied in targeting various frequency-dependent responses in proxy-based reconstructions [36, 37]. The RCS method is expected to preserve any l-f signal up to wavelengths exceeding the lengths of the individual segments used in building the chronologies. On the other hand, the 180-year splines will highlight m-f signal by removing the lowermost frequencies, the most multicentury-timescale variance potentially present in the data. The 30-year spline extracts the h-f signal and will remove in addition much of the multidecadal and all longer scale variances. The three frequency ranges correspond to interannual-to-decadal, interannual-to-multidecadal, and centennial types of variability in the time domain. Furthermore, digital low-pass filters were used to extract the chronology variance with a frequency lower than 10 years [38–40].

We used Fourier analysis to define the spectral content in the chronologies, that is, a spectral description in terms of cycles of varying length, the actual frequencies that generate the original series. The wavelet approach has advantages over the more traditional methods for analyzing potentially nonstationary signals, which have discontinuities and non-periodic characteristics [41–43]. In the complex interactions in climate there are a number of components which tend to damp out the more rapid fluctuations. Thus climate time series with lower frequency/longer period cycles have to contain a greater proportion of the observed variance to achieve the same significance as higher frequency/shorter period features [44]. In the time domain confidence intervals (c.i.) were calculated separately for the l-f and m-f ranges of growth variability using nonparametric bootstrap method (sampling with replacement) [45, 46]. Arithmetic averages (AA) and weighted averages (WA) (see [47] and references therein) were applied in combining regional chronologies.

Simple linear regression models (transfer functions) [12, 48–50] were developed for each of the three versions of the six chronologies and their averages to be used in turn as climate predictors. Individual models were then tested in split period calibration verification, where the total calibration period (1908–2002) was divided into two equal 48-year halves: 1908–1955 and 1955–2002. These subperiods, used for calibration during one period and verification during the other, are referred to as early calibration-late verification and late calibration-early verification (EC-LV and LC-EV, resp.). Both subperiods should produce positive values of reduction of error (RE) [12, 38, 51] and coefficient of efficiency (CE) [12, 51] statistics for the model in order to pass the verification tests. RE and CE values may vary from +1 to −∞, with 0 indicating that the reconstruction model performs no better as a predictor than the calibration (RE) or verification (CE) period mean value. The models are also compared using coefficient of determination ($R^2$) and explained variance ($r^2$).

## 3. Results

*3.1. Periodicity, Trends, and Interannual Shifts in Growth.*
Five of the six l-f series (RCS, no filtering) share significant oscillatory mode at 200–250 years of cycle length (Figure 1) and SXD shows even longer-term features (Figure 1(b)). Most series also have strong multidecadal scale variability, viz. significant peaks at 67 years in SRW and SXD, as well as at 83–125 years in FRW, FXD, and RRW. Although all series have cumulated some concentration of variance at m-f scale, the two western series (SRW and SXD) have 67-year peaks and both of the XD series have a (possibly related) peak at 33 years. In addition the two XD series differ evidently from the others with still distinctly higher frequency content at 0.1–0.3 cpa (Figure 1). Based on these results the common spectral content of these data is concentrated principally on the relatively narrow frequency ranges from 0.005 cpa to 0.5 cpa; that is, periodicities are from 200 to 2 years.

Next, the six l-f time series were low-pass filtered, normalized, and then averaged (c.i. around the mean in Figure 2(a), in 1006–1996, since six years are lost at both ends due to filtering). Correlations between these l-f series are all positive, ranging from 0.81 (FRW-FHI) to 0.21 (FHI-RRW) with a mean value of 0.52. Correlation declines in the three RW series with increasing distance; the *r*-value (from west to east) is 0.55 between SRW and FRW, 0.34 between SRW and RRW, and 0.49 between FRW and RRW. The two density series (SXD and FXD) have high linear association ($r = 0.66$). However their correlation is lower than that between either the two western or central RW and XD series (SRW-SXD, $r = 0.73$ and FRW-FXD, $r = 0.77$).

The mean regional l-f chronology (Figure 2(a)) shows an overall increasing trend (regression of growth on time in 1006–1996; $y = 0.0006x$), reflecting the diverse long-term rates of change in the original series. Producing a distribution of the mean values and then locating the lower and upper bounds of this distribution, five significant (95% c.i.) periods of positive growth in the average series are discerned: 1083–1104, 1159–1178, 1428–1436, 1750–1769, and the longest from 1918 to 2002, separated by periods of poor growth of varying lengths and magnitudes between them. These periods are determined in the regional growth signal within the uncertainty limited by (sources of error) interregional differences (e.g., western, central, and eastern provenance) as well as differences due to proxy type (RW, XD, and HI). The highest values in the second millennium are clearly recorded in the last century (Figure 2(a)). The bootstrapped confidence intervals are nonsymmetric and particularly wide in the first and last centuries.

The six m-f (filtered) series are more synchronous than the l-f series as their average has narrower c.i. and many more significant periods (Figures 2(a) and 2(b)) in addition to having a higher mean correlation ($r = 0.58$). Correlation is the highest between FRW and FXD ($r = 0.74$) as well as between FRW and FHI ($r = 0.74$) and the lowest between SXD and FHI ($r = 0.39$) as well as between SXD and RRW ($r = 0.39$). Twelve distinct and significant (95% c.i.) periods of above average growth are dated (Figure 1(b)): 1083–1101, 1157–1183, 1283–1289, 1411–1447, 1489–1498, 1535–1547, 1559–1572, 1625–1634, 1653–1665, 1750–1767, 1850–1862, and the longest in 1920–1956. The confidence limits indicate wider spread in the original six series in the 13th and 14th centuries than in the rest of the m-f series.

The average of the six h-f (no filtering) series represents the h-f variability in this work (Figure 2(c)). Average correlation between all the six series is 0.41. It is the highest between SRW and FRW ($r = 0.68$) and the lowest between SXD and FHI ($r = 0.22$). In the mean series growth was the lowest in 1601 (−2.6 s.d. units below the mean) and the highest in 1826 (2.4 s.d. units above the mean). The greatest biennial shifts (difference between any two consecutive years) occurred between 1600 and 1601 (3.8 s.d. from 1.2 to −2.6) as well as between 1640 and 1641 (3.8 s.d. from 1.8 to −2.0). In the 20th century, 1903 was unusually low and 1937 was high in the growth index. It is worth noticing that even this h-f series has evident decadal fluctuations.

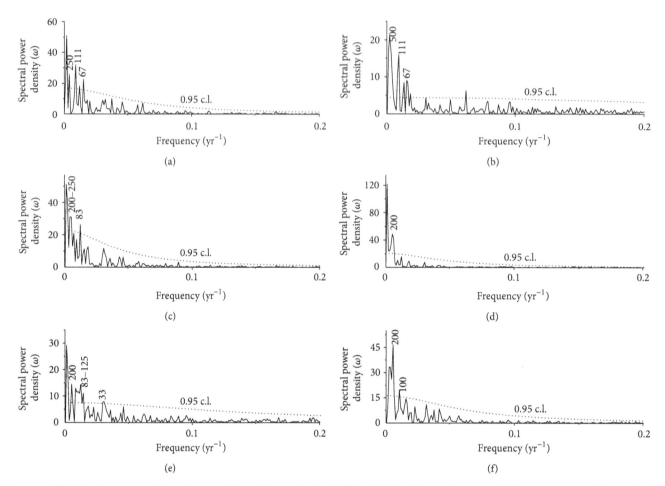

FIGURE 1: The Fourier spectra of the six growth-based chronologies (RCS indexing) of Scots pine. Smooth lines are 0.95 confidence levels, calculated for red noise with AR(1) coefficients $\alpha$: (a) SRW ($\alpha = 0.7$), (b) SXD ($\alpha = 0.20$), (c) FRW ($\alpha = 0.77$), (d) FHI ($\alpha = 0.75$), (e) FXD ($\alpha = 0.43$), and (f) RRW ($\alpha = 0.69$). Cpa is cycles per annum.

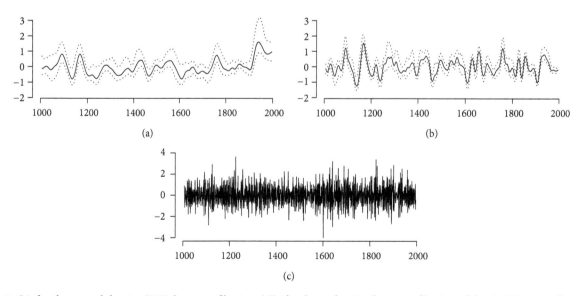

FIGURE 2: C.i. for the mean l-f series (RCS, low-pass filtering; (a)), for the m-f series (low-pass filtering of the six 180-year spline indexed series; (b)) and the average of h-f series (30-year splines, without low-pass filtering; (c)).

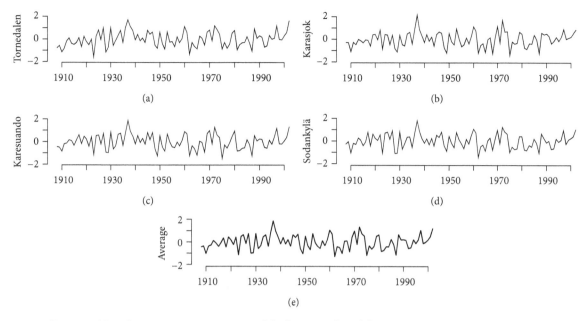

FIGURE 3: Mean June–August temperatures of the four records and their average (z-scores in 1908–2002).

*3.2. Individual Proxies versus July and June–August Temperatures.* Our climatic targets are arithmetic averages of normalized monthly data. July temperatures are first modeled using individual proxies and then June–August temperatures using individuals as well as their various combinations. The strength of targeted regional signal is indicated by correlation of June–August mean temperatures among the four stations (Figure 3), which vary between 0.88 (Tornedalen-Karasjok) and 0.94 (Karesuando-Sodankylä), while the mean correlation is 0.92. This June–August mean has even higher correlation ($r = 0.96$, $r^2 = 0.93$) with the June–August mean of the Bottenviken regional temperature record exclusively from northern Sweden [28], used in calibrations in some previous studies [5, 21]. All four records show, for example, cool conditions in the early 20th century until around 1915, the relative warmth of the 1930s, and a recent warming since the late 1980s. Summer temperature (June–August, normalized scale) varies between −1.32 and 1.88, which here form the limits for interpolation (Figure 3). The mean temperature has a modest positive trend ($y = -0.07 + 0.0014x$) in 1908–2002.

When the six l-f proxies were individually regressed on mean July temperature (Table 1(A)), only FXD and RRW produced positive verification statistics. Although SXD has positive RE values in both periods, the CEs are marginally negative. Among the m-f series (Table 1(B)), both of the western series (SRW and SXD) clearly pass the verification tests, while FRW and FHI do not. FXD and RRW series have improved (higher than in l-f) positive RE and CE values. In the higher frequencies (Table 1(C)), in addition to all three RW series, also FHI passes both tests in both periods. However, both SXD and FXD fall just below zero (−0.02) using the more searching CE statistic in the LC-EV period.

The same proxies were then calibrated against mean June–August temperature. In the l-f range (Table 2(A)), SXD, FXD, and RRW pass the tests. Now SXD produces the highest

RE and CE values in both periods. In the medium frequencies (as previously in the case of July, Table 1(B)), the four series (SRW, SXD, FXD, and RRW) show positive verification performance. FRW and FHI again have negative RE and CE values in the LC-EV period. In the higher frequencies (Table 2(C)), all six series pass the tests. The three RW series have generally lower explained variance ($r^2 \leq 0.21$) as well as lower RE and CE values ($\geq 0.18$) than the two other types of proxies (XD and HI), each of which has $r^2 \geq 0.31$ and both RE and CE $\geq 0.28$.

*3.3. June–August Temperature Signal in l-f, m-f, and h-f Averages.* All six series (their l-f, m-f, and h-f variants) were averaged to a simple mean (AA) and to a weighted mean (WA) and tested as combined predictors of June–August temperature (Table 3(A)–(C)). Simple mean (AA) produced positive verification results (both RE and CE during both periods) only in the higher frequencies (Table 3(B), Figure 2(c)). The failure of l-f and m-f models here as well as in previous comparisons (Tables 1 and 2) is due to excess growth variability as compared to temperature in the LC-EV period, which is most pronounced in FRW and FHI. However, the test values for AA are only slightly weaker than for WA. WA also had positive results in the lowermost frequencies (Table 3(A)). In the medium frequencies, neither method (AA or WA) resulted in an acceptable model (Table 3(B)).

In the next step only those series which individually passed RE and CE tests (in both transfer models for June–August temperatures, Table 2) were weighted and combined, namely, the three l-f series (3RCS; $0.56 * SXD + 0.44 * FXD + 0.14 * RRW$) and four m-f series (4Spline; $0.22 * SRW + 0.54 * SXD + 0.41 * FXD + 0.16 * RRW$). When these were calibrated against June–August temperature, both of them passed the verification tests (Table 3(C)). It should be noted

TABLE 1: Six proxies versus mean July temperatures (average of the four records, no filtering) in 1908–2002. Reduction of error (RE), coefficient of efficiency (CE), and $r^2$ during EC-LV (early calibration in 1908–1955 and late verification in 1955–2002) and LC-EV (late calibration in 1955–2002 and early verification in 1908–1955). Only test results with RE and CE > 0 are included.

| | EC-LV | | | LC-EV | | |
|---|---|---|---|---|---|---|
| | RE | CE | $r^2$ | RE | CE | $r^2$ |
| (A) RCS indexing | | | | | | |
| FXD | 0.29 | 0.21 | 0.35 | 0.10 | 0.02 | 0.14 |
| RRW | 0.25 | 0.17 | 0.19 | 0.21 | 0.14 | 0.16 |
| (B) 180-year spline indexing | | | | | | |
| SRW | 0.30 | 0.22 | 0.29 | 0.24 | 0.18 | 0.32 |
| SXD | 0.26 | 0.18 | 0.25 | 0.15 | 0.08 | 0.12 |
| FXD | 0.35 | 0.28 | 0.36 | 0.14 | 0.07 | 0.15 |
| RRW | 0.27 | 0.19 | 0.21 | 0.23 | 0.17 | 0.18 |
| (C) 30-year spline indexing | | | | | | |
| SRW | 0.27 | 0.19 | 0.30 | 0.4 | 0.34 | 0.43 |
| FRW | 0.29 | 0.21 | 0.39 | 0.13 | 0.05 | 0.24 |
| FHI | 0.52 | 0.47 | 0.57 | 0.23 | 0.17 | 0.23 |
| RRW | 0.18 | 0.09 | 0.25 | 0.27 | 0.20 | 0.33 |

TABLE 2: Six proxies versus mean June–August temperatures in 1908–2002 (no filtering). RE, CE, and $r^2$ during EC-LV (1908–1955 and 1955–2002) and LC-EV (1955–2002 and 1908–1955). Only test results with RE and CE > 0 are included.

| | EC-LV | | | LC-EV | | |
|---|---|---|---|---|---|---|
| | RE | CE | $r^2$ | RE | CE | $r^2$ |
| (A) RCS indexing | | | | | | |
| SXD | 0.53 | 0.53 | 0.56 | 0.53 | 0.53 | 0.56 |
| FXD | 0.43 | 0.42 | 0.44 | 0.42 | 0.42 | 0.44 |
| RRW | 0.1 | 0.1 | 0.11 | 0.15 | 0.15 | 0.17 |
| (B) 180-year spline indexing | | | | | | |
| SRW | 0.17 | 0.17 | 0.21 | 0.07 | 0.07 | 0.28 |
| SXD | 0.47 | 0.47 | 0.54 | 0.45 | 0.45 | 0.57 |
| FXD | 0.35 | 0.35 | 0.44 | 0.27 | 0.26 | 0.43 |
| RRW | 0.12 | 0.12 | 0.13 | 0.15 | 0.15 | 0.18 |
| (C) 30-year spline indexing | | | | | | |
| SRW | 0.18 | 0.17 | 0.21 | 0.14 | 0.13 | 0.20 |
| SXD | 0.49 | 0.49 | 0.50 | 0.41 | 0.41 | 0.42 |
| FRW | 0.14 | 0.14 | 0.16 | 0.11 | 0.10 | 0.15 |
| FHI | 0.46 | 0.46 | 0.48 | 0.28 | 0.28 | 0.31 |
| FXD | 0.39 | 0.39 | 0.40 | 0.37 | 0.37 | 0.38 |
| RRW | 0.16 | 0.16 | 0.17 | 0.14 | 0.14 | 0.15 |

that since sample replication is at least five in 1000–2002 in each of the six series it is thus >15 in 3RCS model, >20 in 4Spline model, >10 in the XD models, and >30 in models where all six series were included.

Because of the persistence in the time series of tree growth they are usually filtered to enhance the signal to be analyzed. It is often recommendable to also reconstruct l-f and h-f variations separately smoothing the proxy and instrumental series prior to calibration [3, 52]. In order to further study the correspondence between these two mean series (3RCS and 4Spline) and the target above decadal scales, the l-f and m-f proxies as well as the June–August temperature were low-pass filtered (10-year smoothing; see [38–40]) and then tested

again (using correspondingly shorter 42-year calibration and verification periods in 1914–1996 due to filtering). The verifications of these models produced positive results for both; however, here the l-f model was superior to the m-f model.

The four successful multiproxy WA model combinations for June–August temperature (Table 3) were recalibrated using the full 95-year calibration period (Figure 4). The mean of all six RCS-indexed series (with the highest weight on SXD and the lowest on FRW and RRW) produced a slightly inferior model as compared to the 3RCS series ($R^2$ = 0.57–0.59, 3RCS in Figure 4(c)). In the m-f scale the only successful combination (4Spline, Figure 4(b)) has

TABLE 3: Simple mean (AA) and weighted mean (WA) of all six series versus mean June–August temperatures (no filtering). The WA of RCS-based (A) and 30-year spline indexed series (B) (see also Figure 4(a)). Only test results with RE and CE > 0 are included. The WAs of those three RCS-indexed series (3RCS; 0.56 * SXD + 0.44 * FXD + 0.14 * RRW) and four 180-year spline indexed series (4Spline; 0.22 * SRW + 0.54 * SXD + 0.41 * FXD + 0.16 * RRW) which individually passed RE and CE tests (see Table 2) versus mean June–August temperatures (C; the full models shown in Figures 4(b) and 4(c)).

| | EC-LV | | | LC-EV | | |
|---|---|---|---|---|---|---|
| | RE | CE | $r^2$ | RE | CE | $r^2$ |
| (A) RCS | | | | | | |
| WA | 0.55 | 0.55 | 0.67 | 0.21 | 0.21 | 0.53 |
| (B) 30-year spline indexed | | | | | | |
| AA | 0.56 | 0.56 | 0.57 | 0.50 | 0.50 | 0.51 |
| WA | 0.65 | 0.65 | 0.65 | 0.58 | 0.57 | 0.58 |
| (C) WAs of selected three RCS and four 180-year spline indexed series | | | | | | |
| 3RCS | 0.58 | 0.58 | 0.61 | 0.53 | 0.52 | 0.58 |
| 4Spline | 0.46 | 0.46 | 0.60 | 0.28 | 0.28 | 0.60 |

intermediate values between l-f and h-f scales. The WA of all six h-f series (30-year spline indexed, Figure 4(a)) shows the best fit with the target ($R^2$ = 0.62). Here the modeled and observed values have particularly synchronous (one- or two-year) peaks in, for example, 1923, 1928–1929, 1949, 1962, and 1979–1980 (Figure 4(c)). The l-f model reproduces, for example, the longer increasing trend from 1908 to 1937 as well as that of the last fifteen years more consistently than the h-f series (Figures 4(a)–4(c)). However, the reproduction of the twin peaks in 1969–1974 is noticeably poorer.

The two density series (SXD and FXD) were further combined (WA) in both the l-f and m-f ranges and calibrated against longer warm season periods (Table 4). Three different periods were used, four months mean temperature from May to August (A), five months mean from April to August (B), and six months mean from April to September (C). All these six proxy combinations (both l-f and m-f) pass the verification tests (RE and CE in both periods: EC-LV and LC-EV) for all three temperature periods. Model fit ($R^2$) recalibrated using all available data varies from 0.48 to 0.58. The weights in WA are rather equal for SXD and FXD, ranging from 0.51 to 0.34, but slightly favoring SXD. The best empirical models (for May–August temperatures, Table 4) for the l-f and m-f bandwidths of the two XD-series were derived by recalibration in 1908–2002 with $R^2$ = 0.58 for the l-f and $R^2$ = 0.53 for the m-f model.

*3.4. Building a Proxy for Subcentury Scale Temperature Variability.* A composite m-f summer temperature proxy was produced using the 4Spline variant (WA of SRW, SXD, FXD, and RRW) (Figure 5(a), calibration in Table 3(C) and Figure 4(b)). Several significant (above 0.99 c.l.) multidecadal characteristics from periods of 33 and 67 years up to a peak exceeding a century (111 years) are highly significant in the Fourier spectrum (Figure 5(c)). The wavelet spectrum shows that the fluctuations spreading over these bandwidths are particularly apparent in the 12th and 17th centuries. On the other hand there is a noticeable lack of significant features

residing in the decadal to bidecadal ranges (Figures 5(b) and 5(c)). In this multidecadal scale the 20th century does not stand out as unusual in the past millennium.

In order to assess the correspondence of amplitude, duration, and timing of recent cycles between Fennoscandia and the North Atlantic the m-f proxy (4Spline based) and AMO were also visually compared (Figure 6). Correlation between the two series (in 1866–1992) is 0.27, rising to 0.49 using 10-year moving averages (MA, Figure 6(a)) and to 0.56 using 20-year MA. Presuming Fennoscandian proxy variability is preceding the AMO cycle (evident in Figure 6(a)) and correcting for (removing) this 10-year time shift the correlation rises to 0.34, 0.73, and 0.82, respectively. The 20-year MA effectively removes all higher frequencies and the relationship seems in agreement particularly during early to mid-20th century. This illustrates that multidecadal variability in Fennoscandian and North Atlantic climates is closely similar in scale and length.

To assess the potential of these data in modelling the AMO (e.g., time stability of such statistical relationship), all six m-f proxies were also compared to annual SST anomalies. Screening the pool of candidate predictors it was also possible to build a transfer model for the AMO. The WA of three series (SRW, FXD, and RRW) was used with the same calibration (1922–1990) as well as verification (1856–1921) periods as in Gray et al. [14]. This model passed the verification trials (RE = 0.05 and CE = 0.05), but model performance is the lowest ($R^2$ = 0.17) in this work.

## 4. Discussion

Five of the six series from different parts of northern Fennoscandia show evidence of a dominant 200–250 years of periodicity. Similar periodicity was found significant in a harmonic decomposition of the average of six central European instrumental as well as stalagmite proxy series from the Austrian Alps in 500–1935 [53]. However, the pronounced minimum during recent centuries appears in our proxies somewhat later (in the early 20th century) as compared to the 1880s noted by Lüdecke et al. [53]. Previously Helama et

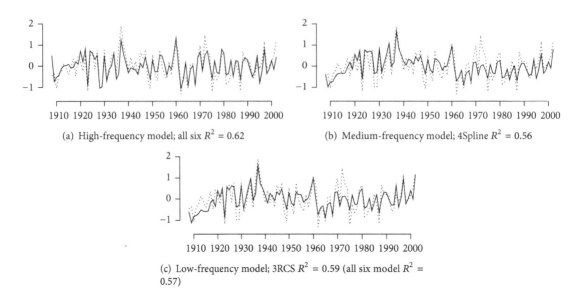

(a) High-frequency model; all six $R^2 = 0.62$

(b) Medium-frequency model; 4Spline $R^2 = 0.56$

(c) Low-frequency model; 3RCS $R^2 = 0.59$ (all six model $R^2 = 0.57$)

FIGURE 4: Final linear regression models on June–August mean as target (dotted line) with various combinations of proxies recalibrated in 1908–2002 (solid line): (a) WA of all six h-f (30-year spline indexed) series, (b) WA of SRW, SXD, FXD, and RRW (four 180-year spline indexed, 4Spline) m-f series, and (c) WA of the SXD, FXD, and RRW (3RCS series).

TABLE 4: Weighted averages of the Swedish and Finnish density series (SXD and FXD) built using two types of indexing (RCS and 180-year splines, Sp) versus a four-month mean temperature (no filtering) from May to August (A), a five-month mean from April to August (B), and a six-month mean from April to September (C). Only test results with RE and CE > 0 are included.

| | EC-LV | | | LC-EV | | |
|---|---|---|---|---|---|---|
| | RE | CE | $r^2$ | RE | CE | $r^2$ |
| RCS_A | 0.55 | 0.54 | 0.56 | 0.55 | 0.55 | 0.57 |
| Sp_A | 0.33 | 0.33 | 0.55 | 0.29 | 0.28 | 0.58 |
| RCS_B | 0.49 | 0.48 | 0.50 | 0.49 | 0.48 | 0.51 |
| Sp_B | 0.26 | 0.25 | 0.47 | 0.25 | 0.25 | 0.50 |
| RCS_C | 0.44 | 0.43 | 0.45 | 0.53 | 0.52 | 0.53 |
| Sp_C | 0.24 | 0.24 | 0.41 | 0.32 | 0.32 | 0.52 |

Weights used in averaging: 0.51 * SXD + 0.47 * FXD (RCS_A); 0.47 * SXD + 0.42 * FXD (SP_A); 0.47 * SXD + 0.41 * FXD (RCS_B); 0.42 * SXD + 0.35 * FXD (Sp_B); 0.46 * SXD + 0.39 * FXD (RCS_C); 0.41 * SXD + 0.34 * FXD (Sp_C).

al. [54] discussed centennial and multidecadal temperature variability over Northern Fennoscandia that bears a potential link to oceanic origins [8, 55] and reviewed the paleoclimatic literature relevant to the topic.

Several individual Fennoscandian tree growth-based proxies have indicated prominent spectral features at about 23, 30, and 90 years, 23–33 years, and 30.8–31.8 and 80.3–87.7 years in Finland [11, 25, 56] as well as relatively time-stable peaks at 32–33 years and at ~55–100 years in Sweden [10]. In an analysis of seven northern hemisphere temperature reconstructions (including, e.g., [15, 57–59]) Ogurtsov et al. [60] reported that they have an unambiguous 60–80-year multidecadal variability in common (AD 1000–1930), which is close to the range of a 67-year cycle indicated by our data. The 4Spline reconstruction without the more or less discrepant secular characteristics (see [17]) clearly records a bimodal structure—c.a. 110-year and 60–70-year variations— as well as a 33-year climatic cycle (similar to the Bruckner cycle). Based on these evidence there exists obvious potential for analyses in subcentury scales and a real possibility to gain

insight into the extent of general natural variability in the subarctic region.

In the lower frequencies of these data (low-pass filtering of the RCS and 180-year spline indexed series) several growth surges and troughs coincide in each group and they were also dated in the averages in the time domain (statistically significant fluctuations; five in l-f and 12 in m-f range). The latest, 37-year period (1920–1956) is the longest continuous multidecadal scale surge and the 85-year period (1918–2002) is the longest l-f surge. The combined m-f chronology is evidently more consistent than the l-f chronology as the more diverging l-f trends are left out. Overall signal strength (measured as mean correlation) between the six series is higher in m-f ($r = 0.58$) than either in l-f ($r = 0.52$) or in h-f series ($r = 0.41$). At least part of the l-f disagreement is possibly due to the (noisy) RCS method, which is usually recommended for large data sets of various age classes of trees for each year, ideally grown under a range of representative ecological conditions [10, 21, 61–64]. Even the ample data bases of SRW and FRW may not fully meet these requirements.

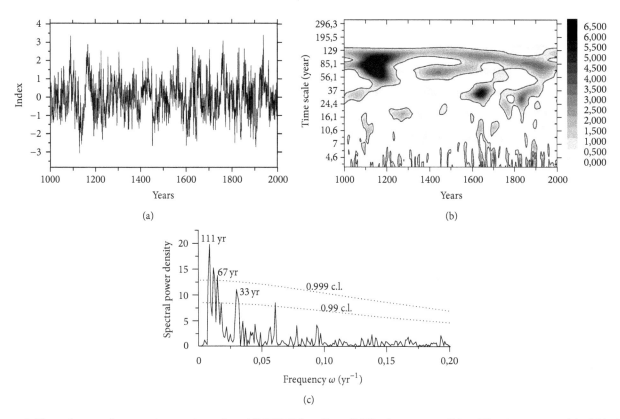

FIGURE 5: The m-f proxy of summer temperatures since AD 1000 ((a), unfiltered). Wavelet spectrum (b) and Fourier spectrum (c) of this time series.

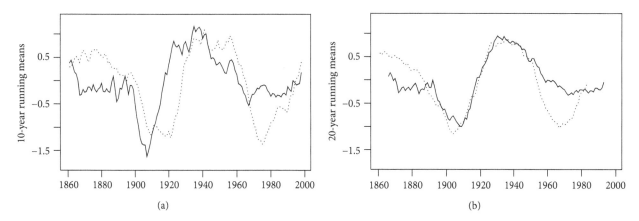

FIGURE 6: Comparison of annual AMO cycle (dash) and the Fennoscandian multidecadal (m-f) summer temperature proxy (solid) during common period using 10-year (a) and 20-year (b) smoothing.

The four instrumental temperature records used in calibrations show remarkable agreement justifying their averaging to a regional mean. July is usually one of the most important factors among monthly temperatures related to various growth parameters from different parts of the region [12, 21, 23, 24, 54]. When the six proxies were individually regressed on mean July temperature, only FXD and RRW passed the verification trials in the l-f range, SRW and SXD in the m-f range, and all but the two XD series in the h-f range.

The growth response of June–August temperatures has often been found time stable and this period has recently been used in temperature reconstructions in the region [2–5, 17]. A simple mean of all six proxy series (AA) works as a verifiable predictor of mean June–August temperature only in the h-f range (30-year spline indexing, Figure 2(c)) in these data. Although all series pass the tests, the SXD, FHI, and FXD have superior individual calibration and verification performance in this context. This is consistent with the results obtained previously by Lindholm et al. [17]. The SXD is obviously the best individual temperature proxy in these three frequency classes, usually most heavily weighted in the various WAs. Otherwise the mean is generally a more powerful predictor of regional temperatures than the individuals. The WA of the six series proved to be a successful predictor of the standard summer period also in the l-f range. Interestingly McCarroll et al. [4] were able to reconstruct June–August temperatures using the simple mean of part of these same data (SRW, SXD, and RRW) in a larger network of nine series. The contrast is plausibly explained mainly by their successful application of different mixed indexing methods (a combination of RCS and splines) as well as inclusion of the coastal areas in both the proxy and climate data and perhaps to a lesser extent by different climate station data as target and different calibration and verification periods.

The use of only a more limited number of l-f and m-f series (those which individually passed the RE and CE tests) produced the best reconstruction models of June–August temperatures. Both types of models (3RCS for l-f and 4Spline for m-f) use (the RW and XD) data from all the three subregions (western, central, and eastern). If the number of predictor variables is further limited to XD series only, it is possible to develop excellent models also for longer summer periods, up to half a year from April to September. Previously this XD quality was associated only with SXD [10, 13, 21] but now we have successfully combined both SXD and FXD for this purpose, since the FXD clearly shares this wide temperature-response time-window property.

The arithmetic averages of l-f and m-f series show some divergence between low-frequency growth variability and mean regional summer temperatures (especially FRW and FHI), leading to overestimations in calibrations (and consequently poor verification) as seen, for example, in Table 2(A)-(B). The phenomenon is often related to the autocorrelation structure of the series and factored out by smoothing (using weighted moving averages), including predictors from years lagging and/or leading current year of growth. Such model structures were compared by Briffa et al. [12], and various combinations of them have frequently been applied in reconstructions including RW in the region [13, 21, 54, 65]. In this work we only compared current year's growth with current temperature. Thus, the rather modest performance of FRW, FHI, and SRW in models of July and June–August temperatures is perhaps not so surprising. The apparently narrow seasonal response pattern of RW series (as compared to broader response time window in XD) generally corresponds with previous results [12, 13, 21, 22, 54, 63].

The autocorrelation structure of a chronology varies through time and depends on the way in which the series have been standardized. Thus, comparisons of indexing methods targeting different (dependent) frequency response are pertinent in dendroclimatological reconstruction work [16, 24, 61]. When the three types (l-f, m-f, and h-f) of reconstruction models are plotted together with the target June–August temperature their differences are not immediately apparent, even between the two extremes—30-year spline and RCS-based series—although the differences in closer detail are distinct and the h-f type reproduces the finer

details more meticulously than the two other types. This is because their differences lie in the only gradually changing l-f characteristics, long-term trends, and periodicities and calibration is necessarily done in a short period.

Comparison of Fennoscandian summer temperature proxy and annual AMO revealed that although the timing of the periods does not match, the amplitude and duration of the cycles do (in roughly 1880–1950). Furthermore Fennoscandian proxy variability was found preceding the SST cycle by about 10 years. Despite the shift in dating, data from all three subregions (SRW, FXD, and RRW) enabled us to build a model for the AMO. The calibration and verification periods were applied according to Gray et al. [14] in their reconstruction of the AMO using a network of 12 proxies including an early version of the FRW (RE = 0.25 in this model as compared to our model with RE = 0.05). The 10-year temporal shift plausibly explains the rather low test values of our AMO model. The comparison also illustrates that the 180-year splines (m-f) are a proper method to capture the relevant frequency range in these data.

As AMO is lagging behind it is obviously not a direct cause of multidecadal oscillation in Fennoscandia. However, the similarities between AMO and proxy may well be related to common origin judged purely by arising periodicities. Such widespread multidecadal temperature variability has been reported from the North Atlantic, the Arctic, and in Europe [9, 53, 55]. In Fennoscandian proxies the recognition of such cycles would help to explain part of the uncertainties in the 20th century trends. Soon after the severe drop at the turn of the (19th and 20th) centuries the highest relative growth (and thus pronounced warming) is usually recorded in 1930s–1940s with a peak in 1937 [2, 4], followed by medium values in 1950s–1970s and a recovery only in the last decades. If the 20th century warming occurred already by midcentury this is somewhat in contrast with the gradual warming expectation as a response to rising $CO_2$ concentration in the atmosphere but corresponds to some similarities in SST and Arctic surface air temperature fluctuations, which show two maxima, in 1930s–1940s and in the last decades [9].

Our m-f reconstruction provides a useful tool for interregional comparisons targeting the subcentury scales. As far as climatic signals can be extracted from more or less noisy proxies it is within reason to pay attention to the targeted frequencies in building the proxies. This will also help to understand the challenge in distinguishing between internally driven ocean-atmosphere-land dynamics and externally forced warming signal. Realizing a dominant multidecadal phenomenon also in Fennoscandian climate helps to explain in this work the higher linear association in m-f than either in l-f or h-f growth variability, although the RCS-based model has advantages over the 180-year splines based model, for example, in the decadal-to-centennial scales as was shown by calibration using separate filtered data sets. It may well also in part explain the rather modest model performance of FRW and FHI in the early 20th century due to excess variance in growth as compared to temperature (LC-EV period divergence). Moreover this frequency band easily relates to interregional (statistical) connection between

Fennoscandian and Arctic climates in particular related to the North Atlantic.

## 5. Conclusion

Six growth-based proxies of Scots pine from the northern Fennoscandian timberline between the Swedish Scandes and the Khibiny Low Mountains were combined to enhance their frequency-dependent properties in growth variability in AD 1000–2002. Five significant periods of good growth were found in the low frequencies and 12 periods in the medium frequencies. Most of the l-f series share significant oscillatory modes in 200–250 years of cycle length and strong multidecadal components. Thus we were able to build a separate m-f reconstruction using the 4Spline model. This series lacks the multicentury frequency bands dominant in the l-f series but has high concentrations of variance with a bimodal structure (ca. 110-year and 60–70-year variations) as well as a 33-year cycle.

Linear transfer models were built for short and long warm seasons using the six proxies in three frequency ranges. When they were individually regressed on mean July temperature only FXD and RRW passed the verification trials in the l-f range, SRW and SXD in the m-f range, and all but the two XD series in the h-f range. A simple mean of all six series works as a successful predictor of mean June–August temperature only in the h-f range. The SXD, FXD, and FHI were superior to the RW. On the other hand, the weighted mean of these same six series provides a good model of the standard summer period also in the l-f range. The use of only a more limited number of l-f and m-f series produced the best reconstruction models of June–August temperatures ($R^2 = 0.56$–$0.62$). Both types of models (three of the RCS series for l-f and four of the 180-year spline indexed series for m-f) use the RW and XD data from all the three subregions, western, central, and eastern. If predictor variables are further limited only to XD series, it is possible to develop excellent models also for longer summer periods, up to half a year from April to September ($R^2 = 0.48$–$0.58$). Previously this XD quality was associated only with SXD.

We can explain the otherwise peculiar general view of the 20th century by presuming that the growth surge and corresponding warming in the early half of the 20th century are part of a widespread dominant cycle taking place along a rising centennial trend (both types are evident in the l-f and m-f series). It is within reason to link the early multidecadal features more closely to natural phenomena and the overall 20th century rise to external forcing. Analyzing different frequency ranges in proxies will potentially help in meeting a major future challenge to separate the externally forced from internally driven variations (both natural and external components integrated in long records) for the detection and attribution of anthropogenic climate change.

## Conflict of Interests

The authors declare that there is no conflict of interests regarding the publication of this paper.

## Acknowledgments

This work was supported by the Academy of Finland (Grant no. SA 138937). M. G. Ogurtsov expresses his thanks to the exchange program between the Russian and Finnish Academies (Project no. 16), to the program of the Presidium of RAS no. 22, and to RFBR Grants 11-02-00755, 13-02-00277, and 13-02-00783 for financial support.

## References

[1] K. R. Briffa, V. V. Shishov, T. M. Melvin et al., "Trends in recent temperature and radial tree growth spanning 2000 years across northwest Eurasia," *Philosophical Transactions of the Royal Society B: Biological Sciences*, vol. 363, no. 1501, pp. 2269–2282, 2008.

[2] U. Büntgen, C. C. Raible, D. Frank et al., "Causes and consequences of past and projected Scandinavian summer temperatures, 500–2100 AD," *PLoS ONE*, vol. 6, no. 9, Article ID e25133, 2011.

[3] J. Esper, U. Büntgen, M. Timonen, and D. C. Frank, "Variability and extremes of northern Scandinavian summer temperatures over the past two millennia," *Global and Planetary Change*, vol. 88-89, pp. 1–9, 2012.

[4] D. McCarroll, N. Loader, R. Jalkanen et al., "A, 1200-year multiproxy record of tree growth and summer temperature at the northern pine forest limit of Europe," *Holocene*, vol. 23, pp. 471–484, 2013.

[5] T. M. Melvin, H. Grudd, and K. R. Briffa, "Potential bias in "updating" tree-ring chronologies using regional curve standardisation: re-processing 1500 years of Torneträsk density and ring-width data," *Holocene*, vol. 23, pp. 364–373, 2013.

[6] I. V. Polyakov, G. V. Alekseev, R. V. Bekryaev et al., "Observationally based assessment of polar amplification of global warming," *Geophysical Research Letters*, vol. 29, no. 18, pp. 25-1–25-4, 2002.

[7] R. A. Kerr, "A North Atlantic climate pacemaker for the centuries," *Science*, vol. 288, no. 5473, pp. 1984–1986, 2000.

[8] M. E. Schlesinger and N. Ramankutty, "An oscillation in the global climate system of period 65–70 years," *Nature*, vol. 367, no. 6465, pp. 723–726, 1994.

[9] I. V. Polyakov, R. V. Bekryaev, G. V. Alekseev et al., "Variability and trends of air temperature and pressure in the maritime arctic, 1875–2000," *Journal of Climatology*, vol. 16, pp. 2067–2077, 2003.

[10] K. R. Briffa, P. D. Jones, T. S. Bartholin et al., "Fennoscandian summers from ad 500: temperature changes on short and long timescales," *Climate Dynamics*, vol. 7, no. 3, pp. 111–119, 1992.

[11] H. H. Lamb, *Climate-Present, Past and Future*, Methuen, London, UK, 1992.

[12] K. R. Briffa, P. D. Jones, J. R. Pilcher, and M. K. Hughes, "Reconstructing summer temperatures in northern Fennoscandinavia back to AD 1700 using tree-ring data from Scots pine," *Arctic & Alpine Research*, vol. 20, no. 4, pp. 385–394, 1988.

[13] K. R. Briffa, T. S. Bartholin, D. Eckstein et al., "A 1,400-year tree-ring record of summer temperatures in Fennoscandia," *Nature*, vol. 346, no. 6283, pp. 434–439, 1990.

[14] S. T. Gray, L. J. Graumlich, J. L. Betancourt, and G. T. Pederson, "A tree-ring based reconstruction of the Atlantic Multidecadal

Oscillation since 1567 A.D.," *Geophysical Research Letters*, vol. 31, no. 12, 2004.

[15] J. Esper, E. R. Cook, and F. H. Schweingruber, "Low-frequency signals in long tree-ring chronologies for reconstructing past temperature variability," *Science*, vol. 295, no. 5563, pp. 2250–2253, 2002.

[16] M. Lindholm, *Reconstruction of past climate from ring-width chronologies of Scots pine (Pinus sylvestris L.) at the northern forest limit in Fennoscandia [Dissertation]*, University of Joensuu, 1996.

[17] M. Lindholm, T. Aalto, H. Grudd, D. McCarroll, M. Ogurtsov, and R. Jalkanen, "Common temperature signal in four well-replicated tree growth series from northern Fennoscandia," *Journal of Quaternary Science*, vol. 27, pp. 828–834, 2012.

[18] M. Lindström, "Northernmost Scandinavia in the geological perspective," *Ecological Bulletin*, vol. 38, pp. 17–37, 1987.

[19] T. Ahti, L. Hämet-Ahti, and J. Jalas, "Vegetation zones and their sections in northwestern Europe," *Annles Botanici Fennici*, vol. 5, pp. 169–211, 1968.

[20] S. Tuhkanen, "A circumboreal system of climatic- phytogeographical regions," *Acta Botanica Fennica*, vol. 127, pp. 1–50, 1984.

[21] H. Grudd, "Torneträsk tree-ring width and density ad 500–2004: a test of climatic sensitivity and a new 1500-year reconstruction of north Fennoscandian summers," *Climate Dynamics*, vol. 31, no. 7-8, pp. 843–857, 2008.

[22] H. Grudd, K. R. Briffa, W. Karlén, T. S. Bartholin, P. D. Jones, and B. Kromer, "A 7400-year tree-ring chronology in northern Swedish Lapland: natural climatic variability expressed on annual to millennial timescales," *Holocene*, vol. 12, no. 6, pp. 657–665, 2002.

[23] M. Lindholm and R. Jalkanen, "Subcentury scale variability in height-increment and tree-ring width chronologies of Scots pine since AD 745 in northern Fennoscandia," *Holocene*, vol. 22, no. 5, pp. 571–577, 2012.

[24] Y. M. Kononov, M. Friedrich, and T. Boettger, "Regional summer temperature reconstruction in the Khibiny Low Mountains (Kola Peninsula, NW Russia) by means of tree-ring width during the last four centuries," *Arctic and Alpine Research*, vol. 41, pp. 460–468, 2009.

[25] M. Lindholm, R. Jalkanen, H. Salminen, T. Aalto, and M. Ogurtsov, "The height-increment record of summer temperature extended over the last millennium in fennoscandia," *Holocene*, vol. 21, no. 2, pp. 319–326, 2011.

[26] M. Lindholm, M. Ogurtsov, T. Aalto, R. Jalkanen, and H. Salminen, "A summer temperature proxy from height increment of Scots pine since 1561 at the northern timberline in Fennoscandia," *Holocene*, vol. 19, no. 8, pp. 1131–1138, 2009.

[27] P. Klingbjer and A. Moberg, "A composite monthly temperature record from Tornedalen in northernn Sweden, 1802–2002," *International Journal of Climatology*, vol. 23, no. 12, pp. 1465–1494, 2003.

[28] H. Alexandersson, "Temperature and precipitation in Sweden 1860–2001," *SMHI Meteorologi*, vol. 104, 2002.

[29] A. Kaplan, M. A. Cane, Y. Kushnir, A. C. Clement, M. B. Blumenthal, and B. Rajagopalan, "Analyses of global sea surface temperature 1856–1991," *Journal of Geophysical Research C: Oceans*, vol. 103, no. 9, pp. 18567–18589, 1998.

[30] K. R. Briffa, P. D. Jones, F. H. Schweingruber, S. G. Shiyatov, and E. R. Cook, "Unusual twentieth-century summer warmth in a 1,000-year temperature record from Siberia," *Nature*, vol. 376, no. 6536, pp. 156–159, 1995.

[31] K. R. Briffa and F. H. Schweingruber, "Recent dendroclimatic evidence of northern and central European summer temperatures," in *Climate Since A.D., 1500*, R. S. Bradley and P. D. Jones, Eds., pp. 366–392, Routledge, London, UK, 1992.

[32] E. R. Cook, B. M. Buckley, R. D. D'Arrigo, and M. J. Peterson, "Warm-season temperatures since 1600 BC reconstructed from Tasmanian tree rings and their relationship to large-scale sea surface temperature anomalies," *Climate Dynamics*, vol. 16, no. 2-3, pp. 79–91, 2000.

[33] E. R. Cook and K. Peters, "The smoothing spline: a new approach to standardizing forest interior tree-ring series for dendroclimatic studies," *Tree-Ring Bulletin*, vol. 41, pp. 45–53, 1981.

[34] E. R. Cook, *A time series analysis approach to tree-ring standardization [Dissertation]*, University of Arizona, 1985.

[35] E. R. Cook, K. R. Briffa, S. Shiyatov, and V. Mazepa, "Tree-ring standardization and growth-trend estimation," in *Methods of Dendrochronology: Applications in the Environmental Science*, E. R. Cook and L. Kairiukstis, Eds., pp. 104–122, Kluwer Academic, Dordrecht, The Netherlands, 1990.

[36] V. Trouet, J. Esper, N. E. Graham, A. Baker, J. D. Scourse, and D. C. Frank, "Persistent positive north atlantic oscillation mode dominated the medieval climate anomaly," *Science*, vol. 324, no. 5923, pp. 78–80, 2009.

[37] H. F. Zhu, X. Q. Fang, X. M. Shao, and Z. Y. Yin, "Tree ring-based February-April temperature reconstruction for Changbai Mountain in Northeast China and its implication for East Asian winter monsoon," *Climate of the Past*, vol. 5, no. 4, pp. 661–666, 2009.

[38] H. C. Fritts, *Tree Rings and Climate*, Academic Press, 1975.

[39] C. W. Stockton and H. C. Fritts, "Conditional probability of occurrence for variations in climate based on width of annual tree-rings in Arizona," *Tree-Ring Bulletin*, vol. 31, pp. 3–24, 1971.

[40] C. Valmore and J. R. LaMarche, "Frequency-dependent relationships between tree-ring series along an ecological gradient and some dendroclimatic implications," *Tree-Ring Bulletin*, vol. 34, pp. 1–20, 1974.

[41] I. Daubechies, "Recent results in wavelet applications," *Journal of Electronic Imaging*, vol. 7, no. 4, pp. 719–724, 1998.

[42] S. Mallat, *A Wavelet Tour of Signal Processing*, Academic Press, 1999.

[43] C. Torrence and G. P. Compo, "A practical guide to wavelet analysis," *Bulletin of the American Meteorological Society*, vol. 79, no. 1, pp. 61–78, 1998.

[44] W. J. Burroughs, *Weather Cycles: Real or Imaginary?* Cambridge University Press, 1994.

[45] P. S. P. Cowpertwait and A. V. Metcalfe, *Introductory Time Series with R*, Springer, 2009.

[46] M. J. Crawley, *The R Book*, John Wiley & Sons, 2009.

[47] D. McCarroll, M. Tuovinen, R. Campbell et al., "A critical evaluation of multi-proxy dendroclimatology in northern Finland," *Journal of Quaternary Science*, vol. 26, no. 1, pp. 7–14, 2011.

[48] H. C. Fritts, "Statistical reconstruction of spatial variations in climate," in *Methods of Dendrochronology: Applications in the*

Six Temperature Proxies of Scots Pine from the Interior of Northern Fennoscandia Combined...

115

*Environmental Science*, E. R. Cook and L. A. Kairiukstis, Eds., pp. 193–210, Kluwer Academic, Dordrecht, The Netherlands, 1990.

[49] H. C. Fritts, T. J. Blasing, B. P. Hayden, and J. E. Kutzbach, "Multivariate techniques for specifying tree-growth and climate relationships and for reconstructing anomalies in paleoclimate," *Journal of Applied Meteorology*, vol. 10, no. 5, pp. 845–864, 1971.

[50] G. R. Lofgren and J. H. Hunt, "Transfer functions," in *Climate From Tree Rings*, M. K. Hughes, P. M. Kelly, J. R. Pilcher, and V. C. LaMarche, Eds., pp. 50–56, Cambridge University Press, 1982.

[51] E. R. Cook, K. R. Briffa, and P. D. Jones, "Spatial regression methods in dendroclimatology: a review and comparison of two techniques," *International Journal of Climatology*, vol. 14, no. 4, pp. 379–402, 1994.

[52] J. Guiot, "Methods of calibration," in *Methods of Dendrochronology: Applications in the Environmental Science*, E. R. Cook and L. A. Kairiukstis, Eds., pp. 165–178, Kluwer Academic, Dordrecht, The Netherlands, 1990.

[53] H. J. Lüdecke, A. Hempelmann, and C. O. Weiss, "Multiperiodic climate dynamics: spectral analysis of long-term instrumental and proxy temperature records," *Climates of the Past*, vol. 9, pp. 447–453, 2013.

[54] S. Helama, M. Timonen, J. Holopainen et al., "Summer temperature variations in Lapland during the Medieval Warm Period and the Little Ice Age relative to natural instability of thermohaline circulation on multi-decadal and multi-centennial scales," *Journal of Quaternary Science*, vol. 24, no. 5, pp. 450–456, 2009.

[55] T. L. Delworth and M. E. Mann, "Observed and simulated multidecadal variability in the Northern Hemisphere," *Climate Dynamics*, vol. 16, no. 9, pp. 661–676, 2000.

[56] M. Timonen, J. Jiang, S. Helama, and K. Mielikäinen, "Significant changes of subseries-means in the Finnish tree-ring index of 7638 years, with comparisons to glaciological evidence from Greenland and Alps," *Quaternary International*, vol. 319, pp. 143–149, 2014.

[57] K. R. Briffa, "Annual climate variability in the Holocene: interpreting the message of ancient trees," *Quaternary Science Reviews*, vol. 19, no. 6, pp. 87–105, 2000.

[58] P. D. Jones, K. R. Briffa, T. P. Barnett, and S. F. B. Tett, "High-resolution palaeoclimatic records for the last millennium: interpretation, integration and comparison with General Circulation Model control-run temperatures," *Holocene*, vol. 8, no. 4, pp. 455–471, 1998.

[59] M. E. Mann, R. S. Bradley, and M. K. Hughes, "Northern hemisphere temperatures during the past millennium: inferences, uncertainties, and limitations," *Geophysical Research Letters*, vol. 26, no. 6, pp. 759–762, 1999.

[60] M. Ogurtsov, M. Lindholm, and R. Jalkanen, "Will a new little ice age begin in the next few decades?" *Applied Physics Research*, vol. 5, pp. 70–77, 2013.

[61] E. R. Cook, K. R. Briffa, D. M. Meko, D. A. Graybill, and G. Funkhouser, "The segment length curse' in long tree-ring chronology development for palaeoclimatic studies," *Holocene*, vol. 5, no. 2, pp. 229–237, 1995.

[62] R. D'Arrigo, R. Wilson, B. Liepert, and P. Cherubini, "On the "Divergence Problem" in Northern Forests: a review of the tree-ring evidence and possible causes," *Global and Planetary Change*, vol. 60, no. 3-4, pp. 289–305, 2008.

[63] J. Esper, D. Frank, U. Büntgen, A. Verstege, R. Hantemirov, and A. V. Kirdyanov, "Trends and uncertainties in Siberian indicators of 20th century warming," *Global Change Biology*, vol. 16, no. 1, pp. 386–398, 2010.

[64] S. Helama, M. Lindholm, M. Timonen, and M. Eronen, "Detection of climate signal in dendrochronological data analysis: a comparison of tree-ring standardization methods," *Theoretical and Applied Climatology*, vol. 79, no. 3-4, pp. 239–254, 2004.

[65] M. Lindholm and M. Eronen, "A reconstruction of midsummer temperatures from ring-widths of scots pine since AD 50 in northern Fennoscandia," *Geografiska Annaler A: Physical Geography*, vol. 82, no. 4, pp. 527–535, 2000.

# 11

# Efficiencies of Inhomogeneity-Detection Algorithms: Comparison of Different Detection Methods and Efficiency Measures

**Peter Domonkos**

*Centre for Climate Change, University of Rovira i Virgili, Campus Terres de l'Ebre, Avenue Remolins 13-15, 43500 Tortosa Tarragona, Spain*

Correspondence should be addressed to Peter Domonkos; peter.domonkos@urv.cat

Academic Editors: S. Feng, L. Makra, and A. P. Trishchenko

Efficiency evaluations for change point Detection methods used in nine major Objective Homogenization Methods (DOHMs) are presented. The evaluations are conducted using ten different simulated datasets and four efficiency measures: detection skill, skill of linear trend estimation, sum of squared error, and a combined efficiency measure. Test datasets applied have a diverse set of inhomogeneity (IH) characteristics and include one dataset that is similar to the monthly benchmark temperature dataset of the European benchmarking effort known by the acronym COST HOME. The performance of DOHMs is highly dependent on the characteristics of test datasets and efficiency measures. Measures of skills differ markedly according to the frequency and mean duration of inhomogeneities and vary with the ratio of IH-magnitudes and background noise. The study focuses on cases when high quality relative time series (i.e., the difference between a candidate and reference series) can be created, but the frequency and intensity of inhomogeneities are high. Results show that in these cases the Caussinus-Mestre method is the most effective, although appreciably good results can also be achieved by the use of several other DOHMs, such as the Multiple Analysis of Series for Homogenisation, Bayes method, Multiple Linear Regression, and the Standard Normal Homogeneity Test.

## 1. Introduction

The underlying climate signal in observed in situ climatic data is often masked either by changes in observational practices, exposure, and instrumentation or by local changes in the environment where the observations are taken. If these changes (called inhomogeneities (IH)) have a significant impact on the statistical characteristics of the observed data, then the time series are inhomogeneous, and their usefulness is limited in assessments of observed climate change. Given that almost any long climate series is potentially inhomogeneous, various techniques have been developed to detect and adjust series where necessary (see, among others, the seminar series of Homogenisation and Quality Control in Climatological Databases, WMO-HMS [1–6]).

There are several options to eliminate the IHs from observed time series. The timing and cause of many potential IHs are documented in network management documents (so-called metadata). Good metadata information greatly facilitates the development of appropriate corrections for inhomogeneous time series, so that the time series can be made more suitable for climate studies. However, metadata is generally incomplete ([7–12], etc.) or at least cannot be assumed to be comprehensive. Thus, the benefit of metadata [13] cannot always be fully exploited in practice, and "even with the best possible metadata, some statistical inhomogeneity detection is advised" [14].

The need for efficient homogenization methods has encouraged researchers to create and use various statistical tools. As a result, nearly twenty statistical homogenization methods are in use. Taking into consideration the number of options available in how these methods are used and the parametric choices within a particular DOHM, the diversity of the applied DOHMs is even greater in practice.

Although several reviewing papers about homogenization methods have been published in the recent years

([13–19], etc.), a realistic evaluation of the advantages and possible disadvantages of homogenization procedures remains elusive. Sometimes even the principles are questioned. For example, in an investigation of real and simulated time series of radiosonde data with some arbitrary selected DOHMs, [20] found that the rate of false detections was usually higher than that of the correct IH detections. While this example is relevant only to the special properties of radiosonde data, results like this may lower confidence more generally in homogenization procedures.

Most homogenization studies focus on the detection and correction of the shifts in the monthly, seasonal, and annual means, since the correction of biases in section means is the most important for the reliable estimation of climate trends and low frequency climate variability. The homogenization of daily data is an even more complex problem as the homogenization on longer time scales, but the number of daily homogenization methods has been growing fast recently ([21–23], etc.), from which the more accurate estimation of extreme value statistics is expected. In this study the background noise of the simulated time series is white noise, and the inhomogeneities in them are changes (biases) of the section means by definition; thus, the results and conclusions are primarily applicable for monthly and annual homogenization.

Detection of IHs is usually conducted via relative homogeneity testing (e.g., the differences between two series are examined for breaks). In that case, detected IHs are supposed to belong to the so-called candidate series, although the chance that they at least partly belong to some reference series usually cannot be ruled out. To keep this risk low, application of statistical homogenization can be recommended only when several time series of the same geographical-climatic region are available [7, 24–27], and many of the spatial correlations are higher than 0.7 [13, 18, 28–32]. These conditions are generally true for the surface air temperature datasets in the extratropical land areas, and in most cases also for precipitation datasets. Monthly and annual temperature time series in Europe and the USA usually have dense networks, and the spatial correlations are often around or above 0.9 for them [12, 33]. In these networks the use of relative time series is advantageous since there is no comparable alternative for eliminating the impacts of local IHs from observed datasets.

Here, we examine the efficiencies of DOHMs as applied to series simulated to represent the test series in a relative homogenization approach. Only objective methods are tested that can be applied in fully automated way. The study does not address the mechanics of creating relative time series (differences or ratios) from raw climate observations nor do we consider the implications of using iteration in detection and correction even though these aspects may significantly impact the ultimate efficiency in practice. Rather, the rationale is that the change point detection components of DOHMs should be analyzed separately from the complete homogenization procedures, (i) because the most efficient detection parts can then likely be paired with spatial comparison and iteration segments of any other homogenization methods, and (ii) considering only the complete homogenization procedure without addressing specific components

tends towards a "black box" approach whereby the advantages and disadvantages of particular elements are difficult to identify.

Although there have been made some comparative examinations aiming to reveal the capability of detecting IHs by different homogenization methods [16, 30, 34–45], all these studies examine some arbitrary selections of DOHMs, and the test datasets used mostly do not have realistic statistical properties. One of the fundamental open questions is the role of similarities and dissimilarities between the statistical properties of real and simulated time series, a topic which has only been discussed by Menne and Williams Jr. [16, 40], Domonkos [41, 42], Titchner et al. [45], and Venema et al. [46]. This topic is intensively discussed in this study relying on the empirical efficiencies from various simulated datasets.

The methodology of the present study mostly follows the rules introduced in [42]. The new lines of investigations in the present study in revealing the performances of DOHMs are as follows: (i) comparisons between test series with randomly positioned change points on the one hand and those with short term, platform shaped biases on the other hand; (ii) comparisons between test series with relatively large shift sizes (as they are in [46]) on the one hand, and those with mostly moderate shift sizes (as which were found empirically in [42]) on the other hand; (iii) the role of empirical autocorrelation in relative time series; (iv) experiments with moving signal-to-noise ratio. After presenting the results, the reality of the test datasets used will be discussed along with some peculiarities of the results, and we will make some comparisons between the blind test results of [46] and our results.

## 2. Methods

*2.1. Concepts and Definitions.* The efficiencies of DOHMs are quantified using ten test datasets as benchmarks. All the simulated time series that comprise the benchmarks were generated to mimic the properties of time series of differences derived by the comparison of one candidate series with some IHs and a reference series of "good quality" (i.e., without IHs). They are referred to as relative time series ($\mathbf{X}$). Specifically, each dataset comprises $N$ relative time series of $n$ year length as follows:

$$\mathbf{X}_p = \left[ x_{p,1}, x_{p,2}, \ldots x_{p,n} \right]^T, \quad p = 1, 2, \ldots N. \quad (1)$$

In this study $n = 100$ and $N = 10,000$. All the time series contain a standard white noise process ($\mathbf{W}$) whose standard deviation equals 1, as well as a term for cumulated effects of IHs, named also station effect ($\mathbf{H}$).

The origin of the noise is the natural fluctuation of spatial differences of climatic elements. The relative time series generally do not contain low frequency noise, because in a particular observing network of the same climatic region, the low frequency climatic changes tend to be common. However, this assumption is not exactly true; so two of the test datasets were simulated to have characteristics with slight deviations from this rule (see Section 2.3).

Each element of time series (the index $p$ will not be in use hereafter) can be expressed as a sum of the interstation noise ($w$) and station-specific IH effects ($h$) as follows:

$$x_i = w_i + h_i, \quad i = [1, 2, \ldots n]. \tag{2}$$

**X** always represents raw time series (before homogenization), while **U** represents homogenized time series. If homogenization is perfect, then $\mathbf{U} \equiv \mathbf{W}$.

Always the first moment (section-average) of the time series is biased by the imposed IHs in the simulations. Three types of IHs are used, namely, (i) change point (sudden shift), (ii) trend (gradual change), and (iii) platform (pair of sudden shifts). These are defined in more detail as follows.

    (i) Change point: if $h_{i+1} \neq h_i$ and the change is not a part of a gradual change (cf. (iii)), then a change point type IH exists at time $i$.

    (ii) Trend: gradual change of $h$ over a period $[j, k] (1 \leq j < k \leq n)$. The artificial trends are always linear ($h_{i+1} - h_i = h_i - h_{i-1}$ for each $i$ ($i \in [j + 1, k - 1]$)), and their minimum duration is 5 years.

    (iii) Platform: a pair of change points of the same magnitude, but with different signs. If $h_{i+1} \neq h_i$ and $\exists k$, ($k \in [i + 1, n]$) for which $h_{k+1} - h_k = -(h_{i+1} - h_i)$, then a platform exists whose first year is $i + 1$ and last year is $k$. When $k - i$ represents a relatively short time period (say, $k - i \leq 10$), platforms are also referred to as short-term IHs. From this point of view, outliers are also platforms with 1 year duration.

Note that although from the combination of IHs type (i) and type (ii), IH of any shape could be constructed; type (iii) is included as a distinct type in the simulation process, because an earlier study [42] showed that the shifts of successive change points often have the opposite signs.

The absolute value of an IH is considered to be its "magnitude", while the magnitude with the sign is considered to be its "size." During the simulation, IH magnitudes ($m$) and other statistical characteristics are expressed with their ratio to the standard deviation of the white noise ($s_e$), while detected magnitudes ($m^*$) are expressed with their ratio to the estimated standard deviation of the white noise ($s_e^*$) as follows:

$$s_e^* = \sqrt{1 - R^2} \cdot s_T \quad \text{if } R > 0,$$
$$s_e^* = s_T \qquad\qquad \text{if } R \leq 0. \tag{3}$$

In (3), $R$ denotes 1-year lag autocorrelation, and $s_T$ means the empirical standard deviation of the time series. The application of the unit $s_e^*$ follows from the fact that during the detection process $s_e$ is known only for simulated time series, while for relative time series from real observations this characteristic is unknown. In contrast, $s_e^*$ can easily be calculated for any time series. $s_e^*$ is usually higher than $s_e$ but never higher than $s_T$. Thus $s_e^*$ is a better estimation of $s_e$ than $s_T$ would be.

The detection processes are always paired with a standard adjustment procedure (SA) in this study. Let us suppose that

$Q$ change points have been detected with timings $t_1, t_2, \ldots t_Q$, as well as $t_0 = 0$ and $t_{Q+1} = n$ by definition. The segment between adjacent change points $t_k$ and $t_{k+1}$ is denoted by $K$ and segment means with upper stroke. For the SA, $m^*(t_k)$ is calculated as the difference of the adjacent segment means around $t_k$. If there is a detected trend with $e_K$ annual change for section $K$, it is taken into account in the calculation according to

$$m^*(t_k) = \overline{x_K} - \overline{x_{K-1}}$$
$$\qquad - 0.5 \left(e_K (t_{k+1} - t_k) + e_{K-1} (t_k - t_{k-1})\right). \tag{4}$$

Then the adjustment of $m^*$ is applied for all $x_i$ of $i \leq t_k$. Derivation of $m^*$ and SA is applied for trend IHs with the same logic, but taking into account the gradualness. Note that the estimation of $m^*$ in the various homogenization approaches may differ from the SA used here; however, the uniformity of adjustment technique applied in this study is essential for testing the performance of detection parts separately from other properties of homogenization methods.

Finally, in the discussion below, $f$ stands for the average frequency of IHs (i.e., the number of events in 100 years), while $s$ equals the standard deviation of the IH sizes.

*2.2. DOHMs Examined.* The nine DOHMs that we examine (Table 1) are widely used in climatology. All they are objective methods, hence they can be applied automatically to find IHs in time series. Only one nonparametric method, the Wilcoxon Rank Sum test, was selected because the efficiencies of nonparametric methods are more limited (see, e.g., [37]).

The significance thresholds used attempt to ensure the 0.05 rate first type error (FTE) in pure white noise processes. The values are generally taken from the reference studies with some exceptions. For Bay, the Caussinus-Lyazrhi criterion [47] is applied, while for MLR and tts the thresholds are calculated with Monte-Carlo technique. In C-M and MAS some kind of joint detection of multiple IHs is applied, while in tts and in the second phase of E-P single change points are searched in subsections of the time series. The cutting algorithm [35] is applied when reference studies do not give other proposal for treating multiple IHs (i.e., for Bay, MLR, SNH, SNT, WRS, and in the first phase of E-P). In the cutting algorithm only one change point can be detected at a specific step, but cutting the time series into two parts at the timing of detected change points, multiple IHs can be detected. Subsection examination and cutting algorithm need the definition of the minimum length of subperiods for searching change points in them, it is 10 years in this study after Easterling and Peterson [35] and Moberg and Alexandersson [48]. The shortest period between two adjacent detected change points is 5 years for such methods. In SNT the detected IHs are always trends when the estimated duration of change is at least 5 years and always change points in the reverse case. The version of MLR used here differs in one more detail from the original description, that is, only 1-year-lag autocorrelations are considered in calculating FTE (instead of all lags between 1 year and 3 year). This modification has no substantial effect on the performance of MLR (not shown).

TABLE 1: DOHMs with their abbreviations used in the study.

| DOHM | Abbreviation | Reference |
| --- | --- | --- |
| Bayes method | Bay | Ducré-Robitaille et al. 2003 [37] |
| Caussinus-Mestre method (PRODIGE) | C-M | Caussinus and Mestre 2004 [53] |
| Easterling-Peterson method | E-P | Easterling and Peterson 1995 [35] |
| Multiple Analysis of Series for Homogenisation | MAS | Szentimrey 1999 [67] |
| Multiple Linear Regression | MLR | Vincent 1998 [68] |
| Standard Normal Homogeneity Test for shifts only | SNH | Alexandersson 1986 [69] |
| Standard Normal Homogeneity Test for shifts and trends | SNT | Alexandersson and Moberg 1997 [70] |
| $t$-test | tts | Ducré-Robitaille et al. 2003 [37] |
| Wilcoxon Rank Sum test | WRS | Wilcoxon 1945 [71] |

A uniform prefiltering of outliers is applied before the use of any DOHM. Anomalies from the average of the time series are considered to be outliers if their absolute values are higher than 4 standard deviations of the time series elements. This threshold is often used in practice (e.g., [9, 25, 49]). Detected outliers are replaced with an anomaly value of zero.

*2.3. Test Datasets.* Obviously, the higher the resemblance between the simulated and real statistical properties, the higher the confidence that the assessed efficiencies based on simulated datasets are valid for real climatic datasets. Unfortunately, the exact statistical properties of IHs occurring in real climatic time series are not known. In the benchmark surrogated dataset of the COST HOME project (hereafter: Benchmark, [46]) the mean frequency of IHs is 5 per 100 years in artificial test datasets. It is based on the experience that the frequency of detected IHs in long climatic time series is approximately 5 per 100 years [13, 50], and considering that the IHs with low magnitudes are more frequent than those with high magnitudes [16, 40, 42], the IH sizes have normal distribution with 0 expected value. However, it is only a rough approach to the true properties of observational time series; see more discussion about this problem in Section 4.1 of this study and in [42], as well.

The mathematical description of ten test datasets is presented below, together with some ideas about the motivation of creating them. When no specification for IH-type is given, the type is always change point. Trends are included in three datasets only, but in those three their role is much greater (about 25% of long-term biases) than in the Benchmark (2% of all IHs).

(A) CH1B0 refers to 1 Big Change point without any restriction for R (0). In this dataset, exactly one IH is included in each time series. Its timing ($j$) is 40 or 60, and $m = 3$. In this simple case it is easy to demonstrate the time function of station effect as

$$h_i = 0, \quad \text{if } i \leq j,$$

$$h_i = 3, \quad \text{if } i > j, \tag{5}$$

$$1 \leq i < n, \quad j = 40 \quad \text{or} \quad j = 60.$$

(B) CH5B0: The intension of this dataset is to mimic the Benchmark (note that $f = 5$ and $s = 0.8°C$

are the key characteristics for the raw time series of the Benchmark). In the simulation process of this dataset, the probability of the introduction of a new change point was 0.05 at each year. Thus the average number of change points in time series is 5, and the mean frequency per time series has binomial distribution. The sizes are normally distributed with a mean of zero and $s$ of 3.5. The value of $s$ came from the estimation that $0.8°C$ is likely 3-4 times higher than the typical $s_e$ for true relative time series in dense observing networks.

(C) PF5B0 was generated in the same way as CH5B0, but instead of individual change points platforms were introduced to the time series, again with probability of 0.05 at each year. The length of the platform has uniform distribution between 1 year and 10 years. The sizes of IHs have normal distribution with a mean of zero and $s$ of 3.5.

(D) HUSTR: this dataset is the "Hungarian standard," because its characteristics were provided by an empirical procedure in which the statistical properties of the detected IHs in test datasets were approached via series of experiments to the same properties for true relative time series [42]. Those relative time series were constructed from observed temperature series in Hungary. The time series of HUSTR include rather complex structures of randomly distributed IHs of different types (change points, platforms, and trends) and magnitudes, as well as noise that differs from white noise. In this dataset the number of IHs is high, and short-term platforms are particularly frequent. Some IH-sizes are large, but the majority of them is small. $R \geq 0.4$ in each time series. The full description of its generation is presented in [42]. The change point frequency and the standard deviation of the change point magnitudes are $f = 31.1$ and $s = 1.20$.

(E) HUST0: like HUSTR, but without any restriction for R, $f = 30.3$ and $s = 1.02$.

For the following four datasets $s$ was set to be similar to the s of HUSTR and HUST0.

(F) CH5S0: the same as CH5B0, but $s = 1$.
(G) PF5S0: the same as PF5B0, but $s = 1$.

(H) CH5SR: it was generated in the same way as CH5S0, but only time series with $R \geq 0.4$ were retained.

(I) PF5SR: it was generated in the same way as PF5S0, but only time series with $R \geq 0.4$ were retained.

(J) CHPF0: "compromise dataset." The term compromise is used because this dataset contains both long-term IHs and short-term platforms, as well as a few trend IHs. The frequency of IHs ($f$ = 10.4) is higher than in the Benchmark, but much lower than in the Hungarian standard. During its simulation a new IH is introduced with 7% probability at each year, more specifically with 3, 3, and 1% probability for change points, platforms, and trends, respectively. Data were simulated from 50 years before the starting point of time series until 50 years after the end of time series for ensuring the temporal uniformity of occurrences of IHs within the examined 100 years. The duration of trends has even distribution between 5 and 99 years, while that of the platforms is the same as in PF5B0. This dataset is examined with moving $s$.

Table 2 summarizes the properties of the IHs in the ten test datasets. The diversity of the shown characteristics serves well the objective of the paper, that is, to find conclusions that are not related to the specific IH-properties of a given test dataset.

*2.4. Measures of Efficiency.* Four kinds of measures are examined: (a) detection skill, (b) skill of linear trend estimation, (c) sum of squared errors (SSE), and (d) combined maximal bias (CMB).

Let the sum of correct detections, that of false detections, and the total number of change points be denoted by $S_R$, $S_F$, and $S$, respectively. Although these concepts are clear in case of one or a few of fairly large IHs, their occurrences are not easy to be identified in complex structures. Therefore, the concepts "change point," "correct detection," and "false detection" must be defined for a quantitative and objective evaluation. The application of some arbitrary parameters is unavoidable for these definitions.

A *true change point* exists in time series **X** at year $j$ ($3 \leq j \leq n-3$), if

$$\frac{1}{k}\left|\sum_{i=j-k+1}^{j} x_i - \sum_{i=j+1}^{j+k} x_i\right| \geq 2, \quad \text{for each } k \text{ of } k = \{1, 2, 3\}.$$
(6)

Equation (6) means that change points with magnitude ($m^*$) at least 2 are considered only, and the shift of this magnitude must be apparent comparing each symmetric half-window pairs, up to window width of 6 years.

*Correct detection:* there is a detected change point at year $j$ with $m^* \geq 1.5$, and a true change point with a shift of the same sign as the detected IH has, really exists in section $[j-1, j+1]$ of **X**.

*False detection:* there is a detected change point at year $j$ with $m^* \geq 1.5$, but no true change with the same direction occurs at all, taking into account any of the possible comparisons of section means for symmetric half windows around $j$ up to window width of 6 years in **X**. There is no minimum threshold here for the magnitudes of true changes, only their signs are considered.

(a) *Detection skill* ($E_D$):

$$E_D = \frac{S_R - S_F}{S}.$$
(7)

Pieces of the detection result that do not meet with the conditions of either the correct detection or the false detection are not taken into account in the calculation of the detection skill. For perfect detection $E_D = 1$. In case of half of the detected change points are false, $E_D = 0$. Note that in datasets with very few change points ($S$ is small) $E_D$ can easily be negative.

For the following three measures error terms will be defined first, following them the way of their conversion to efficiency measures.

(b) *Error of linear trend estimation:* linear trends are fitted to the time series with minimizing the SSE between the trend line and the annual values of time series. The fitting is accomplished both for **U** and **W**, and the one for **W** is considered to be perfect. The differences between these two slopes are error terms. The procedure was accomplished for the whole (100 year long) time series, as well as for the last 50 years of the series. Thereafter the arithmetical average of these two errors is taken.

(c) *Sum of squared errors* (SSE):

$$\text{SSE} = \sum_{i=1}^{n} (u_i - w_i)^2.$$
(8)

(d) *Combined maximal bias* (CMB): This measure evaluates the maximum difference between **U** and **W**, but in a way where detections with time-lapse error only are considered as partially right detections. When true IHs of **X** are detected right but with some time lapse, in CMB the detection is considered good, but a penalty term is applied for the time lapse. The penalization depends on the size of the time-lapse. Following this idea and comparing the annual values of **U** and **W**, the annual series of a combined error term **B** (the combination of size errors and time lapses) can be calculated. Naturally, when $u_i = w_i$, $b_i = 0$. The below formula shows the case, when $u_i > w_i$ (the reverse case is handled with the same logic rules):

$$b_i = \min_k \left(g_k + u_i - \min\left(u_i, \max_j\left(w_j\right)\right)\right),$$
(9)

where $g_k$ is the penalty term of $k$-year lapse:

$$g_k = \exp\left(c_1\left(k - c_2\right)\right) - c_3,$$

$$c_1 = 0.369, \qquad c_2 = 3.297,$$

$$c_3 = 0.2962; \quad j \in \left[j_1, j_2\right],$$

$$j_1 = \max\left(1, i - k\right),$$

$$j_2 = \min\left(i + k, n\right),$$

$$k = [0, 1, 2, \ldots, 15].$$
(10)

TABLE 2: Properties of the test datasets.

| | $f(CH)$ | $f(trend)$ | $f(PFa)$ | $f(PFb)$ | $f'$ | $f$ | $s$ | $R$ |
|---|---|---|---|---|---|---|---|---|
| CH1B0 | 1.0 | 0 | 0 | 0 | 1.0 | 1.0 | 0 | — |
| CH5B0 | 5.0 | 0 | 0 | 0 | 5.0 | 5.0 | 3.50 | — |
| PF5B0 | 0 | 0 | 2.5 | 2.5 | 5.0 | 9.2 | 3.50 | — |
| HUSTR | 3.1 | 2.3 | 5.4 | 17.9 | 16.2 | 31.1 | 1.20 | $R \geq 0.4$ |
| HUST0 | 2.9 | 2.4 | 5.1 | 17.9 | 15.5 | 30.3 | 1.02 | — |
| CH5S0 | 5.0 | 0 | 0 | 0 | 5.0 | 5.0 | 1.00 | — |
| PF5S0 | 0 | 0 | 2.5 | 2.5 | 5.0 | 9.2 | 1.00 | — |
| CH5SR | 5.8 | 0 | 0 | 0 | 5.8 | 5.8 | 1.15 | $R \geq 0.4$ |
| PF5SR | 0 | 0 | 3.9 | 2.7 | 7.8 | 12.2 | 1.34 | $R \geq 0.4$ |
| CHPF0 | 3.3 | 1.5 | 1.4 | 1.4 | 7.6 | 10.4 | 0.0...4.0 | — |

$f(CH)$: frequency of change-points that are not parts of platforms, $f(trend)$: frequency of trends, $f(Pfa)$: frequency of platforms with longer than 5 years duration, $f(Pfb)$: frequency of platforms with maximum 5 years duration, $f'$: frequency of all but Pfb IHs, $f$: frequency of all IHs, $s$: standard deviation of IH-sizes, and $R$: restriction for the autocorrelation. Each frequency characteristic is shown as number per time series. Frequency characteristics show the frequency of introduced IHs during the generation, except for column $f$, where the empirical total frequencies are presented. ($f$ is often slightly lower than the frequency of all introduced IHs, due to superposition of IHs or stretch-out over the end of the time series.)

In (9), the term $\max(w_j)$ indicates that each homogenized value ($u_i$) is compared with a true value for which the bias is minimal in the $2k$ wide window around $i$. This optimization is repeated applying different window width, but the penalty for time lapse exponentially increases with $k$. With the present parameterization the penalty for 3-year (4-year) lapse is 0.6 (1.0). After having $b_i$ for each $i$, CMB is calculated as the difference between the extremes of $b_i$ values as

$$CMB = \max_{i,j} \left| b_i - b_j \right|, \quad i, j \in [1, 2, \ldots, n]. \quad (11)$$

The transformation of the error terms in (7), (8), and (9) to efficiency measures is as follows. Let the error terms and the efficiencies be denoted by $q$ and $E$, respectively; then the general form of the connection between the error terms and efficiencies is given by

$$E = \frac{q(\mathbf{X}) - q(\mathbf{U})}{q(\mathbf{X})}. \quad (12)$$

In this way the maximal achievable value of $E$ is always 1, and the sign of $E$ shows whether a homogenization has resulted in quality improvement or not. Efficiency measures of trend detection, SSE, and CMB are denoted by $E_T$, $E_S$, and $E_C$, respectively.

The described four efficiency measures characterize DOHMs in different ways. Only $E_D$ evaluates directly the detection results, while the other measures can be applied on adjusted time series. However, the usefulness of $E_D$ is limited by the facts that (i) the calculation of $E_D$ contains arbitrary parameters and (ii) the general purpose of homogenization is to achieve the highest reliability of trends and other characteristics of variability that are present in true time series [20], and not the identification of change points. Therefore the use of efficiency characteristics showing the performance in preserving or reconstructing the true climatic characteristics of time series such as $E_T$, $E_S$, and $E_C$ is preferred rather than that of the detection skill [46, 51]. When $E_T$, $E_S$, and $E_C$ are applied on evaluating DOHMs, the meanings of

the obtained characteristics slightly differ from those for whole homogenization methods. The similarities and differences are characterized by the following peculiarities of time series homogenization. (i) The application of SA is optimal when the reference series is homogeneous. (ii) Although reference series are seldom homogeneous when real time series are homogenized, the bias in candidate series is often substantially larger than that in reference series. Therefore the good performance of DOHM+SA is a necessary, although not satisfactory, requirement from any homogenization method. (iii) The lack of the detection of IHs leaves the station effect to be inhomogeneous between two adjacent detected IHs on the one hand, while false detections shorten the sections between adjacent IHs unnecessarily reducing the sub samples for calculating sub section means on the other hand. The mentioned two problems reduce efficiencies, any kind of correction method is applied. The potential impact of these errors on the final homogenization results can be quantified by the remaining SSE after homogenization when the remaining SSE is influenced by detection errors only.

## 3. Results

*3.1. Case Studies.* Figure 1 presents the efficiencies showing the four kinds of characteristics in four distinct sections. The presentation starts with the characteristics of $E_S$ followed by those of $E_T$, $E_C$, and $E_D$, in this order.

The simplest case is when only 1 change point is included in each time series. These time series can be treated most easily, but, unfortunately, the appearance of this case is not common in climatic datasets. Figure 1 shows that for CH1B0 all the DOHMs perform well, except for tts. The weakness of tts arises from its low detection power. As tts examines parts of time series separately, a relatively strict significance threshold has to be applied to keep the rate of FTE low, but it results in relatively poor detection power.

The results show that DOHMs usually perform well also for CH5B0. The largest IHs of CH5B0 can relatively easily be identified, owing to their characteristic magnitude

| $E_S$ (SSE) | Bay | C-M | E-P | MAS | MLR | SNH | SNT | tts | WRS |
|---|---|---|---|---|---|---|---|---|---|
| CH1B0 | 98.6 | 97 | 95.8 | 98.6 | 98.9 | 98.7 | 98.6 | (43.1) | 98.8 |
| CH5B0 | 97.3 | 97.4 | 94.3 | 97.2 | 94.2 | 97 | 95.2 | 74.8 | 94.5 |
| PF5B0 | (26.3) | 60.3 | 31.6 | 58.7 | 48.6 | (20.5) | (11.1) | (−24) | (−15) |
| HUSTR | 76.7 | 78.8 | 73.2 | 77.9 | 76.8 | 76.4 | 75 | (39.3) | 73.3 |
| HUST0 | 64.3 | 65 | 61.8 | 62.1 | 64.7 | 63.7 | 62.4 | 53.1 | 60.9 |
| CH5S0 | 85 | 82.4 | 73.4 | 80 | 78.4 | 84.4 | 83.8 | (18.1) | 83.2 |
| PF5S0 | −51 | −55 | −66 | −56 | −56 | −54 | −56 | −75 | −61 |
| CH5SR | 91.9 | 90.2 | 83.8 | 89.8 | 88.5 | 91.4 | 90.9 | (22.7) | 90.7 |
| PF5SR | 5.3 | 23.5 | 11.4 | 19.5 | 20.3 | 1.6 | −4 | (−48) | (−17) |
| $E_T$ (Trend) | Bay | C-M | E-P | MAS | MLR | SNH | SNT | tts | WRS |
| CH1B0 | 93 | 92.7 | 86.7 | 93.5 | 93.5 | 93.1 | 92.9 | (39.1) | 93.2 |
| CH5B0 | 92.9 | 93.7 | 89.3 | 92.5 | 89.8 | 92.5 | 89.3 | (57.3) | 91 |
| PF5B0 | 55.5 | 74.2 | 63.1 | 71.8 | 69.4 | 52.5 | 47.3 | (23.5) | (38.8) |
| HUSTR | 73.3 | 76.7 | 69.4 | 74 | 74.4 | 72.3 | 69.8 | (22.9) | 71.3 |
| HUST0 | 59.2 | 62.7 | 56.7 | 56.5 | 59.6 | 58.9 | 57.8 | 44.1 | 57.7 |
| CH5S0 | 73.7 | 75.2 | 60.7 | 68.7 | 66.6 | 72.9 | 71.2 | (9.4) | 72.5 |
| PF5S0 | 11 | 20.3 | 3.2 | 12.9 | 11.6 | 10.1 | 9.1 | −1 | 7.3 |
| CH5SR | 83.7 | 84.5 | 73.7 | 81.5 | 79.9 | 83.2 | 81.1 | (14.4) | 82.8 |
| PF5SR | 42.4 | 57.6 | 48.8 | 53.4 | 53.6 | 38.5 | 34.6 | (8.1) | 33.7 |
| $E_C$ (CMB) | Bay | C-M | E-P | MAS | MLR | SNH | SNT | tts | WRS |
| CH1B0 | 91.2 | 87.2 | 72.6 | 86.4 | 92 | 91.3 | 88.3 | (38.5) | 91.8 |
| CH5B0 | 81.1 | 81.4 | 75 | 80 | 72.8 | 81.1 | 72.4 | (41.6) | 77.5 |
| PF5B0 | 20.9 | 35.3 | 24.5 | 31.9 | 24 | 19.5 | 16.5 | (−6) | 5.9 |
| HUSTR | 45.5 | 44.8 | 42 | 44.2 | 45.2 | 45.5 | 45.3 | (14.8) | 43.8 |
| HUST0 | 34.7 | 32.1 | 31 | 26.1 | 34.1 | 34.8 | 34.4 | 26.2 | 33.6 |
| CH5S0 | 46.2 | 40.5 | 31.1 | 35.6 | 40.3 | 45.9 | 44.7 | (6.7) | 45.1 |
| PF5S0 | 1.7 | −2 | −9 | −6 | 1.2 | 1.2 | 1 | −4 | −1 |
| CH5SR | 59.1 | 55.1 | 47.4 | 52.7 | 53.7 | 58.7 | 56.7 | (10.4) | 57.8 |
| PF5SR | 22 | 26.7 | 25.8 | 26.5 | 26.7 | 20.6 | 20.1 | 1.2 | 14 |
| $E_D$ (Detection) | Bay | C-M | E-P | MAS | MLR | SNH | SNT | tts | WRS |
| CH1B0 | 96.1 | 91.6 | 90.4 | 84 | 96.2 | 96.2 | 88.2 | (39) | 94.7 |
| CH5B0 | 86.8 | 92.7 | 81.8 | 88.8 | 67.8 | 86.1 | (54.4) | (46.5) | 74.9 |
| PF5B0 | (44.6) | 92.6 | 62.6 | 88.2 | 62.7 | (42.5) | (40.1) | (18.2) | (18.4) |
| HUSTR | 54.8 | 79.3 | 53.1 | 77.1 | 58.7 | 52.1 | (49.1) | (24) | (39.7) |
| HUST0 | (25.8) | 60 | 25.3 | 56.5 | 34.2 | (29) | (27.2) | (9.5) | (20) |
| CH5S0 | 3.1 | (−133) | (−109) | (−210) | (−13) | 1.4 | 26.8 | 13.5 | −1 |
| PF5S0 | 17.9 | 18.7 | −7 | (−17) | 21.1 | 18.6 | 18.2 | 5.8 | 5.5 |
| CH5SR | 23.5 | (−47) | (−24) | (−85) | (6) | 23.2 | 46.2 | 20.4 | (−11.8) |
| PF5SR | 43.8 | 62.6 | 56.6 | 65.1 | 55.5 | 42.2 | 41.3 | (12.9) | (17.3) |

FIGURE 1: Four efficiency characteristics for each DOHM with each test dataset. Bold and underlined: best DOHM of a particular test (BDOHM hereafter), bold: less than 5% lag behind BDOHM, italics: more than 15% lag behind BDOHM, and italics with brackets: more than 30% lag behind BDOHM. Lightness of cell background improves with growing efficiency.

and duration. Yet the average performance is usually lower for CH5B0 than for CH1B0. The highest performances are produced by C-M, but Bay, MAS, and SNH are nearly as good, and almost all DOHMs have an efficiency above 50%. $E_S$ values are particularly high; they are usually well above 90%. The detection skill is relatively poor for SNT and MLR, its likely reason is that using these DOHMs sometimes trends are identified instead of two or more change points.

The only difference between PF5B0 and CH5B0 is that time series in PF5B0 include short-term platforms instead of long-term IHs. The frequency and the magnitude distribution of the events are unchanged (although the number of change points is doubled in this way, since a platform consists of two change points). This change of dataset properties has dramatic effect on the performances of DOHMs. For PF5B0 the majority of the efficiencies calculated are lower than 50%, although they are still positive with few exceptions. C-M and MAS have considerably higher efficiencies than the other DOHMs have; MLR shows the third best performance, while tts and WRS produced the poorest results.

Examining the results of the Hungarian standard (HUSTR) one can see that the performances of the DOHMs are generally better than for PF5B0, but poorer than for CH5B0. C-M has the best results again, but the advantage of C-M and MAS is much smaller than for dataset PF5B0; moreover, Bay and MLR have similar or slightly better performance than MAS has for $E_T$ and $E_C$. Considerably poorer results occur only with tts.

When the simulation method of the standard dataset is applied without limiting autocorrelation values (HUST0), the performances are poorer than for HUSTR, but the mean difference is moderated. The order of the skills does not change much either. In detection skill the C-M and MAS are much better than the other DOHMs, while for other skills the differences are small.

The results for CH5S0 show a spectacularly big difference in comparison with the previously analyzed results, namely, the detection skill of C-M and MAS is not the best in this case, but just the opposite relation can be seen. While most of the DOHMs perform near zero detection skill, those of MAS, C-M, and E-P are strongly negative. To understand this phenomenon, consider that (i) in CH5S0 the frequency of IHs of considerable size ($m^* > 2$) is very low, it is only 0.064 per 100 years on average; (ii) MAS, C-M, and E-P often have higher false alarm rate than the other DOHMs have (not shown). While in case of usual frequency of large IHs the high power of detection overbalances the drawback of relatively high false alarm rate; it does not operate for cases of rare IH occurrences. Interestingly, in spite of the large negative skills in $E_D$, the performances of C-M and MAS are still good for $E_S$, $E_T$, and $E_C$, and in trend detection skill C-M is the best.

Very different results were obtained using PF5S0. In this case most of the efficiencies scattered around zero, while SSE-reduction is not achieved by any of the DOHMs. This is the kind of results that would be good to avoid in the practical application of homogenization methods. Although the skills in preserving long-term linear trends are still slightly positive, the $E_S$ results show the disruption of short-term variability. An interesting feature is the very big difference between the results of CH5S0 and PF5S0 (similarly, as between CH5B0 and PF5B0), since the magnitude distribution of the inserted IHs is the same for the two datasets. In PF5S0 the mean frequency of IHs with $m^* > 2$ is 0.166 per 100 years which is although significantly higher than that for CH5S0, it still indicates that most of the time series are free from IHs of large magnitudes. It can be seen that the change of the IH-form from solely shifts to short-term platforms resulted in the complete cessation of negative detection skills of C-M and MAS, but, on the other hand, dominant $E_S$ values are dropped 140% approximately (from +80% to −60% in most of the methods).

The results with CH5SR and PF5SR prove that limiting the autocorrelation at 0.4 has a significant positive impact on the efficiencies. Although $E_D$ values of MAS, C-M, and E-P are still negative, its importance seems to be minor relative to the high efficiencies in $E_S$ and $E_T$. Large negative efficiencies related to platform type IHs of small size, as like with PF5S0, are not present with PF5SR except for tts, and most of the efficiencies are significantly positive.

The general differences between the performances of individual DOHMs when they are applied to the same test datasets in the nine case studies are as follows. Apart from tts the DOHMs tend to show similar efficiencies concerning $E_T$, $E_S$, and $E_C$, while the differences are generally larger in detection skill. C-M has generally the highest performance, except when there is only one IH in the time series or when the signal to noise ratio is very unfavorable. The skill of C-M is particularly good in preserving long-term climatic trends, which is given by the fact that the $E_T$ of C-M is always positive, and it is always the highest in comparison with the $E_T$ values of other DOHMs with the only exception of CH1B0. MAS is most often the second best DOHM, while the following three places in the rank order are for Bay, MLR, and SNH. Bay often, SNH several times produced better results than MLR, but the performance of MLR seems to be more uniform for different datasets than that of Bay and SNH. tts (WRS) performs markedly (slightly) poorer than the other DOHMs examined.

*3.2. Sensitivity to IH Magnitudes.* Experiments with compromise-form datasets (CHPF0) were performed applying increasing $s$ from 0 to 4, and the remaining SSE after homogenization were calculated in relation to the background noise ($s_e^*$). In Figure 2 the results of tts are not included, because the SSE of tts are excessively large. For large signal to noise ratio ($s > 2$) the rank order is similar to what was dominant in the case studies: C-M has the best performance, followed by MAS, Bay, SNH, and MLR. For moderate signal to noise ratio ($1 < s < 2$) C-M is still the best; it is followed by Bay, but all the performances have little variation, that is, they hardly depend on the choice of the DOHM. When the signal to noise ratio is small ($s < 1$), the C-M is not the best, but SSE are generally small with any of the examined DOHMs, except with E-P they are slightly larger.

## 4. Discussion

*4.1. Appropriateness of Test Datasets.* The creation of test datasets with realistic statistical properties needs the knowledge of the relation between the characteristics of detected and true IHs. This relation is examined applying various DOHMs on two test datasets (Figure 3). Although the detected frequency ($f^*$) depends on the DOHM applied, the difference from the true frequency is always negative, because very small IHs cannot be detected by any of the DOHMs. This negative difference is even more striking when the IHs are short-term platforms (Figure 3(b)). From these results it is clear that the direct observation of the "best" IH-properties through the examination of observational datasets is not possible. The properties of Benchmark were set by expert decisions of experienced homogenizers, but when time series contain large number of hardly detectable IHs, experts' estimations might be biased. When HUSTR was constructed,

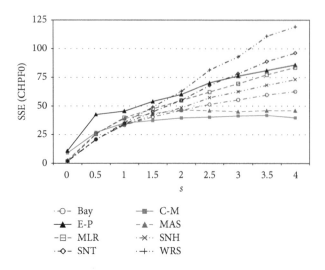

FIGURE 2: Remaining sum of squared errors (SSE) after the homogenization of CHPF0 with various DOHMs and using ideal reference series. The unit of SSE is the estimated standard deviation of background noise ($s_e^*$).

a distinct approach to the assessment of IH-properties was applied relative to the Benchmark. The main idea was to find IH-populations for which the detection results are similar to the detection results from real climatic time series. This approach could provide more reliable assessments of true IH properties than earlier studies, but further examination is probably needed because the very high frequency of detected short-term platforms in HUSTR may have origins other than IHs, for example, the natural variability of spatial temperature-gradients.

Figure 4 shows the magnitude distributions for change point type IHs in three datasets. It can be seen that the amount of small IHs is the highest in HUSTR, though it must be noted that there is no way to be assured about the reality of the amount of very small IHs ($m^* < 1$), because they have little impact on the detection results. For $m^* > 2$ the relation between CH5B0 and HUSTR is reversed, that is, CH5B0 contains more medium-size and large IHs than HUSTR does. Note that the differences are large, since the scale is logarithmic. Figure 4 illustrates also that the general failure with homogenizing PF5S0 is due to the lack of large IHs, beyond the short duration of IH caused biases in that dataset.

Relying on [42], the use of HUSTR-like test datasets can be favored. Taking into account that an unknown portion of IHs in HUSTR might have other sources than true IHs, datasets with smaller frequency of platforms than in HUSTR but with higher frequency of them than in the Benchmark could be closer to the reality, and CHPF0 is a realization of this idea. Anyhow, the author does not state that all kinds of climatic time series could be well represented with one or two test datasets of specific properties, but he suggests that the use of different kind test datasets, especially ones including large number of short-term IHs, is essentially important in testing efficiencies of IH detection algorithms.

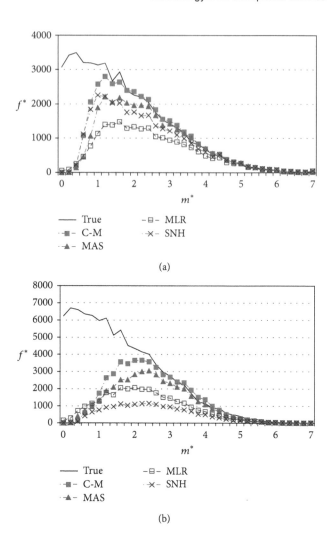

FIGURE 3: Magnitude distribution of true and detected IHs in test datasets, (a) (upper) CH5B0 and (b) (bottom) PF5B0. Frequencies ($f^*$) are shown using an arbitrary unit.

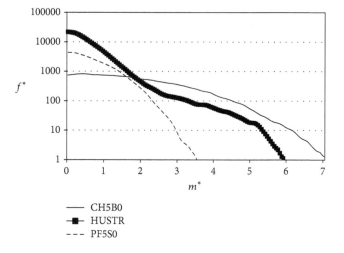

FIGURE 4: Magnitude distribution of IHs in three test datasets. Frequencies ($f^*$) are shown using an arbitrary unit.

*4.2. Efficiency of DOHMs.* The most important finding is that the performance of DOHMs strongly depends on the properties of the used test datasets, and thus the results are varied. (A similarly high diversity of results using simulated radiosonde datasets is described in [45].)

Nevertheless, some consistencies are evident in the results shown here. For example, efficiencies are always higher (a) for datasets with random sequences with only change points than for those with short-term platforms, (b) for datasets with time series of high autocorrelation than for datasets without restriction of $R$, and (c) for datasets with large IHs compared to those with low spread of IH-magnitudes. The apparent diversity of efficiencies is likely realistic; first because the frequency and intensity of IHs are diverse in practice, and second, because the networks of observed data have different spatial correlations causing strong diversity of signal/noise ratios.

In some earlier studies [41, 42] it was found that DOHMs with joint detection of multiple IHs perform better than other DOHMs. In this respect the present study confirms the earlier findings. The results presented here show that C-M is generally the best DOHM, but there are several other DOHMs whose performances are usually not much poorer. DOHMs of the best performances include joint detection of multiple IHs (C-M and MAS) or cutting algorithm (Bay, MLR, and SNH). By contrast, sequential detection of IHs using moving windows cannot be recommended which is proven by the fact that the performance of tts is always substantially poorer than using DOHMs with joint detection or cutting algorithm. The cause of the failure with tts is that the use of time windows results in distinct decisions about the significances of individual IHs, disregarding the data in other parts of the time series. Note that E-P also includes examinations for sections of limited length, and likely it does not effect positively its final performance. The traditional way of correcting inhomogeneous series is to apply the bias term on the candidate series with the opposite sign as it was detected in the relative time series. In examinations with limited window width the biases are usually assessed also with such windows [9, 35], and the adjustments are derived from such assessments. However, as the temporal coherence of such bias estimations is poor, it often results in particularly poor final performance in preserving the true climatic variability of time series, in spite of the fact that the detection skill of change points can be better than applying the SA (not shown). Examinations of arbitrary separated sections can be useful when the final objective is to make decisions about the significance of one or few potential IHs, but not for the task of general statistical homogenization. Among DOHMs with cutting algorithm the nonparametric WRS performed the poorest.

All the results show that the performance of MAS is very similar to that of C-M. Note that the performance of MAS is not better with its originally suggested correction technique than with SA (not shown), at least in the model task of this study (i.e., in case of homogeneous reference series).

An interesting feature of Figure 2 is that when time series are homogenized with C-M or MAS the increase of SSE with rising $s$ stops around $s = 2$. It promises that applying one

of these DOHMs, the remaining uncertainty of homogenized time series can be assessed, since that does not depend on the magnitudes of IHs in the raw dataset. However, it should be noted that that problem is still complex, since the remaining SSE depends on the frequency, shape, and temporal structure of IHs.

In an earlier study [41], moving parameter examinations were performed for DOHMs, in which significance thresholds and the shortest period allowed between two adjacent change points were varied in homogenizing HUSTR. The main findings of that study are that when time series are presumed to have large IHs, the optimal significance thresholds are lower than those which are generally applied, and the optimal shortest period is 2-3 years for C-M and MAS (1 year in this study) and 3–5 years (most often 4 years) for other DOHMs (5 years in this study). On the other hand, that study also showed that the optimization of parameters does not have robust effect on the performance of DOHMs. In this study we did not apply the parameters optimized on HUSTR by Domonkos [41] because the true optimums obviously depend on the properties of test datasets.

Most of the examined test datasets do not contain trend type IHs. MLR and SNT might be expected to perform better when time series contain more trends. However, the variation of trend frequency in HUSTR (not shown) indicated that raising the trend ratio within long-term IHs hardly influences the rank-order of efficiencies. Even with a 70% trend ratio, the MLR and SNT did not perform better than C-M, the only exception being that the $E_C$ of SNT became to be the best among DOHMs. On the other hand, the performance of SNT is generally slightly poorer than that of C-M, MAS, Bay, MLR, and SNH.

Most DOHMs are applied with 0.05–010 FTE except E-P. In E-P the significance thresholds that are suggested by Easterling and Peterson [35] are applied. Lund and Reeves [52] presented the mathematical explanation why these thresholds are not restrictive enough for keeping FTE low. However, the application of stricter significance thresholds substantially worsens the performance of E-P in $E_S$, $E_T$, and $E_C$ (not shown); therefore, we did not change the original parameterization.

Our results indicate that DOHMs with joint detection of multiple IHs, and particularly C-M, perform better than the other DOHMs except when the signal to noise ratio is too low. C-M is a maximum likelihood method, and the step function fitting in that aims at the minimization of the residual SSE [53]. The number of steps is determined by the Caussinus-Lyazrhi criterion, which is based on the information theory [47]. However, the methodological development of DOHMs has not been terminated with the creation of C-M. A further improvement of performance is expected from the network-wide joint detection of IHs [54] and from the harmonization of work on monthly and annual time scales [55], not mentioning here the problem of daily data homogenization which requires the development of distinct methods relative to the homogenization of annual and monthly data.

While in the COST HOME project full homogenization methods were tested, in this study only the detection parts are examined. The comparison of the results show that the efficiencies of complete methods are usually much lower [46] than the efficiencies of detection methods with idealized reference series. This difference indicates that the time series comparison, the calculation of correction terms, as well as the treatment of data-gaps and outliers may also be substantial sources of homogenization errors. This finding confirms the need for further investigations into strategies for minimizing the risk of negative efficiencies. Optimally the best segments of homogenization methods should be put together (i.e., the best detection method with the best correction method, etc.), and the performance of the best methods should be repeatedly checked applying test datasets of varied properties.

### 4.3. Reduction of the Risk of Negative Efficiency.

Given that almost all results indicate positive efficiency, the expectation that the application of DOHMs generally results in improvement in the quality of observed climatic time series is realistic. However, under certain conditions, alterations from the natural climatic characteristics are relatively frequent (e.g., with PF5S0). It is important to note that obviously no improvement can be achieved using any DOHM on a homogeneous time series; thus the fact that test datasets with negative efficiencies of DOHMs for them can be constructed is not a discouraging indication in itself. On the other hand, a general expectation from DOHMs is that their procedures should indicate with fairly high probability if there is no realistic chance to make quality improvement on time series. The application of DOHMs for time series with low signal to noise ratio is problematic because of the increasing proportion of false alarms and the possible disruption of the natural variability structure shown by raw observed time series. For time series comprising pure white noise, the traditional expectation of preserving time series without adjustments is 90–95%. Figure 5 shows the FTE values for the examined DOHMs. It can be seen that 100 year long series of white noise are not subjected to adjustments with 89–96% probability, except for E-P the ratio is 82% only. When time series contain IHs, but the signal to noise ratio is insufficient for their correct detection, these series should be treated in the same way as homogeneous series. Unfortunately, in case of PF5S0, 25–50% of the series are adjusted (the ratio depends on DOHM; see Figure 6), and their quality is worsened, since the detection results are generally poor due to the low signal-to-noise ratio. The signal to noise ratio can be low due to the very small size of IHs, short duration of biases, high level of background noise, or the existence of other noise than white noise (see also [20, 56]). The frequent adjustment of time series in low signal to noise conditions is a kind of over-homogenization. For instance, when the dataset properties are changed from CH5S0 to PF5S0, it resulted in the mean decline of $E_S$ from +80% to −60%, which might seem to be shocking at the first glance. Notwithstanding, the degree of disruption by over-homogenization is usually much less serious in absolute scale. Examining more the differences between CH5S0 and PF5S0, the mean SSE of raw time series in $s_e$ unit (which is common for all datasets) is 254.4 for CH5S0, but only 26.8 for PF5S0. Calculating with +80% $E_S$ for CH5S0 and −60% $E_S$ for PF5S0, the remaining mean SSE after adjustments is 50.9 (42.8) for

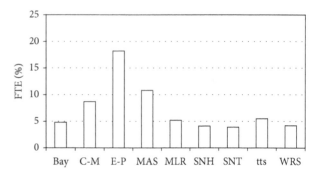

FIGURE 5: First type errors of DOHMs in pure white noise.

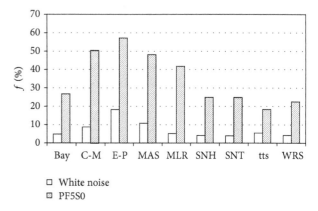

FIGURE 6: Ratios of qualification "nonhomogeneous" for time series of pure white noise and those for PF5S0.

CH5S0 (PF5S0). The latter results show that the possible over-homogenization usually has little true impact on the quality of observed time series, but one has to keep in mind that the undesired impacts can be great when the background noise of relative time series is high (e.g., due to relatively low spatial correlations in network), as well as additional error-terms should be added due to the imperfectness of reference time series, a problem that is not examined in this study.

A crucial question is that how often low signal to noise ratio occurs in relative time series of true observed datasets; because if its occurrence is frequent, the problem of over-homogenization can be serious. Unfortunately, short-term IHs are likely frequent in true observed time series, because they can easily be produced (i) when the cause of the IH is temporal, (ii) when technical problems are discovered with some delay, and the elimination of the problem is not paired with the backward correction of the data, and (iii) if two shifts of the same direction and significant magnitude are consecutives, the chance that the bias will be realized and corrected increases; therefore, the probability of two consecutive shifts with the opposite directions is higher than that with the same direction.

Another problem is that homogenizations are often step-by-step procedures, searching and correcting inhomogeneous time series from the ones with the highest biases proceeding towards the ones with smaller biases. Results of Figure 6 indicate that the chance that such homogenization procedures do not stop at the best stage is unfavorably high.

There are several options for minimizing the chance of over-homogenization. The intensive use of metadata information in the homogenization process is a widely applied alternative ([18, 29, 57–63], etc.). In spite of the fact that metadata usually do not provide quantitative information [58], the building of metadata information into the numerical evaluation process is one of the most promising ways of improving efficiency of homogenization methods [30, 40, 64–66]. Another option of possible developments is to think out again the parameterization of DOHMs. When some introductory pieces of information (e.g., autocorrelation) clearly indicate that a time series under examination is inhomogeneous, conventional significance thresholds often turn out to be too strict [41]. On the other hand, the problem of potential over-homogenization indicates that traditional significance thresholds may be too low for deciding the usefulness of applying adjustments. The results of this study show that the over-homogenization effects cease when 0.4 minimum threshold is applied for the autocorrelation of relative time series. The author does not suggest that it is an optimal solution, but he points to the fact that some problems exist around the basic mathematical model applied in homogenization procedures, that is, relative time series cannot be modeled well by one or several randomly situated IHs plus white noise. Some homogenization procedures apply a kind of moderation of correction terms to reduce the chance of over-homogenization. Examples for such treatments are the USHCN homogenization method in which adjustments are applied only when the data of at least three nearby stations concordantly indicate the existence and sign of a local shift in the candidate series [40], as well as Multiple Analysis of Series for Homogenization [67] in which always the lowest threshold of the confidence interval of shift-size is applied in particular adjustments. Naturally, a moderation of correction terms like these might sometimes cause under-homogenization (i.e., lack of adjustments application or too small adjustments). Thus the proper way of the minimization of over-homogenization is still an open question.

## 5. Conclusions

In this study the efficiency for detection algorithms of nine DOHMs is tested using ten different test datasets and four measures of efficiency. The statistical properties of the examined datasets are varied; one of them is similar to the benchmark surrogated dataset of COST HOME project (CH5B0), another is derived empirically to reproduce statistical characteristics of IH-detection results from real climatic time series (HUSTR), and a third one forms a compromise around the halfway between the Benchmark-like CH5B0 and the Hungarian standard (CHPF0). The diversity of test dataset properties and the different kinds of efficiency measures are necessary because the efficiency of homogenization strongly depends on the properties of real observed time series (which are truly diverse) and on the preferred purpose(s) of the homogenization. Our main findings are as follows.

(i) The application of DOHMs is generally beneficial for the quality of time series. When DOHMs are used for time series containing at least one large IH relative to the noise level, the bias of adjusted time series is usually smaller than half of that in raw time series. The mean error of linear trend estimation can be reduced by 75% in the Hungarian standard and by 90% in the Benchmark dataset with the application of DOHMs. Note that this estimation is valid only when all the errors out of the detection-segment (i.e., time series comparison, assessment of correction terms, and treatment of data gaps and outliers) are insignificantly small.

(ii) Short-term, platform-shaped IHs are much more difficult to detect precisely than randomly scattered solely IHs are, particularly when the IH-magnitudes are small. Differences between efficiencies for different kinds of datasets are often larger than those for different DOHMs.

(iii) Results of different efficiency measures often strongly differ, indicating different skills even by applying the same DOHM and for the same test dataset.

(iv) Efficiencies of individual DOHMs do not usually have the same rank order when different efficiency measures or results from different test datasets are compared, although several relationships seem to be stable. The results presented show that C-M is often the most effective DOHM, although the performances of further 4-5 DOHMs (MAS, Bay, MLR, and SNH) are still not much poorer. DOHMs capable of detecting both change points and trends (MLR and SNT) have no better performance than C-M has, even when time series contain trends. The efficiency of SNT is usually slightly lower than that of the earlier version of Standard Normal Homogeneity Test (SNH), and it is not among the best five. DOHMs have to treat multiple IHs either with cutting algorithm or (optimally) through the joint detection of all IHs. tts does not apply any of these techniques, and thus it is not capable of preserving real climatic characteristics of time series. Consequently, DOHMs with sequential tests cannot be recommended for homogenizing time series.

(v) Observed climatic time series cannot be modeled well with the composition of a white noise process plus one or several randomly scattered change points. Not considering the differences between such simple models and the true world may result in unnecessary disruption in the real climatic characteristics of time series. On the other hand, the results show that the degree of the potential disruption is small, at least when some conservative conditions of DOHM application are given, that is, when relative time series of low noise level can be built from the time series of networks of high spatial correlations.

## Acknowledgments

The author thanks Matthew Menne, Manola Brunet, Phil Jones, and three anonymous reviewers for their useful comments in improving the clearness of the paper. The research was supported by the European projects COST ES0601 and EURO4M FP7-SPACE-2009-1/242093 and by the Spanish project "Tourism, Environment and Politics" ECO 2010-18158.

## References

[1] WMO Hungarian Meteorological Service, *Proceedings of the First Seminar for Homogenization of Surface Climatological Data*, Edited by S. Szalai, Hungarian Meteorlogical Service, Budapest, Hungary, 144 pages, 1996.

[2] WMO Hungarian Meteorological Service, *Proceedings of the Second Seminar for Homogenization of Surface Climatological Data*, Edited by S. Szalai, T. Szentimrey, and Cs. Szinell, WCDMP-41, WMO-TD 932. WMO, Geneva, Switzerland, 214 pages, 1999.

[3] WMO Hungarian Meteorological Service, *Third Seminar for Homogenization and Quality Control in Climatological Databases*, Edited by S. Szalai, Hungarian Meteorological Service, 2001, http://www.met.hu/.

[4] WMO Hungarian Meteorological Service, *Fourth Seminar for Homogenization and Quality Control in Climatological Databases*, Edited by S. Szalai, WCDMP-56, WMO-TD, 1236, WMO, Geneva, Switzerland, 243 pages, 2004.

[5] WMO Hungarian Meteorological Service, *Proceedings of the Fifth Seminar for Homogenization and Quality Control in Climatological Databases*, Edited by M. Lakatos, T. Szentimrey, Z. Bihari, and S. Szalai, WCDMP-No. 71, WMO-TD, 1493, WMO, Geneva, Switzerland, 203 pages, 2008.

[6] WMO Hungarian Meteorological Service, *Proceedings of the Sixth Seminar for Homogenization and Quality Control in Climatological Databases*, Edited by M. Lakatos, T. Szentimrey, Z. Bihari, and S. Szalai, WCDMP-No. 76, WMO-TD., 1576, WMO, Geneva, Switzerland, 116 pages, 2010.

[7] M. Brunet, O. Saladié, P. Jones et al., "The development of a new dataset of Spanish daily adjusted temperature series (SDATS) (1850–2003)," *International Journal of Climatology*, vol. 26, no. 13, pp. 1777–1802, 2006.

[8] D. P. Rayner, "Wind run changes: the dominant factor affecting pan evaporation trends in Australia," *Journal of Climate*, vol. 20, no. 14, pp. 3379–3394, 2007.

[9] M. Staudt, M. J. Esteban-Parra, and Y. Castro-Díez, "Homogenization of long-term monthly Spanish temperature data," *International Journal of Climatology*, vol. 27, no. 13, pp. 1809–1823, 2007.

[10] M. Rusticucci and M. Renom, "Variability and trends in indices of quality-controlled daily temperature extremes in Uruguay," *International Journal of Climatology*, vol. 28, no. 8, pp. 1083–1095, 2008.

[11] S. C. Sherwood, C. L. Meyer, R. J. Allen, and H. A. Titchner, "Robust tropospheric warming revealed by iteratively homogenized radiosonde data," *Journal of Climate*, vol. 21, no. 20, pp. 5336–5350, 2008.

[12] M. J. Menne, C. N. Williams Jr., and R. S. Vose, "The U.S. Historical Climatology Network monthly temperature data, version 2," *Bulletin of the American Meteorological Society*, vol. 90, pp. 993–1007, 2009.

[13] I. Auer, R. Böhm, A. Jurković et al., "A new instrumental precipitation dataset for the greater Alpine region for the period 1800–2002," *International Journal of Climatology*, vol. 25, pp. 139–166, 2005.

[14] E. Aguilar, I. Auer, M. Brunet, T. C. Peterson, and J. Wieringa, "WMO Guidelines on climate metadata and homogenization," WCDMP 53, WMO, Geneva, Switzerland, 2003.

[15] T. C. Peterson, D. R. Easterling, and T. R. Karl, "Homogeneity adjustments of in situ atmospheric climate data: a review," *International Journal of Climatology*, vol. 18, pp. 1493–1517, 1998.

[16] M. J. Menne and C. N. Williams Jr., "Detection of undocumented changepoints using multiple test statistics and composite reference series," *Journal of Climate*, vol. 18, no. 20, pp. 4271–4286, 2005.

[17] J. Reeves, J. Chen, X. L. Wang, R. Lund, and X. Lu, "A review and comparison of change-point detection techniques for climate data," *Journal of Applied Meteorology and Climatology*, vol. 46, pp. 900–915, 2007.

[18] M. Brunet, O. Saladié, P. Jones et al., "A case-study/guidance on the development of long-term daily adjusted temperature datasets," Tech. Rep. WC-DMP-66/WMO-TD-1425, WMO, Geneva, Switzerland, 2008.

[19] B. Trewin, "A daily homogenized temperature data set for Australia," *International Journal of Climatology*, vol. 33, pp. 1510–1529, 2013.

[20] S. C. Sherwood, "Simultaneous detection of climate change and observing biases in a network with incomplete sampling," *Journal of Climate*, vol. 20, no. 15, pp. 4047–4062, 2007.

[21] P. M. Della-Marta and H. Wanner, "A method of homogenizing the extremes and mean of daily temperature measurements," *Journal of Climate*, vol. 19, no. 17, pp. 4179–4197, 2006.

[22] X. L. Wang, H. Chen, Y. Wu, Y. Feng, and Q. Pu, "New techniques for the detection and adjustment of shifts in daily precipitation data series," *Journal of Applied Meteorology and Climatology*, vol. 49, no. 12, pp. 2416–2436, 2010.

[23] Z. Li, Z. Yan, L. Cao, and P. Jones, "Adjusting inhomogeneous daily temperature variability using wavelet analysis," *International Journal of Climatology*, 2013.

[24] V. Alexandrov, M. Schneider, E. Koleva, and J.-M. Moisselin, "Climate variability and change in Bulgaria during the 20th century," *Theoretical and Applied Climatology*, vol. 79, no. 3-4, pp. 133–149, 2004.

[25] F. G. Kuglitsch, A. Toreti, E. Xoplaki, P. M. Della-Marta, J. Luterbacher, and H. Wanner, "Homogenization of daily maximum temperature series in the Mediterranean," *Journal of Geophysical Research D*, vol. 114, no. 15, Article ID D15108, 2009.

[26] A. C. Costa and A. Soares, "Trends in extreme precipitation indices derived from a daily rainfall database for the South of Portugal," *International Journal of Climatology*, vol. 29, no. 13, pp. 1956–1975, 2009.

[27] Z. Yan, Z. Li, Q. Li, and P. Jones, "Effects of site change and urbanisation in the Beijing temperature series 1977–2006," *International Journal of Climatology*, vol. 30, no. 8, pp. 1226–1234, 2010.

[28] T. C. Peterson and D. R. Easterling, "Creation of homogeneous composite climatological reference series," *International Journal of Climatology*, vol. 14, no. 6, pp. 671–679, 1994.

[29] M. Begert, T. Schlegel, and W. Kirchhofer, "Homogeneous temperature and precipitation series of Switzerland from 1864 to 2000," *International Journal of Climatology*, vol. 25, no. 1, pp. 65–80, 2005.

[30] A. T. DeGaetano, "Attributes of several methods for detecting discontinuities in mean temperature series," *Journal of Climate*, vol. 19, no. 5, pp. 838–853, 2006.

[31] J. C. Gonzalez-Hidalgo, J.-A. Lopez-Bustins, P. Štepánek, J. Martin-Vide, and M. de Luis, "Monthly precipitation trends on the Mediterranean fringe of the Iberian Peninsula during the second-half of the twentieth century (1951–2000)," *International Journal of Climatology*, vol. 29, no. 10, pp. 1415–1429, 2009.

[32] S. M. Vicente-Serrano, S. Beguería, J. I. López-Moreno, M. A. García-Vera, and P. Štepánek, "A complete daily precipitation database for northeast Spain: reconstruction, quality control, and homogeneity," *International Journal of Climatology*, vol. 30, pp. 1146–1163, 2010.

[33] P. Domonkos and P. Štepánek, "Statistical characteristics of detectable inhomogeneities in observed meteorological time series," *Studia Geophysica et Geodaetica*, vol. 53, pp. 239–260, 2009.

[34] T. A. Buishand, "Some methods for testing the homogeneity of rainfall records," *Journal of Hydrology*, vol. 58, no. 1-2, pp. 11–27, 1982.

[35] D. R. Easterling and T. C. Peterson, "A new method for detecting undocumented discontinuities in climatological time series," *International Journal of Climatology*, vol. 15, no. 4, pp. 369–377, 1995.

[36] J. R. Lanzante, "Resistant, robust and non-parametric techniques for the analysis of climate data: theory and examples, including applications to historical radiosonde station data," *International Journal of Climatology*, vol. 16, no. 11, pp. 1197–1226, 1996.

[37] J. F. Ducré-Robitaille, L. A. Vincent, and G. Boulet, "Comparison of techniques for detection of discontinuities in temperature series," *International Journal of Climatology*, vol. 23, pp. 1087–1101, 2003.

[38] M. Syrakova, "Homogeneity analysis of climatological time series—experiments and problems," *Időjárás*, vol. 107, pp. 31–48, 2003.

[39] G. Drogue, O. Mestre, L. Hoffmann, J.-F. Iffly, and L. Pfister, "Recent warming in a small region with semi-oceanic climate, 1949–1998: what is the ground truth?" *Theoretical and Applied Climatology*, vol. 81, no. 1-2, pp. 1–10, 2005.

[40] M. J. Menne and C. N. Williams Jr., "Homogenization of temperature series via pairwise comparisons," *Journal of Climate*, vol. 22, no. 7, pp. 1700–1717, 2009.

[41] P. Domonkos, "Testing of homogenisation methods: purposes, tools and problems of implementation," in *Proceedings of the 5th Seminar and Quality Control in Climatological Databases*, WCDMP-No. 71, WMO-TD, 1493, pp. 126–145, WMO, 2008.

[42] P. Domonkos, "Efficiency evaluation for detecting inhomogeneities by objective homogenisation methods," *Theoretical and Applied Climatology*, vol. 105, pp. 455–467, 2011.

[43] P. G. F. Gérard-Marchant, D. E. Stooksbury, and L. Seymour, "Methods for starting the detection of undocumented multiple changepoints," *Journal of Climate*, vol. 21, no. 18, pp. 4887–4899, 2008.

[44] C. Beaulieu, O. Seidou, T. B. M. J. Ouarda, X. Zhang, G. Boulet, and A. Yagouti, "Intercomparison of homogenization techniques for precipitation data," *Water Resources Research*, vol. 44, no. 2, Article ID W02425, 2008.

[45] H. A. Titchner, P. W. Thorne, M. P. McCarthy, S. F. B. Tett, L. Haimberger, and D. E. Parker, "Critically reassessing tropospheric temperature trends from radiosondes using realistic validation experiments," *Journal of Climate*, vol. 22, no. 3, pp. 465–485, 2009.

[46] V. Venema, O. Mestre, E. Aguilar et al., "Benchmarking monthly homogenization algorithms," *Climate of the Past*, vol. 8, pp. 89–115, 2012.

[47] H. Caussinus and F. Lyazrhi, "Choosing a linear model with a random number of change-points and outliers," *Annals of the Institute of Statistical Mathematics*, vol. 49, no. 4, pp. 761–775, 1997.

[48] A. Moberg and H. Alexandersson, "Homogenization of Swedish temperature data—part II: homogenized gridded air temperature compared with a subset of global gridded air temperature since 1861," *International Journal of Climatology*, vol. 17, no. 1, pp. 35–54, 1997.

[49] L. A. Vincent, T. C. Peterson, V. R. Barros et al., "Observed trends in indices of daily temperature extremes in South America 1960–2000," *Journal of Climate*, vol. 18, pp. 5011–5023, 2005.

[50] T. C. Peterson, "Assessment of urban versus rural in situ surface temperatures in the contiguous United States: no difference found," *Journal of Climate*, vol. 16, pp. 2941–2959, 2003.

[51] P. Domonkos, "Homogenising time series: beliefs, dogmas and facts," *Advances in Science and Research*, vol. 6, pp. 167–172, 2011.

[52] R. Lund and J. Reeves, "Detection of undocumented change-points: a revision of the two-phase regression model," *Journal of Climate*, vol. 15, no. 17, pp. 2547–2554, 2002.

[53] H. Caussinus and O. Mestre, "Detection and correction of artificial shifts in climate series," *Journal of the Royal Statistical Society C*, vol. 53, no. 3, pp. 405–425, 2004.

[54] F. Picard, E. Lebarbier, M. Hoebeke, G. Rigaill, B. Thiam, and S. Robin, "Joint segmentation, calling, and normalization of multiple CGH profiles," *Biostatistics*, vol. 12, no. 3, pp. 413–428, 2011.

[55] P. Domonkos, "Adapted caussinus-mestre algorithm for networks of temperature series (ACMANT)," *International Journal of Geosciences*, vol. 2, pp. 293–309, 2011.

[56] V. C. Slonosky and E. Graham, "Canadian pressure observations and circulation variability: links to air temperature," *International Journal of Climatology*, vol. 25, no. 11, pp. 1473–1492, 2005.

[57] E. Aguilar, T. C. Peterson, P. Ramírez et al., "Changes in precipitation and temperature extremes in Central America and northern South America, 1961–2003," *Journal of Geophysical Research*, vol. 110, Article ID D23107, 2005.

[58] M. Brunetti, M. Maugeri, F. Monti, and T. Nanni, "Temperature and precipitation variability in Italy in the last two centuries from homogenised instrumental time series," *International Journal of Climatology*, vol. 26, no. 3, pp. 345–381, 2006.

[59] D. Camuffo, C. Cocheo, and G. Sturaro, "Corrections of systematic errors, data homogenisation and climatic analysis of the Padova pressure series (1725–1999)," *Climatic Change*, vol. 78, no. 2-4, pp. 493–514, 2006.

[60] R. Brázdil, K. Chromá, P. Dobrovolný, and R. Tolasz, "Climate fluctuations in the Czech Republic during the period 1961–2005," *International Journal of Climatology*, vol. 29, pp. 223–242, 2009.

[61] Q. Li, H. Zhang, J. I. Chen, W. Li, X. Liu, and P. Jones, "A mainland china homogenized historical temperature dataset of 1951–2004," *Bulletin of the American Meteorological Society*, vol. 90, no. 8, pp. 1062–1065, 2009.

[62] M. Türkes, T. Koç, and F. Saris, "Spatiotemporal variability of precipitation total series over Turkey," *International Journal of Climatology*, vol. 29, pp. 1056–1074, 2009.

[63] M. Syrakova and M. Stefanova, "Homogenization of Bulgarian temperature series," *International Journal of Climatology*, vol. 29, no. 12, pp. 1835–1849, 2009.

[64] J. R. Christy, W. B. Norris, K. Redmond, and K. P. Gallo, "Methodology and results of calculating central California surface temperature trends: evidence of human-induced climate change?" *Journal of Climate*, vol. 19, no. 4, pp. 548–563, 2006.

[65] L. Haimberger, "Homogenization of radiosonde temperature time series using innovation statistics," *Journal of Climate*, vol. 20, no. 7, pp. 1377–1403, 2007.

[66] M. P. McCarthy, H. A. Titchner, P. W. Thorne, S. F. B. Tett, L. Haimberger, and D. E. Parker, "Assessing bias and uncertainty in the HadAT-adjusted radiosonde climate record," *Journal of Climate*, vol. 21, no. 4, pp. 817–832, 2008.

[67] T. Szentimrey, "Multiple Analysis of Series for Homogenization (MASH)," in *Proceedings of the 2nd Seminar for Homogenization of Surface Climatological Data*, WCDMP 41, WMO-TD 962, pp. 27–46, WMO, 1999.

[68] L. A. Vincent, "A technique for the identification of inhomogeneities in Canadian temperature series," *Journal of Climate*, vol. 11, no. 5, pp. 1094–1104, 1998.

[69] H. Alexandersson, "A homogeneity test applied to precipitation data," *Journal of Climatology*, vol. 6, no. 6, pp. 661–675, 1986.

[70] H. Alexandersson and A. Moberg, "Homogemzation of Swedish temperature data—part I: homogeneity test for linear trends," *International Journal of Climatology*, vol. 17, no. 1, pp. 25–34, 1997.

[71] F. Wilcoxon, "Individual comparisons by ranking methods," *Biometric Bulletin*, vol. 1, pp. 80–83, 1945.

# 12

# Temporal and Spatial Variability of Air Temperatures in Estonia during 1756–2014

**Agu Eensaar**

*Centre of Real Sciences, Tallinn University of Applied Sciences, Pärnu Maantee 62, 10135 Tallinn, Estonia*

Correspondence should be addressed to Agu Eensaar; agu.eensaar@gmail.com

Academic Editor: Pedro Ribera

The change in the statistical and temporal parameters of air temperatures in the Estonian cities, that is, Tallinn and Tartu, was analyzed for two centuries. The results showed that the change of air temperature in Estonia exceeded 0.5°C per ten years for the time 1979–2012. For the longer period, that is, 1880–2012, the average annual rise in the air temperature was within the range of 0.1°C per ten years. The analysis of frequency distributions of the average annual air temperatures and Welch's $t$-test demonstrated the considerable rise in air temperature (the significance level of 0.05) in Estonia, which took place in 1901–2014 and was witnessed only in the months from November to April. However, no significant rise in air temperature was detected in Estonia from May to October.

## 1. Introduction

Climate change has a considerable impact on the natural environment and human activities that can differ by regions. The relevance of the issue is based on the fact that the Intergovernmental Panel on Climate Change (IPCC) was established in 1988 by the World Meteorological Organization (WMO) and the United Nations Environment Programme (UNEP) with the aim to provide a comprehensive assessment of all aspects of climate change [1].

To forecast climate change and its impact, the peculiarities in the changes of various components related to the climate need to be examined. One of the significant parameters characterizing climate change is the air temperature. Global warming, verified by recent studies [2], is one of the major global concerns. In addition, it is also essential to study the changes in air temperature at local or regional level. Several studies have been dedicated towards the examination of climate change in the Baltic Sea regions [3–5]. The major cities in Estonia include Tallinn, the capital city, with ca. 430 thousand residents, and Tartu, with ca. 100 thousand residents (Figure 1). Therefore, it is significantly important to examine the peculiarities of the climate change of these cities. The present study aims at investigating the trends and variability in air temperatures in Estonian cities, such as Tallinn and Tartu, during 1756–2014.

## 2. Data and Methods

The use of instrumental measurements that were organized by a network of systematically arranged weather stations started in Estonia in the middle of the 19th century. Prior to this period, the meteorological observations were performed by individuals. Instrumental observations during the 18th century and the first-half of the 19th century in Estonia were performed using various measuring instruments. In addition, no uniform observation rules were followed during that period [6]. Therefore, it is difficult to estimate the accuracy of earlier observations. So far, the first known regular meteorological observations in Tallinn were performed in the years 1777–1779 by an artillerist named Breckling [6]. Professor Carpov of Schola cathedralis Tallinnensis (Tallinn Cathedral School) performed meteorological observations in the years 1779–1800 [7]. Later, Ivanov, Rickers, Schiefer-decker, and other naval officers of the Port of Tallinn acted as single observers [8]. There are also additional miscellaneous indirect data concerning the meteorological situation of the earlier period [9]. Observations, made by enthusiasts, differ

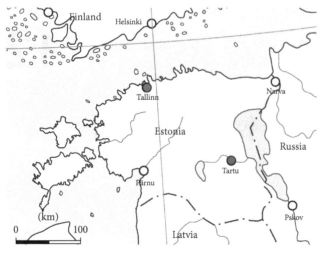

FIGURE 1: Map of Estonia.

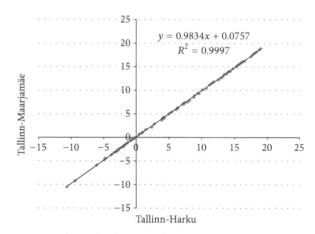

FIGURE 2: Relationship between the average monthly air temperatures (°C) in Tallinn-Maarjamäe and Tallinn-Harku.

considerably in their quality; yet after critical analysis this data appears more precise than any reconstruction made on the basis of indirect data. Due to the unavailability of local regular observations, Tarand [10] reconstructed the average monthly air temperatures in Tallinn for the period of 1756–1804 using the air temperature data of the nearby cities (i.e., St. Petersburg, Russia; Stockholm, Sweden; Loviisa and Porvoo, Finland). When comparing the data, taking into consideration the different level of industrialization in these cities, it has been concluded that the trend of global warming in the Nordic region is 0.8°C per 100 years and the urban heat island effect of most of the cities is 0.4°C per 100 years. In addition to that, Tarand et al. [8] has established that there is a great likelihood that many of the air temperature observations concerning summer and dating from the second half of the 18th century and the beginning of the 19th century have been influenced by the direct solar radiation effect common only to the countries in the Nordic region.

The ice data of the Port of Tallinn were also used as the essential supporting data. In the period of 1805–1849, simultaneous observations were taken at two locations in Tallinn. Neither of the two basic air temperature measuring areas in Estonia (Tallinn and Tartu) succeeded in organizing observations at the same location for a longer period. Thus, the air temperature has been measured at seven different locations in Tallinn and nine locations in Tartu. Based on parallel observations, Tarand et al. [8] found the location corrections, taking into account the effects of mesoclimate on the results of temperature measurements of meteostations located on the territories of Tallinn and Tartu. He used the location corrections and results from the earlier and later systematic meteorological observations at several observation points for the reconstruction of a time-series of average monthly air temperatures in Tallinn in the years 1756–2000, which was reduced to the Tallinn-Maarjamäe observation point (N 59°27′24″; E 24°48′50″). For the time being, the air temperatures in Tallinn were measured at the Tallinn-Harku meteorological station of the Estonian Weather Service (N 59°23′53″; E 24°36′10″), which

later published the results of meteorological observations in yearbooks, the latest being in 2013 [11]. The data of Estonian weather observations are stored in the databases of Statistics Estonia (SE) (http://www.stat.ee/en), which collects and stores data according to the international classifications and methods. At present, the meteorological station in Maarjamäe is no longer operative; hence, there is a lack of direct measuring data of air temperatures. The parallel data of air temperatures of Tallinn-Maarjamäe and Tallinn-Harku during 1992–2000 was used to determine the mutual relationship between the average monthly air temperatures of the observation points (Figure 2). Based on the relationship, the temperatures measured in the years 2001–2014 at the Tallinn-Harku meteorological station were reduced to the Tallinn-Maarjamäe observation point. This provided an opportunity to extend Tarand's et al. [8] time-series of average monthly temperatures until the end of 2014.

The air temperature data of Tartu were recorded from 1821 onwards using J. F. Parrot's observations [12, 13]. By using the location corrections found for different observation points and the results of meteorological observations performed at various observation points in Tartu, Tarand et al. [8] reconstructed a time-series of average monthly air temperatures in Tartu for the years 1821–1999, which was reduced to the Tartu-Tõravere meteorological station (N 58°15′51″; E 26°27′41″). Presently, the meteorological observations are taken at the Tõravere meteorological station of the Estonian Weather Service. Therefore, no correction factors have been used while extending Tarand's et al. [8] time-series of temperatures in Tartu until to 2014.

The changes in air temperature were analyzed in Tallinn during the years 1756–2014 and in Tartu during 1821–2014. The average monthly air temperatures analyzed were found to originate from the monograph [8] to which the average air temperatures (until the year 2014) found from the data available in the SE database of measurements of the Estonian Weather Service were added.

The meteorological conditions are formed due to a multitude of factors that make it difficult to determine a cause-and-effect relationship, which is only possible in the case of

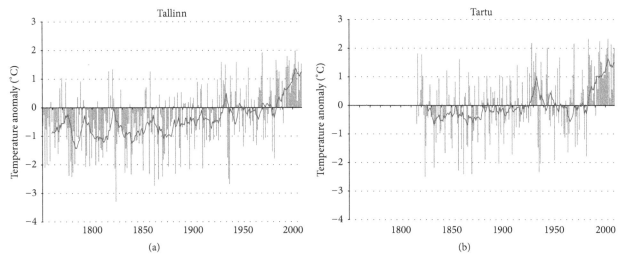

FIGURE 3: Mean temperature annual anomalies (difference in relation to the 1961–1990 average): (a) Tallinn in years 1756–2014; (b) Tartu in years 1821–2014. The decadal moving average is presented by the red line.

a specific situation. However, every specific meteorological situation may be regarded as a realization of every possible meteorological condition at a given location that has some probability of occurrence. In the course of time, given the large variability of air temperatures, statistical methods are used in this research for finding out the parameters that characterize the change in air temperature [14].

## 3. Results and Discussion

*3.1. Average Annual Air Temperatures.* Firstly, the change of air temperatures over a long period is of interest. The average annual air temperature, represented by $t_y$, was calculated as the weighted average of the average monthly temperatures:

$$t_y = \frac{\sum_i n_i t_i}{\sum_i n_i}, \tag{1}$$

where $i$ is the consecutive number of the month, $n_i$ is the number of days in the month $i$, and $t_i$ is the average temperature of the month $i$.

The average annual temperature (1) depends on the locality and fluctuates significantly with time. Therefore, in order to compare the variability and trends of the temperatures of various localities, it is important to examine the fluctuations in relation to the average value of a certain period. Figure 3 shows the fluctuations (anomalies) of the average annual air temperatures in relation to the average air temperatures of 1961–1990. The figure also shows the moving decade averages of the temperature fluctuations.

Figure 3 confirms that the average annual temperatures in Tallinn and Tartu vary to a great extent. The annual temperatures have changed (in relation to 1961–1990 average) from −3.3°C to +2.0°C in Tallinn and from −2.5°C to +2.0°C in Tartu. In the past few years, the air temperatures in both Tallinn and Tartu revealed a strong rising trend.

There is a rather strong correlative relationship between the average annual air temperatures in Tallinn and Tartu

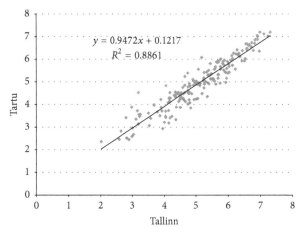

FIGURE 4: The relationship between the average annual air temperatures (°C) in Tallinn and Tartu for the years 1821–2014.

(Figure 4). Thereby the value of linear correlation coefficient is 0.94.

Due to their greater variability, the annual average air temperatures can be considered as random variables characterized by a distribution function. The average air temperature values occur with greater probability and are more extreme with a smaller probability. The change of the mean value and variance of the distribution function of air temperature for longer periods is a variable characteristic of climate change. When calculating the frequency distributions, we have divided the analyzed data into three groups by the centuries: 18th, 19th, and 20th centuries together with the 21st century.

Figure 5 presents the frequency distributions (in percentage) of average annual air temperatures in Tallinn and Tartu in different periods. The Jarque-Bera test was used to test the normality of the data.

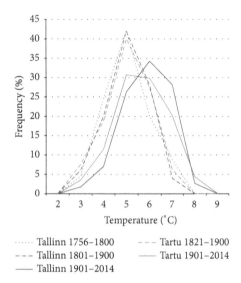

FIGURE 5: Frequency distributions of average annual air temperatures in Tartu and Tallinn in various periods per 1°C intervals.

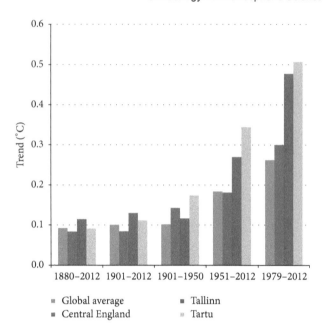

FIGURE 6: Annual average air temperature trends per decade.

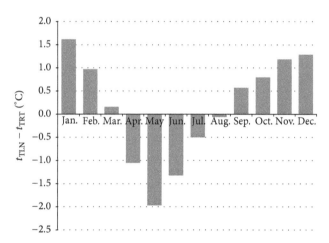

FIGURE 7: The difference between the average air temperatures in 1821–2014 in Tallinn compared to Tartu ($t_{TLN} - t_{TRT}$).

Figure 5 shows the average annual air temperature data that follows the normal distribution with 95% level of confidence. It appears that no significant change has taken place in Tallinn during 1801–1900 compared to the earlier period of 1756–1800. However, the distributions have shifted nearly 1°C in the positive direction in both Tallinn and Tartu during 1901–2014 compared to the earlier periods. This means that the average annual air temperature has risen considerably during the later periods.

In order to assess the nature of the change of Estonian average annual air temperatures, the trends in the changes in Tallinn and Tartu for 10 years per five customarily compared periods and the method of least squares [13] were considered as the basis of the calculations. The trends of global land surface air temperature [2] and air temperature data from Central England [15, 16] were used as references (Figure 6).

It becomes evident from Figure 6 that the rise of the average annual air temperature in Estonia has been faster than it has been global, especially in the later periods, reaching even 0.5°C per ten years. At the same time, the rise in the longer period of 1880–2012 is still closer to the global trend, remaining within the range of 0.1°C per ten years.

*3.2. Average Monthly Air Temperatures.* Since Tallinn is situated on the sea and Tartu lies inland, the differences in air temperature during the year are rather large. It becomes evident from Figure 7 that the average monthly temperature in Tallinn in autumn and winter (from September to March) is higher than that of Tartu, with a difference reaching up to 1.6°C. In spring and summer (from April to August), the average monthly air temperature in Tartu is higher up to 2°C.

Despite the geographical proximity, the temporal course of air temperatures is different in Tallinn and Tartu. Therefore, they both need to be considered when analyzing the Estonian temperature regime.

The fast change of air temperature in Estonia poses a threat to its natural resources and, thus, its features need to be examined in detail. The average monthly air temperature also depends on many circumstances and can be regarded as a random variable and treated statistically. Every meteorological situation has its causes and a course that results from it; however, due to the complexity of the cause-and-effect relationship, such kind of analysis is difficult to perform.

Figure 8 presents the frequency distribution of the occurrence of all average monthly air temperatures in Tallinn (1756–2014) and Tartu (1821–2014). The percentages are calculated for 1°C intervals. The frequency distributions of all monthly air temperatures in Central England in the years 1756–2014 that are calculated on the basis of a previous study's [11] data are used for reference basis. It is evident that the frequency of occurrence of the average monthly temperatures

TABLE 1: Statistical indicators of the average air temperatures (°C) in Tallinn by periods of time.

| Indicator | Period | Jan. | Feb. | Mar. | Apr. | May | Jun. | Jul. | Aug. | Sep. | Oct. | Nov. | Dec. | Year |
|---|---|---|---|---|---|---|---|---|---|---|---|---|---|---|
| Median | 1756–1800 | −6.6 | −5.4 | −3.0 | 2.2 | 7.4 | 13.2 | 16.2 | 15.3 | 11.4 | 6.2 | 1.2 | −4.1 | 4.6 |
| | 1801–1900 | −5.6 | −6.0 | −3.2 | 1.9 | 7.9 | 13.3 | 16.4 | 15.2 | 11.3 | 6.0 | 0.6 | −2.9 | 4.6 |
| | 1901–2014 | −3.8 | −4.8 | −1.6 | 3.4 | 9.0 | 13.5 | 16.6 | 15.7 | 11.4 | 6.4 | 1.4 | −2.2 | 5.4 |
| | 1756–2014 | −5.3 | −5.4 | −2.7 | 2.5 | 8.4 | 13.4 | 16.5 | 15.4 | 11.4 | 6.2 | 1.1 | −2.6 | 4.8 |
| Average | 1756–1800 | −6.6 | −5.9 | −4.0 | 2.0 | 7.5 | 13.4 | 16.5 | 15.6 | 11.7 | 6.1 | 1.0 | −3.5 | 4.5 |
| | 1801–1900 | −5.9 | −6.1 | −3.4 | 1.9 | 7.8 | 13.3 | 16.3 | 15.5 | 11.3 | 5.9 | 0.6 | −3.4 | 4.5 |
| | 1901–2014 | −4.5 | −5.1 | −2.0 | 3.4 | 8.9 | 13.7 | 16.8 | 15.6 | 11.4 | 6.3 | 1.4 | −2.2 | 5.4 |
| | 1756–2014 | −5.4 | −5.6 | −2.9 | 2.6 | 8.2 | 13.5 | 16.6 | 15.6 | 11.4 | 6.1 | 1.0 | −2.9 | 4.9 |
| St. dev. | 1756–1800 | 3.2 | 3.4 | 3.1 | 1.7 | 1.6 | 1.3 | 1.5 | 1.3 | 1.5 | 2.1 | 2.3 | 2.9 | 0.9 |
| | 1756–2014 | 3.6 | 3.6 | 2.8 | 1.9 | 1.9 | 1.6 | 1.6 | 1.5 | 1.4 | 1.8 | 2.1 | 2.9 | 1.1 |
| | 1801–1900 | 3.7 | 3.5 | 2.6 | 1.8 | 1.9 | 1.7 | 1.5 | 1.6 | 1.3 | 1.7 | 2.0 | 3.0 | 1.0 |
| | 1901–2014 | 3.4 | 3.7 | 2.7 | 1.8 | 1.7 | 1.6 | 1.6 | 1.4 | 1.4 | 1.8 | 2.0 | 2.8 | 1.0 |
| Max. | 1756–1800 | 0.2 | 0.4 | 1.0 | 5.4 | 11.6 | 17.0 | 19.8 | 18.7 | 15.4 | 10.2 | 5.8 | 1.8 | 6.4 |
| | 1801–1900 | 0.6 | 0.3 | 2.1 | 6.6 | 12.6 | 16.8 | 20.3 | 20.5 | 14.6 | 9.9 | 4.7 | 1.3 | 6.7 |
| | 1901–2014 | 1.8 | 2.5 | 3.0 | 7.1 | 12.9 | 17.5 | 21.2 | 19.6 | 14.9 | 9.6 | 5.3 | 4.4 | 7.3 |
| | 1756–2014 | 1.8 | 2.5 | 3.0 | 7.1 | 12.9 | 17.5 | 21.2 | 20.5 | 15.4 | 10.2 | 5.8 | 4.4 | 7.3 |
| Min. | 1756–1800 | −13.4 | −16.1 | −9.9 | −2.8 | 4.8 | 11.2 | 13.2 | 13.1 | 9.4 | 1.2 | −6.5 | −13.0 | 2.9 |
| | 1801–1900 | −14.5 | −16.6 | −9.7 | −2.4 | 1.9 | 9.0 | 12.5 | 12.2 | 8.7 | 0.7 | −4.3 | −12.4 | 2.0 |
| | 1901–2014 | −15.2 | −14.3 | −9.9 | −1.1 | 4.4 | 9.8 | 13.2 | 12.6 | 7.1 | 1.8 | −4.8 | −9.2 | 2.6 |
| | 1756–2014 | −15.2 | −16.6 | −9.9 | −2.8 | 1.9 | 9.0 | 12.5 | 12.2 | 7.1 | 0.7 | −6.5 | −13.0 | 2.0 |

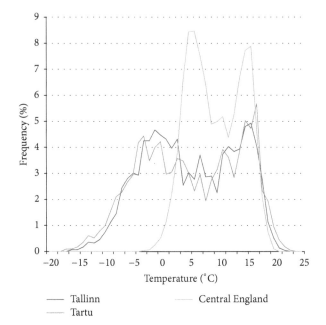

FIGURE 8: Frequency distribution of all the average monthly air temperatures in the years 1756–2014.

Firstly, we are interested in how the average monthly air temperature has changed over the time since the average annual air temperature forms as the weighted average of the average monthly air temperatures. Table 1 presents the statistical indicators (median, average, standard deviation, minimum value, and maximum value) of the average air temperatures in Tallinn, and Table 2 presents those in Tartu in various periods (centuries).

It becomes evident from Tables 1 and 2 that the changes in air temperatures are not homogeneous in all the months during the periods of observation. Thus, for example, the average temperature in January in different centuries has risen from −6.6°C to −4.5°C in Tallinn and from −7.4°C to −6.1°C in Tartu. However, the average air temperatures in September have fallen from 11.7°C to 11.4°C in Tallinn and from 10.9°C to 10.8°C in Tartu. The average monthly air temperatures in Estonia vary to a great extent, changing in the range up to 20°C. The variability of air temperatures in different months is worthy of closer examination. The main issue of interest lies with the frequency distributions of air temperature (occurrence frequencies of various temperatures) in different months and how and whether these distributions have changed over centuries. We observed the frequency distributions of average monthly temperatures both in Tallinn and Tartu in various periods (centuries) in order to find out the nature of change of the air temperature.

Figure 9 presents the frequency distributions of average monthly air temperatures in Tallinn and Tartu. In the case of Tallinn, the distributions of the air temperatures in January and February and those of December and March virtually coincide. In the case of Tartu, the distributions of the air temperatures in January and February and those of May and

above 17°C in Estonia and that in Central England practically coincide. However, in Estonia, the air temperature changes to a much wider extent than in Central England with a significant difference concerning lower temperatures. In Estonia, the low air temperatures occur at a quite high frequency, hence indicating no significant difference in the situations of Tallinn and Tartu.

TABLE 2: Statistical indicators of the average air temperatures (°C) in Tartu by periods of time.

| Indicator | Period | Jan. | Feb. | Mar. | Apr. | May | Jun. | Jul. | Aug. | Sep. | Oct. | Nov. | Dec. | Year |
|---|---|---|---|---|---|---|---|---|---|---|---|---|---|---|
| Median | 1821–1900 | −7.1 | −6.6 | −3.4 | 3.1 | 10.3 | 15.2 | 17.1 | 15.4 | 10.9 | 5.2 | −0.8 | −4.4 | 4.5 |
| | 1901–2014 | −5.3 | −6.0 | −1.8 | 4.3 | 10.7 | 14.6 | 16.9 | 15.7 | 10.8 | 5.5 | 0.5 | −3.4 | 4.9 |
| | 1821–2014 | −6.4 | −6.3 | −2.6 | 3.7 | 10.5 | 14.9 | 16.9 | 15.6 | 10.8 | 5.4 | −0.1 | −3.7 | 4.8 |
| Average | 1821–1900 | −7.4 | −6.9 | −3.3 | 3.2 | 10.1 | 15.2 | 17.2 | 15.6 | 10.9 | 5.3 | −0.7 | −4.8 | 4.5 |
| | 1901–2014 | −6.1 | −6.2 | −2.4 | 4.4 | 10.7 | 14.7 | 17.1 | 15.6 | 10.8 | 5.4 | 0.3 | −3.2 | 4.9 |
| | 1821–2014 | −6.6 | −6.5 | −2.7 | 3.9 | 10.5 | 14.9 | 17.1 | 15.6 | 10.8 | 5.3 | −0.1 | −3.9 | 4.8 |
| St. dev. | 1821–1900 | 3.9 | 3.5 | 2.7 | 1.9 | 2.2 | 1.6 | 1.9 | 1.5 | 1.5 | 1.8 | 2.3 | 3.0 | 1.0 |
| | 1901–2014 | 4.0 | 3.8 | 3.0 | 2.1 | 2.1 | 1.7 | 1.9 | 1.5 | 1.6 | 1.9 | 2.3 | 3.2 | 1.1 |
| | 1821–2014 | 3.9 | 4.0 | 3.1 | 2.1 | 2.0 | 1.7 | 1.9 | 1.5 | 1.6 | 1.9 | 2.3 | 3.3 | 1.1 |
| Max. | 1821–1900 | 0.0 | 0.0 | 1.9 | 7.7 | 15.1 | 19.0 | 21.9 | 19.8 | 14.7 | 8.4 | 4.3 | 0.3 | 6.6 |
| | 1901–2014 | 1.7 | 2.4 | 4.4 | 9.3 | 18.3 | 19.2 | 22.6 | 20.2 | 14.7 | 8.9 | 4.2 | 3.3 | 7.2 |
| | 1821–2014 | 1.7 | 2.4 | 4.4 | 9.3 | 18.3 | 19.2 | 22.6 | 20.2 | 14.7 | 8.9 | 4.3 | 3.3 | 7.2 |
| Min. | 1821–1900 | −17.0 | −17.6 | −9.6 | −0.5 | 3.8 | 10.6 | 11.0 | 12.4 | 7.3 | −0.3 | −5.5 | −14.2 | 2.3 |
| | 1901–2014 | −16.6 | −15.5 | −10.8 | −0.9 | 6.4 | 10.9 | 10.6 | 12.4 | 6.6 | 0.0 | −6.8 | −13.4 | 2.5 |
| | 1821–2014 | −17.0 | −17.6 | −10.8 | −0.9 | 3.8 | 10.6 | 10.6 | 12.4 | 6.6 | −0.3 | −6.8 | −14.2 | 2.3 |

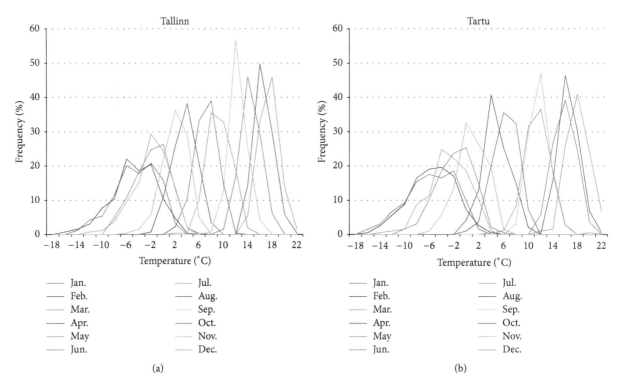

FIGURE 9: Frequency distributions of the average monthly air temperatures: (a) Tallinn, 1756–2014; (b) Tartu, 1821–2014.

September are similar; in addition, the distributions of the air temperatures in August and June are also quite similar.

Subsequently, we examined how the frequency distributions of the average monthly air temperatures have changed over the centuries. Figure 10 represents the frequency distributions of the average air temperatures both in Tallinn and Tartu in different periods month-wise. First, we tried to establish a change in the average situation by visual analysis of the charts. The change in the average air temperature is expressed by a change in the frequency distribution of the temperature.

In the case of January, a shift in the frequency distributions of air temperature toward higher temperatures is evident both in Tartu and in Tallinn (Figure 10(a)). In February, the tendency of a rise in air temperature is no longer strongly perceivable (Figure 10(b)). In March, a shift of about 2°C toward the temperature rise is visible, especially in the years 1901–2014 (Figure 10(c)). The average air temperatures of

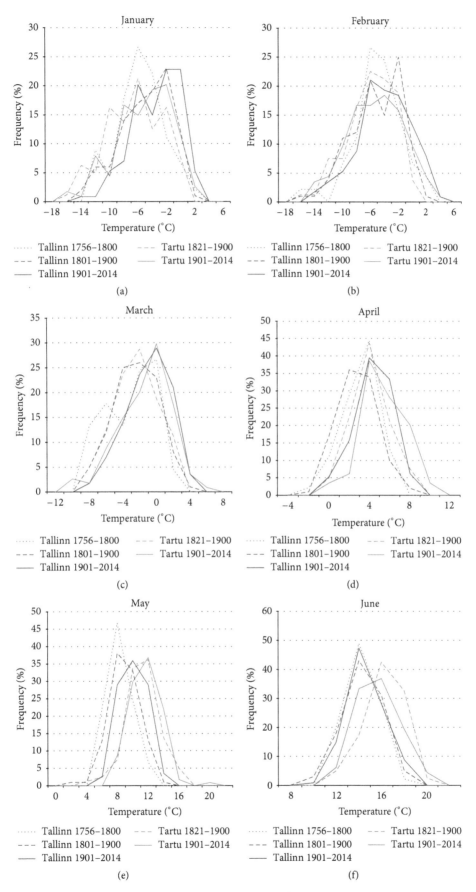

Figure 10: Continued.

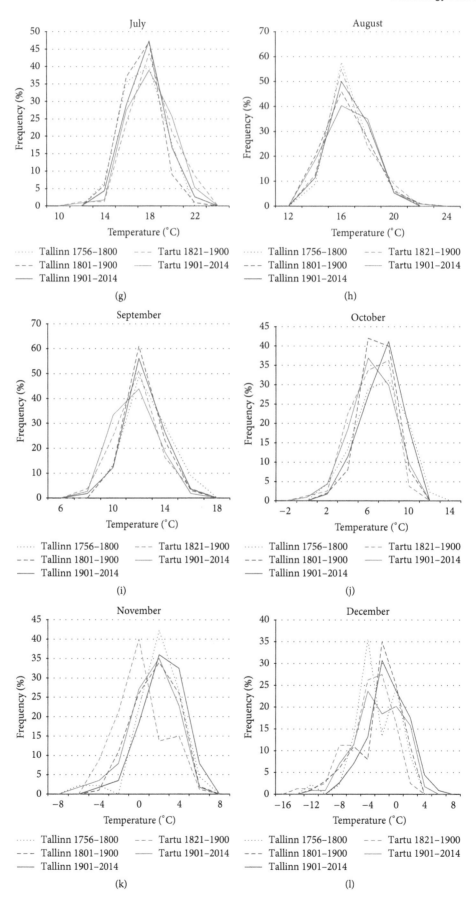

FIGURE 10: Frequency distributions of the average air temperatures in Tallinn and Tartu in different months: (a) January; (b) February; (c) March; (d) April; (e) May; (f) June; (g) July; (h) August; (i) September; (j) October; (k) November; (l) December.

April (Figure 10(d)) have become warmer in the years 1901–2014 compared to the previous periods. At the same time, the air temperatures in Tallinn in the 19th century were lower than the 18th century. In May, the air temperature in Tallinn has significantly risen in the past centuries while in Tartu, the rise is insignificant (Figure 10(e)). In June (Figure 10(f)), the average air temperatures in Tallinn have not changed in the past few centuries. However, in Tartu, the average air temperature in May has reduced. In July, the long-term changes in the air temperatures both in Tallinn and Tartu are insignificant (Figure 10(g)). In addition, in August (Figure 10(h)) and September (Figure 10(i)), no considerable changes have taken place in the average air temperatures. In October, the frequency distributions show a difference between the average air temperatures in Tallinn and Tartu, but there are only minor changes in the frequency distributions of the average air temperatures of various centuries (Figure 10(j)). In November, a significant shift in the frequency distributions of the air temperature in Tartu towards the positive temperatures is perceivable while the change is rather small in the case of Tallinn (Figure 10(k)). In the frequency distributions of the air temperatures in December, a minor shift toward the temperature rise is visible both in Tartu and in Tallinn (Figure 10(l)).

A visual assessment of the frequency distributions allows the assessment of the general nature of changes that have taken place; however, it gives no information on whether the changes hold any statistical relevance. It becomes evident from Figure 10 that the distribution of the average monthly temperatures can be regarded as close to a normal distribution in the first approximation. Thus, the methods of verification of statistical hypotheses dealing with random variables with normal distribution can be used.

In order to check the statistical relevance of the change of the average temperatures of various periods, Welch's $t$-test was used [14]. Let us assume that the air temperature of a period has a normal distribution characterized by average temperature $\mu_1$ and the temperature dispersion $\sigma_1^2$. The average temperature for another period compared is $\mu_2$ and the temperature dispersion is $\sigma_2^2$. The values of the average temperatures and dispersions are unknown to us. For the period $n_1$ and the other period $n_2$, we have the measured values of temperature; these values can be regarded as a sample from a time-series of distribution of temperatures. The $t$-test helps to determine whether the difference between the average temperatures $\bar{x}_1$ and $\bar{x}_2$ obtained from the measured values is statistically relevant or not.

Let us formulate a zero hypothesis that states the average temperatures for the compared periods to be not risen, $H_0: \mu_1 - \mu_2 \leq 0$. An alternative hypothesis is that the average temperatures have risen, $H_1: \mu_1 - \mu_2 > 0$. For calculation of the test statistic, we used the below equation:

$$t_0 = \frac{\bar{x}_1 - \bar{x}_2}{\sqrt{s_1^2/n_1 + s_2^2/n_2}}, \tag{2}$$

where $s_1$ and $s_2$ are standard deviations of measured temperatures (samples).

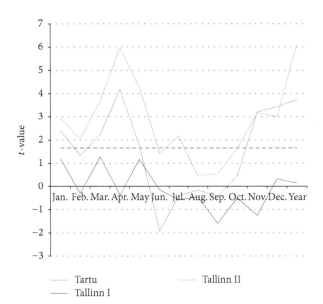

FIGURE 11: A verified $t$-test of the hypothesis of the air temperature change in Tallinn and Tartu. Significance level 0.05. The zero hypothesis (average temperature has not risen) is valid in the critical area of the parameter (below the dashed line). Comparison periods of the average temperatures: Tartu: 1901–2014 in relation to the period of 1821–1900; Tallinn I: the average temperatures of the period of 1801–1900 in relation to the period of 1756–1800; Tallinn II: the average temperatures of the period of 1901–2014 in relation to the period of 1801–1900.

A zero hypothesis can be rejected if $t_0 > t_{\alpha,\nu}$. $t_{\alpha,\nu}$ is a value of Student's $t$-function corresponding to significance level or probability of error $\alpha$, and the number of degrees of freedom $\nu$. The number of degrees of freedom $\nu$ is calculated with

$$\nu = \frac{\left(s_1^2/n_1 + s_2^2/n_2\right)^2}{\left(s_1^2/n_1\right)^2/(n_1-1) + \left(s_2^2/n_2\right)^2/(n_2-1)}. \tag{3}$$

By taking the value of significance level $\alpha = 0.05$, it was determined whether the average monthly and yearly air temperatures in Tartu and Tallinn had risen. The calculations demonstrate (Figure 11) that the air temperatures in Tallinn in the 19th century have not been raised as compared to the 18th century. The changes in the 20th century (together with the beginning of the 21st century), however, are considerable compared to those of the 19th century. The temperature rise is considerable from November to May, except in February, when the temperature rise was evident; however, in Tartu, (inland) the temperature rise has no significant value. From June to October, the zero hypothesis stated zero temperature change compared to the 18th century that remained valid in the case of Tartu. In June, a significant fall in the average temperature was observed. However, in the case of Tallinn, no significant temperature rise has occurred in June and from August to October. At the same time, the rise in the average annual temperatures in relation to the 19th century is statistically relevant (significance level, $\alpha = 0.05$).

## 4. Conclusions

By using a long time-series of average monthly air temperatures in Tallinn and Tartu, as reconstructed by Tarand et al. [8] and supplemented by the results of measurement from the Estonian Weather Service, the change and variations of air temperatures for various periods of time were analyzed. The temperature changes for a short period may not be extrapolated to a longer period as shown in Figure 3, where several periods have existed for a longer time during which the average annual air temperature has suddenly risen and then fallen.

When analyzing the change in the air temperature, it should be taken into account that the temperature changes in a particular locality do not coincide with the global change in air temperature, which are not homogeneous either throughout the year.

Average yearly and monthly air temperatures in Estonia vary to greater extent during long-term periods (Table 2). Hence, the temperatures are observed as random values in this research. Random value can be characterized with frequency distribution beside its average. Based on the frequency distribution, it is possible to estimate the occurrence of different air temperatures.

The changes of average air temperatures may be caused by specific systematic causes as well as by random factors (as a random combination of several common influences). By using Welch's $t$-test of statistical hypothesis, it has been checked whether the long-term period average monthly and yearly air temperatures changes can be explained by unequal variances (null hypothesis). We can speak about the increase of average air temperatures in cases when the empirical values of $t$-statistics of air temperatures measured at observations are higher than the critical value of $t$-statistics. In cases when the empirical value of $t$-statistics is lower than the critical value, we cannot confirm the statement that average air temperatures have risen. As the result of the analysis, it has become evident that the average yearly air temperature in Tallinn as well as Tartu has risen (significance level 0.005). At the same time it cannot be said about the average monthly air temperatures all of months. By analyzing the frequencies of occurrence of the average monthly air temperatures for various periods, it was found that, in Estonia, a significant rise in air temperature in the past century in the coastal city of Tallinn can be observed in autumn and spring, whereas in midwinter, in February, the rise is minimal. Welch's $t$-test shows (Figure 11) that in Estonian inland (Tartu), a rise in air temperature in the past century cannot be observed in the period from May to October (significance level 0.05).

Our findings show that the air temperature has risen in Estonia in winter and early spring; however, in the summer, no significant temperature rise has taken place. According to the hypothesis, albedo is one of the essential factors affecting the ground's thermal balance and thereby the air temperature. The conditions of Estonia depend substantially on the icing of the Baltic Sea and the presence of the snow cover. The correlation coefficients based on the data from 1962–2002 between the mean air temperature of the period December–February and the maximum ice extent in the Baltic Sea are −0.93 (Tallinn) and −0.90 (Tartu) [17]. The correlation coefficients between the mean air temperature of the period November–March and winter snow cover duration are −0.88 (Tallinn) and −0.83 (Tartu) [17]. Thus, global warming in the conditions of Estonia can boost the temperature regime by changing the ice and snow conditions. This would also explain the rising trends of air temperature in Estonia as presented in Figure 6. The follow-up research may contribute to the clarification of this matter.

## Conflict of Interests

The author declares that there is no conflict of interests regarding the publication of this paper.

## References

[1] IPCC, "Climate change 2001: the scientific basis," in *Contribution of Working Group I to the Third Assessment Report of the Intergovernmental Panel on Climate Change*, J. T. Houghton, Y. Ding, D. J. Griggs et al., Eds., p. 881, Cambridge University Press, Cambridge, UK, 2001.

[2] D. L. Hartmann, A. M. G. Klein Tank, M. Rusticucci et al., "Observations: atmosphere and surface," in *Climate Change 2013: The Physical Science Basis. Contribution of Working Group I to the Fifth Assessment Report of the Intergovernmental Panel on Climate Change*, T. F. Stocker, D. Qin, G.-K. Plattner et al., Eds., pp. 159–254, Cambridge University Press, Cambridge, UK, 2013.

[3] A. Tarand and A. Eensaar, "Air temperature," in *Country Case Study on Climate Change Impacts and Adaptation Assessments in the Republic of Estonia*, pp. 17–21, Stockholm Environment Institute-Tallinn, Tallinn, Estonia, 1998.

[4] A. Tarand and P. Ø. Nordli, "The tallinn temperature series reconstructed back half a millennium by use of proxy data," *Climatic Change*, vol. 48, no. 1, pp. 189–199, 2001.

[5] J. Jaagus and K. Mändla, "Climate change scenarios for Estonia based on climate models from the IPCC Fourth Assessment Report," *Estonian Journal of Earth Sciences*, vol. 63, no. 3, pp. 166–180, 2014.

[6] A. Tarand, "Meteoroloogilised vaatlused Eestis enne 1850. Aastat. Teaduse ajaloo küsimusi Eestist, VIII," Tallinn, Estonia, pp. 30–50, 1992.

[7] A. Tarand, 200 aastat professor C. L. Carpovi meteoroloogilistest vaatlustest, EGS Aastaraamat, Tallinn, Estonia, 1979.

[8] A. Tarand, J. Jaagus, and A. Kallis, *Eesti kliima minevikus ja tänapäeval*, Tartu Ülikooli Kirjastus, Tartu, Estonia, 2013.

[9] A. Tarand and P. Kuiv, "The beginning of the rye harvest—a proxi indicator of summer climate in the Baltic area," in *European Paleoclimate and Man, 8*, pp. 61–72, Gustav Fischer, Stuttgart, Germany, 1994.

[10] A. Tarand, *Tallinnas mõõdetud õhutemperatuuri aegrida*, vol. 93 of *Publicationes Geophysicales Universitatis Tartuensis*, 2003.

[11] *Eesti Meteoroloogia Aastaraamat 2013*, Keskkonnaagentuur, Tallinn, Estonia, 2014.

[12] H. Meitern, *Esimesed kestvamad ilmavaatlused Tartus 1821–1834*, Teaduse Ajaloo Lehekülgi Eestist VIII, Tallinn, Estonia, 1992.

[13] O. Kärner and H. Meitern, *Õhutemperatuuri muutustest Tartus viimase 200 aasta jooksul*, vol. 50, Publicationes Geophysicales Universitatis Tartuensis, 2006.

[14] C. Montgomery Douglas and C. R. Runge George, *Applied Statistics and Probability for Engineers*, E Wiley & Sons, New York, NY, USA, 5th edition, 2011.

[15] D. E. Parker, T. P. Legg, and C. K. Folland, "A new daily central England temperature series, 1772–1991," *International Journal of Climatology*, vol. 12, pp. 317–342, 1992.

[16] Met Office Hadley Centre Central England Temperature Data, http://www.metoffice.gov.uk/hadobs/hadcet/data/download .html.

[17] H. Tooming and J. Kadaja, Eds., *Handbook of Estonian Snow Cover*, Estonian Meteorological and Hydrological Institute, Tallinn, Estonia, 2006.

# 13

# Comparative Study of M5 Model Tree and Artificial Neural Network in Estimating Reference Evapotranspiration Using MODIS Products

**Armin Alipour, Jalal Yarahmadi, and Maryam Mahdavi**

*Department of Irrigation and Drainage Engineering, College of Aburaihan, University of Tehran, P.O. Box 33955-159, Pakdasht, Tehran, Iran*

Correspondence should be addressed to Armin Alipour; a.alipour@ut.ac.ir

Academic Editor: Ines Alvarez

Reference evapotranspiration ($\text{ET}_O$) is one of the major parameters affecting hydrological cycle. Use of satellite images can be very helpful to compensate for lack of reliable weather data. This study aimed to determine $\text{ET}_O$ using land surface temperature (LST) data acquired from MODIS sensor. LST data were considered as inputs of two data-driven models including artificial neural network (ANN) and M5 model tree to estimate $\text{ET}_O$ values and their results were compared with calculated $\text{ET}_O$ by FAO-Penman-Monteith (FAO-PM) equation. Climatic data of five weather stations in Khuzestan province, which is located in the southeastern Iran, were employed in order to calculate $\text{ET}_O$. LST data extracted from corresponding points of MODIS images were used in training of ANN and M5 model tree. Among study stations, three stations (Amirkabir, Farabi, and Gazali) were selected for creating the models and two stations (Khazaei and Shoeybie) for testing. In Khazaei station, the coefficient of determination ($R^2$) values for comparison between calculated $\text{ET}_O$ by FAO-PM and estimated $\text{ET}_O$ by ANN and M5 tree model were 0.79 and 0.80, respectively. In a similar manner, $R^2$ values for Shoeybie station were 0.86 and 0.85. In general, the results showed that both models can properly estimate $\text{ET}_O$ by means of LST data derived from MODIS sensor.

## 1. Introduction

Decline in availability of water for agriculture is one of the most serious challenges facing human life that has affected agricultural production in some arid and semiarid regions around the world. Determining future demands of water for agriculture section includes computation of several factors such as runoff, groundwater, precipitation, and evapotranspiration (ET) [1]. ET is identified as the combination of two different processes including evaporation from the soil surface and crop transpiration [2].

Accurate and reliable estimates of ET are necessary to determine temporal variations in irrigation requirement, improve allocation of water resources, and evaluate the effect of changes in land use and crop patterns on the water balance [3]. Considering difficulties in direct measurement of ET [4], this parameter is estimated through reference evapotranspiration ($\text{ET}_O$) and crop coefficient ($K_c$) for a specific crop [5].

Therefore, calculation of $\text{ET}_c$ (evapotranspiration of the given plant) and subsequently crop water requirement as irrigation water depend on $\text{ET}_O$ estimates. $\text{ET}_O$ is defined as the rate of ET from a reference surface in such a way that the surface is assumed to be covered with a hypothetical grass with specific characteristics [2]. A large number of methods have been offered in order to model $\text{ET}_O$ [2, 6, 7]; the majority of these methods are extremely complicated and rely on weather data such as temperature, solar radiation, wind speed, and air humidity [1, 8]. $\text{ET}_O$ is principally calculated by physically based equations (e.g., Penman-Monteith (PM) equation) or by empirical relationships between meteorological variables.

The FAO-Penman-Monteith (FAO-PM) method is currently recommended as an accepted standard technique for computing $\text{ET}_O$. This method employs a variety of complex equations, which are based on weather data, including air temperature, humidity, radiation, and wind speed [9]. The requirement of various weather data, which are not mainly available in many regions, is the major limitation of this method [10]; in addition, it seems that the reordered data are not accurate enough, especially in developing countries [11].

From the viewpoint of temporal and spatial variations of climatic data, weather stations density (the number of stations per unit of area) is not generally adequate for demonstrating spatiotemporal variations. Today, remote sensing techniques, which gather land surface information from a broader geographic scope, are known as an effective tool and methodology for extracting the ground parameters that can be used to estimate ET at regional scales [12]. To this end, in the basins that may face shortage of measured data, remote sensing data can considerably supply needed information such as albedo, leaf area index, and land surface temperature (LST) [13]. Recent advances in remote sensing techniques and enhancement of temporal, spatial, and radiometric resolution of satellite images have produced a range of useful products. These products can be widely used in various research areas, especially water and environment. Many studies have been conducted in estimating $ET_O$ as an important element of hydrological cycle. In this regard, it can be mentioned in the studies of Guerschman et al. [14]; Maeda et al. [4]; Yang et al. [12]; Cleugh et al. [15]; Tian et al. [16]; Hankerson et al. [17]; Li et al. [18].

Recently, some data-based techniques such as artificial neural networks (ANNs), support vector machine, and M5 model tree have been broadly utilized in solving practical problems. Mentioned techniques can simulate processes in which there is not an exact solution, and it is merely possible to approach near-optimal solution [19]. In these methods, a new knowledge is not produced about the process of a problem. In fact, thanks to the proper recognition of the problem process in selecting the input and output variables, caused by the use of modern regression techniques, this helps to better fit the data [20]. Data-driven techniques can efficiently model nonlinear and complicated processes such as ET that its precise calculation depends on numerous weather data and interactions between them [21].

Recently, by the developing new machine learning methods, application of new models in the estimation of $ET_O$ has been increased. Kişi and Öztürk [22] studied the adaptive neurofuzzy inference system (ANFIS) technique in estimation of $ET_O$ in Los Angeles, California, and showed that it could be successfully considered in the $ET_O$ estimation. Kişi and Çimen [23] applied SVM in modeling $ET_O$ in central California and concluded that SVR model could be embedded as a module for estimating $ET_O$ data in hydrological modeling studies. Tabari et al. [24] investigated the potential of SVM, ANFIS, multiple linear regression (MLR), and multiple nonlinear regression (MNLR) for estimating in a semiarid highland environment in Iran. They found that the SVM and ANFIS had better performance in $ET_O$ modeling compared to the others. Besides, several studies have shown high performance of ANNs in modeling ET against statistical methods [8, 25–29]. The substantial benefit of ANN methods, compared to conventional methods, is the ability of solving problems that are difficult to formalize [30]. Hou et al. [10] studied ANNs in simulating $ET_O$ values and demonstrated that the models reflected a noticeable efficiency. Landeras

et al. [21] compared calculated $ET_O$ by ANN to the $ET_O$ values of radiation and thermal models and declared that ANN can provide more accurate results than empirical models.

Due to the simple geometric structure of model tree and providing fast computing set of IF-THEN rules, which are easily understandable [31], the model tree has been utilized in different scientific studies. Pal and Deswal [1] in a study evaluated the capability of M5 model tree to model $ET_O$ in daily scale, and the obtained results showed that M5 model tree can be properly used in this context. Rahimikhoob et al. [11] conducted a study to assess the performance of M5 model tree in reaching $ET_O$ values in a semiarid region of Iran by means of $K_p$ values of pan evaporation. They asserted that this model shows more satisfactory performance than other approaches in estimating $ET_O$ in the study region. Sattari et al. [32] compared ANN and M5 model tree in determining $ET_O$ and concluded that both ANN model and M5 model tree can properly estimate $ET_O$. ANN suggested a better performance; however, M5 model tree offered simple linear regressions that can simply calculate $ET_O$.

The main aim of this study is to evaluate the ability of M5 model tree and ANN model in estimating daily $ET_O$ using LST values obtained from MODerate Resolution Imaging Spectroradiometer (MODIS)/Terra sensor. The accuracy of calculated $ET_O$ by means of M5 model tree and ANN model has been investigated by comparing with the $ET_O$ amounts of FAO-PM method.

## 2. Materials and Methods

*2.1. Study Area and Data.* Farms of Sugarcane Development Project, which are located in Khuzestan province, Iran, were considered as the study area in the present research. Khuzestan province is stretched in southeastern part of the country, with the area of approximately $63,238 \, km^2$ between longitudes of $47°38'$ E to $50°32'$ E and latitudes of $29°57'$ N to $33°00'$ N. This province is located adjoining the boundary of Iraq from the west and Persian Gulf from the south. Khuzestan province according to de Martonne climate classification has a semiarid climate. The average temperature during the winter season is nearly $9.14°C$, and occasionally its minimum falls down a few degrees below $0°C$. The average temperature in the summers is approximately $31.2°C$, whereas its maximum sometimes exceeds $50°C$. Required climatic variables in order to estimate $ET_O$ in this study were collected from weather stations, namely, Amirkabir, Farabi, Gazali, Khazaei, and Shoeybie for the years of 2006 and 2007 in the periods when satellite images existed. The meteorological data consisted of daily observations of maximum and minimum temperature, sunshine duration, relative humidity, and wind speed which have been used in estimating $ET_O$ by FAO-PM equation. Figure 1 illustrates the location of weather stations in the study area.

*2.2. MODIS Images.* In the current study, data of daytime LST ($LST_D$) and nighttime LST ($LST_N$) obtained from the MODIS sensor were used as inputs to the M5 model tree and ANN models. The MODIS sensor, which is on board

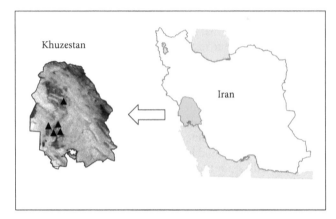

FIGURE 1: Location of weather stations in the study area.

of the Terra and Aqua satellites, is able to provide images in 36 spectral bands between 0.62 and 14.385 $\mu$m. Spatial resolutions of the sensor are 250 m, 500 m, and 1 km in several bands. In the thermal-infrared (TIR), MODIS sensors have 1 km spatial resolution [19]. This sensor performs a range of daily regular observations on the sea, land, and atmosphere in the global scale; moreover, it has an extensive and continuous spectral and spatial coverage. MODIS sensor was launched by the Terra and Aqua satellites in 1999 and 2002, respectively. These satellites both have a sun-synchronous, near polar, circular orbit; Terra satellite everyday crosses over the equator from north to south, while Aqua satellite has an inverse movement from south to north. United States National Aeronautics and Space Administration (NASA) is responsible for processing and providing MODIS sensor's data. NASA supplies the data in a large variety of products applicable to land, ocean, and atmosphere uses [4]. The data and products of MODIS sensor are available on the NASA's website (http://modis.gsfc.nasa.gov/). The images selected for this study are the level three (L3) data of the MODIS sensor (MOD11A1), which are related to Terra satellite. This product provides daytime and nighttime LST data collected on a 1 km spatial resolution and gridded in the sinusoidal projection system [33]. The MODIS sensor LST data is acquired from two TIR channels, that is, 31 (10.78–11.28 $\mu$m) and 32 (11.77–12.27 $\mu$m) using the split-window algorithm [34].

A total of 428 images of the MODIS sensor associated with MOD11A1 production from the Terra satellite in the time periods of 2006 and 2007 covering the study area were retrieved from the Land Processes Distributed Active Archive Center (LP DAAC) (http://lpdaac.usgs.gov). Cloudless sky and fair weather were the reason for selecting these pictures. Changing coordinate systems from the sinusoidal to UTM (datum WGS84) was performed using MODIS Reprojection Tool (MRT). Afterward, locations of study stations were projected on the images and then $LST_D$ and $LST_N$ were extracted from the corresponding pixels of the images. Extracting LST values in the desired stations was done by means of Hawth's Analysis Tools (HAT) in ArcGIS 9.3 software. Unit of LST values extracted from the images was Kelvin, so before

utilizing them as model input, these values were converted to Celsius.

In the present study, whole LST values extracted from MODIS images in three stations, including Amirkabir, Farabi, and Gazali, were used to construct and train the ANN and M5 model tree, and two other stations, Khazaei and Shoeybie, were applied to test mentioned models.

### 2.3. Artificial Neural Network (ANN).
ANNs are known as powerful machine-learning techniques, which are extensively used for numerical prediction and classification [35]. ANNs are also operative tools to model nonlinear systems and need less amount of data as inputs than customary mathematical approaches [27]. A neural network is based on processing a set of data that its analysis procedure has been inspired by biological neural systems. Neural networks consist of an interconnected group of neurons placed near to each other in layers that implement data processing through neurons and their connections. ANNs included one input layer, an output layer, and one or more hidden layers; neurons of each layer are joined to all neurons of the previous layer by means of weighted connections. The input vectors are received by neurons of the input layer and the values can be conveyed to the next layer of processing elements across connections. The number of neurons in each hidden layer is determined by the user. This process continues up to the output layer. ANN modeling is carried out in two stages including training and testing. In the training stage, after getting input data, neural network makes attempts to convert the inputs into desired outputs. Connecting weights of network neurons are determined in this stage; these weights in the testing stage are examined using various datasets.

A multilayer perceptron (MLP) ANN with one input layer, one hidden layer, and one output layer was used to estimate $ET_O$ from $LST_D$, $LST_N$, mean LST ($LST_M$), and day of year (DOY) data as inputs. The transfer function in the networks was log-sigmoid for this study. The accuracy of the networks was evaluated for each epoch in the training through mean squared error (MSE). The backpropagation (BP) algorithm was employed to train MLP neural network. Levenberg-Marquardt (LM), a second-order nonlinear optimization technique, was used with an early stopping criterion to improve the network training speed and efficiency. The target error for the training of networks was equal to $10^{-4}$. With the specification of this target error, the number of iteration for all ANN models was placed to 1000.

### 2.4. M5 Model Tree.
M5 model tree was introduced by Quinlan in 1992 [36]. The model is established according to a binary decision tree in which there are linear regression functions in the terminal node (leaf), which sets a relationship between independent and dependent variables [11]. Tree-based models are made according to a divide-and-conquer method for constructing a relationship between independent and dependent variables [37]. The tree models can be also used for qualitative and quantitative data [9].

Building the M5 model tree involves two separate steps [36, 38]. The first one consists of dividing data into subsets

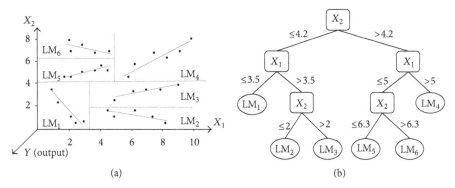

FIGURE 2: (a) Example of M5 model tree, a splitting the input space $x_1 \times x_2$ by M5 model tree algorithm and (b) diagram of model tree with six linear regression models at the leaves [11].

to construct the model tree. The dividing criterion is based on the standard deviation of the subset values that reach a node as a measure of error value in that node and additionally calculating the expected reduction in the error as a result of testing each attribute at that node. Equation of standard deviation reduction (SDR) is given as follows [1]:

$$SDR = SD(T) - \sum \frac{T_i}{T} \times SD(T_i), \qquad (1)$$

where $T$ is defined as a set of examples that reaches the node; $T_i$ demonstrates the subset of examples that have the $i$th output of the potential set, and SD is the standard deviation [39]. Due to branching process, data in child nodes (subtree or smaller nodes) have fewer SD than parent nodes (greater nodes). After examining all possible structures, a structure would be picked out that has the maximum expected error reduction. This dividing often creates a great tree-like structure that leads to overfit structure. In order to avoid overfitting, in the second step the overgrown tree is pruned, and pruned subtrees are replaced with linear regression functions. Figure 2(a) illustrates splitting the input space $x_1 \times x_2$ (independent variables) into six linear regression functions at the leaves (labeled LM1 through LM6) by M5 model tree algorithm. General form of the model is $y = a_0 + a_1 x_1 + a_2 x_2$, in which $a_0$, $a_1$, and $a_2$ are linear regression constants. Figure 2(b) shows branches relations in the form of a tree diagram [19].

*2.5. FAO-PM Equation.* FAO-PM equation, a physically based method, is widely used in the calculation of $ET_O$ and is a reliable method in the estimation of $ET_O$ in various climate conditions [2, 40, 41]. Moreover, FAO proposed this equation as a standard method for estimating $ET_O$. Although the most precise method for estimating $ET_O$ is to employ lysimeter data, considering the absence of such data in the study region, assessment of ANN and M5 model tree performance was made by FAO-PM equation. The equation for estimating daily $ET_O$ has been presented as follows [2]:

$$ET_O = \frac{0.484\Delta(R_n - G) + \gamma(900/(T_a + 273))U_2(e_s - e_a)}{\Delta + \gamma(1 + 0.34U_2)}, \qquad (2)$$

where $ET_O$ is the reference evapotranspiration (mm d$^{-1}$), $\Delta$ is the slope of the saturation vapor pressure curve (kPa °C$^{-1}$) at the daily mean air temperature (°C), $R_n$ and $G$ are the net solar radiation and soil heat flux density (MJ m$^{-2}$ d$^{-1}$), $\gamma$ is the psychrometric constant (kPa °C$^{-1}$), $T$ is the daily mean temperature (°C), $U_2$ is the mean wind speed (ms$^{-1}$), $e_s$ is the saturation vapor pressure (kPa), and $e_a$ is the actual vapor pressure (kPa) [2].

In current study, the daily values of $\Delta$, $R_n$, $e_s$, and $e_a$ were calculated using the equation given by Allen et al. [2]. For calculation of $R_n$, it is needed to calculate daily solar radiation ($R_s$). Allen et al. [2] proposed $R_s$ to be calculated using the Angstrom formula, which relates solar radiation to extraterrestrial radiation ($R_a$) and relative sunshine duration. The formula has the following form:

$$R_s = \left(a + b\frac{n}{N}\right)R_a, \qquad (3)$$

where $n$ and $N$ are, respectively, the actual daily sunshine duration and daily maximum possible sunshine duration, $R_s$ and $R_a$ are, respectively, the daily global solar radiation (MJ m$^{-2}$ d$^{-1}$) and the daily extraterrestrial solar radiation (MJ m$^{-2}$ d$^{-1}$) on a horizontal surface, and $a$ and $b$ are empirical coefficients which depend on location. The values of $a$ and $b$ coefficients are considered as default values, 0.25 and 0.5, respectively, proposed by Allen et al. [2].

*2.6. Statistical Analyses.* In this study, $ET_O$ results obtained from FAO-PM equation were utilized as reference values and the results achieved by ANN and M5 model tree were taken into account as estimated values. The comparison of ANN model and M5 model tree values with FAO-PM equation was made using statistical indices such as coefficient of determination ($R^2$), root mean square error (RMSE), and mean absolute error (MAE), in addition to plotting graphs. These indices are defined as follows:

$$R^2 = \left(\frac{\sum_{i=1}^{n}(x_i - \bar{x})(y_i - \bar{y})}{\sqrt{\sum_{i=1}^{n}(x_i - \bar{x})^2 \sum_{i=1}^{n}(y_i - \bar{y})^2}}\right)^2$$

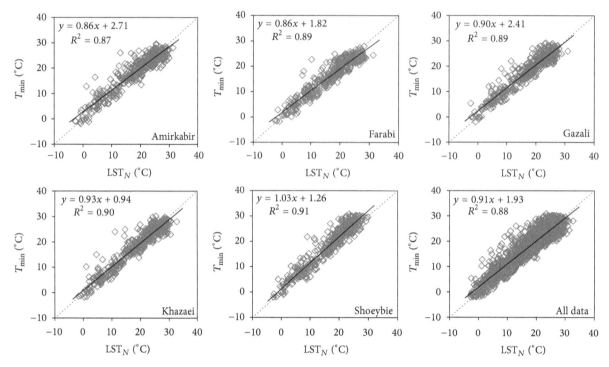

FIGURE 3: Scatter plot of $T_{min}$ against $LST_N$.

$$RMSE = \sqrt{\frac{1}{n}\sum_{i=1}^{n}(x_i - y_i)^2}$$

$$MAE = \frac{1}{n}\sum_{i=1}^{n}|x_i - y_i|,$$

$$(4)$$

where $x_i$ is $ET_O$ computed by FAO-PM, $\overline{x}$ denotes the average amount of computed $ET_O$ computed by FAO-PM, $y_i$ is $ET_O$ computed by ANN/M5, and $\overline{y}$ is the average computed $ET_O$ amounts by last models.

## 3. Results and Discussion

In the first stage of the study, before calculating the $ET_O$, the correlation between LST obtained from MODIS sensor and air temperature was determined. Accordingly, scatter plots of LST values provided by MODIS images versus air temperature are presented for different stations in Figures 3 and 4. As it can be perceived in Figure 3, there is a high correlation between the $LST_N$ and minimum air temperature ($T_{min}$). The greatest $R^2$ was observed between $LST_N$ and $T_{min}$ in Shoeybie station, which was equal to 0.91; the minimum $R^2$ was seen in the Amirkabir station with the value of 0.87. Furthermore, for all stations, $R^2$ of two aforementioned parameters was entirely assessed at 0.88. In Figure 3, symmetrical and appropriate distribution of points around the line of best fit (line 1:1) indicates a strong correlation between the two parameters. In Figure 4, relation between the $LST_D$ and maximum air temperature ($T_{max}$) displays that the greatest $R^2$ between $LST_D$ and $T_{max}$ is related to Amirkabir and Khazaei stations

($R^2 = 0.91$); the minimum amount of this coefficient is associated with Gazali station ($R^2 = 0.81$). For all study stations, $R^2$ was generally equal to 0.78. Overall, it can be concluded that there is a strong relationship between these two parameters. The scatter of points around the line of best fit in the Figure 4 depicts an overestimation of $LST_D$ over $T_{max}$. This overestimation is more pronounced for the temperatures over 30°C.

For all examined datasets, the correlation between $LST_N$ and $T_{min}$ is higher compared to the correlation between $LST_D$ and $T_{max}$ for thermal products of MODIS sensor. Moreover, previous studies concerning estimation of air temperature have similarly confirmed that correlation between $LST_N$ and $T_{min}$ is greater than correlation between $LST_D$ and $T_{max}$ [12, 19].

In the second stage, in order to calculate $ET_O$ using LST values extracted from the MODIS sensor ($LST_D$ and $LST_N$), ANN models and M5 model tree were employed. In addition to $LST_D$ and $LST_N$, two parameters including mean LST ($LST_M$) and day of year (DOY) were considered as inputs for generating model in order to increase accuracy of $ET_O$ estimates. Thus, in producing ANN model and M5 model tree, four parameters including $LST_D$, $LST_N$, DOY, and $LST_M$ were taken into consideration as input and $ET_O$ calculated from weather data using FAO-PM equation as output.

For generating ANN model, training data (whole data of Amirkabir, Farabi, and Gazali stations) were applied, and the number of nodes in the hidden layer was obtained using trial and error method. The number of nodes in the hidden layer was considered varying from 1 to 12; based on error statistical parameters, a network with 3 nodes in the hidden layer indicated the lowest error in comparison of $ET_O$ values

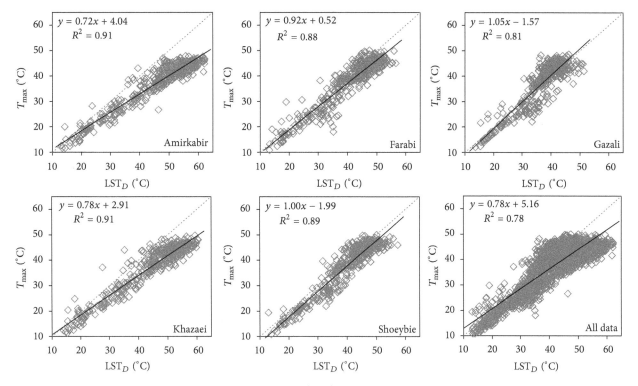

FIGURE 4: Scatter plot of $T_{max}$ against $LST_D$.

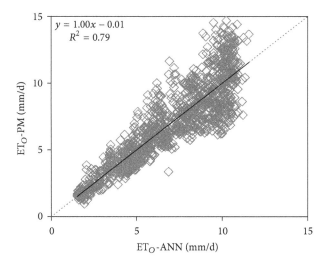

FIGURE 5: Scatter plot of estimated $ET_O$ using ANN against calculated $ET_O$ by FAO-PM equation for training data.

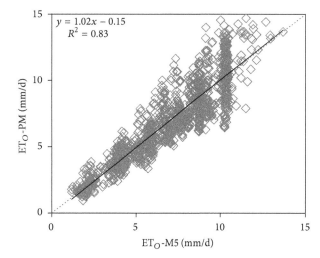

FIGURE 6: Scatter plot of estimated $ET_O$ using M5 model tree against calculated $ET_O$ by FAO-PM equation for training data.

calculated by FAO-PM method. Table 1 presents the values of error parameters for the neural network with the structure of three nodes in the hidden layer. In Figure 5, scatter diagram of calculated $ET_O$ values through ANN for all training data has been plotted against $ET_O$ values obtained by FAO-PM equation. As shown in Figure 3, $R^2 = 0.79$ demonstrates that the ANN has provided an acceptable determination coefficient in approximating $ET_O$. In this figure, the points density increases around the line 1 : 1 in large amounts of $ET_O$; that is, it represents the overestimation or underestimation of the data that are situated in this area.

TABLE 1: Error statistics in estimating $ET_O$ for model tree and ANN methods using training data.

| Model | RMSE (mm/d) | MAE (mm/d) |
|-------|-------------|------------|
| ANN | 1.40 | 1.05 |
| M5 | 1.27 | 0.96 |

For generating M5 model tree, WEKA 3.7 software [42] was used considering the same data applied to the training ANN model. Algorithm 1 shows M5 model tree produced

LSTn <= 19.43:
  LSTms <= 15.55: LM1
  LSTms > 15.55:
    DOY <= 282.5:
      DOY <= 81.5: LM2
      DOY > 81.5:
        DOY <= 106:
          LSTd <= 31.2: LM3
          LSTd > 31.2: LM4
        DOY > 106:
          DOY <= 207:
            LSTn <= 17.72:
              DOY <= 129: LM5
              DOY > 129: LM6
            LSTn > 17.72: LM7
          DOY > 207:
            DOY <= 272.5: LM8
            DOY > 272.5: LM9
    DOY > 282.5:
      DOY <= 311.5: LM10
      DOY > 311.5: LM11
LSTn > 19.43:
  DOY <= 227.5:
    DOY <= 159.5:
      DOY <= 147.5: LM12
      DOY > 147.5: LM13
    DOY > 159.5:
      DOY <= 198.5:
        DOY <= 167.5: LM14
        DOY > 167.5: LM15
      DOY > 198.5:
        DOY <= 216.5:
          DOY <= 206.5: LM16
          DOY > 206.5: LM17
        DOY > 216.5: LM18
  DOY > 227.5: LM19

LM num: 1
$$Pm = 0.0147 * Tn + 0.1048 * Tms - 0.0007 * DOY + 0.8963$$
LM num: 2
$$Pm = 0.0349 * Tn + 0.0312 * Tms + 0.028 * DOY + 1.3107$$
LM num: 3
$$Pm = 0.046 * Tn + 0.081 * Tms - 0.0134 * DOY + 3.9221$$
LM num: 4
$$Pm = 0.0719 * Tn + 0.0642 * Tms - 0.0088 * DOY + 3.8864$$
LM num: 5
$$Pm = 0.0705 * Tn + 0.0073 * Td + 0.0301 * Tms + 0.0003 * DOY + 4.6615$$
LM num: 6
$$Pm = 0.201 * Tn + 0.0147 * Td + 0.0301 * Tms + 0.0003 * DOY + 2.5252$$
LM num: 7
$$Pm = 0.118 * Tn + 0.0301 * Tms + 0.0026 * DOY + 4.7479$$
LM num: 8
$$Pm = 0.0529 * Tn + 0.0301 * Tms - 0.0087 * DOY + 6.254$$
LM num: 9
$$Pm = 0.0529 * Tn - 0.0348 * Tms - 0.0116 * DOY + 8.4773$$
LM num: 10
$$Pm = 0.0328 * Tn + 0.0201 * Tms - 0.0012 * DOY + 3.367$$
LM num: 11
$$Pm = 0.1437 * Tn + 0.0201 * Tms - 0.0012 * DOY + 1.7414$$
LM num: 12
$$Pm = -0.1076 * Tn + 0.0018 * Tms - 0.0393 * DOY + 15.9787$$
LM num: 13
$$Pm = 0.0297 * Tn + 0.0018 * Tms + 0.0032 * DOY + 8.0036$$
LM num: 14
$$Pm = 0.3445 * Tn + 0.0769 * Td + 0.0018 * Tms - 0.1664 * DOY + 26.3097$$
LM num: 15
$$Pm = 0.0316 * Tn + 0.0018 * Tms - 0.006 * DOY + 10.5743$$
LM num: 16
$$Pm = -0.0175 * Tn + 0.0018 * Tms + 0.0225 * DOY + 4.5799$$
LM num: 17
$$Pm = -0.206 * Tn + 0.0018 * Tms + 0.0185 * DOY + 11.1645$$
LM num: 18
$$Pm = 0.0015 * Tn + 0.0018 * Tms - 0.0046 * DOY + 9.6864$$
LM num: 19
$$Pm = 0.022 * Tn + 0.0018 * Tms - 0.0387 * DOY + 16.0569$$

ALGORITHM 1: Tree-structured regression model generated by M5 algorithm using training data.

by the software and established linear regression relations. Afterward, a scatter plot of estimated $ET_O$ by M5 model tree versus calculated $ET_O$ by FAO-PM equation has been presented for training data (Figure 6); $R^2$ value equal to 0.83 indicates proper agreement. The distribution of points around the best fit line represents higher accuracy of this method in the estimation of low $ET_O$ values; in the great values of $ET_O$ the error amount increases as well.

The comparison of error statistics concerning developed models by ANN model and M5 model tree in estimating $ET_O$ using training data can be drawn in Table 1. As seen, M5 model tree, which has shown RMSE and MAE values, respectively, equal to 1.27 and 0.96 mm d$^{-1}$, has resulted in higher accuracy than ANN in estimating $ET_O$.

For testing the methods of ANN and M5 model tree, data of Khazaei and Shoeybie stations have been used. In Figure 7, the scatter plot of calculated $ET_O$ through two models of ANN and M5 model tree versus FAO-PM values for the mentioned stations have been shown. $R^2$ values associated with calculated $ET_O$ by ANN against FAO-PM equation $ET_O$ values for Khazaei and Shoeybie stations were 0.79 and 0.86, respectively. Additionally, the same $R^2$ values related to M5 model tree for aforementioned stations were 0.80 and 0.85, respectively. In both methods, $ET_O$ values were properly distributed around line 1 : 1. For smaller amounts of $ET_O$, there was been a better compatibility between the calculated values by two models and FAO-PM values compared to greater amounts of $ET_O$. In other words, underestimation or overestimation of $ET_O$ is related to warm seasons further than cold seasons. The evolutions of calculated $ET_O$ values over time have been illustrated for two test stations in Figure 8. As perceived in this graph, the maximum amounts of $ET_O$ by these two models were not entirely compatible with FAO-PM values. This could be due to the lack of consideration of some

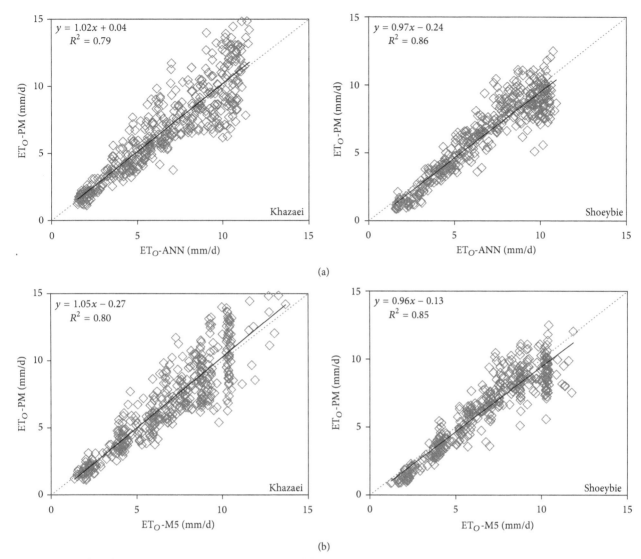

FIGURE 7: Scatter plot of estimated $ET_O$ using ANN and M5 model tree against calculated $ET_O$ by FAO-PM method, (a) ANN method, and (b) M5 method.

weather parameters affecting $ET_O$ rate (e.g., wind speed, solar radiation, and relative humidity) as inputs of ANN model and M5 model tree.

In Table 2, error statistics of $ET_O$ estimation by ANN and M5 model tree have been presented for two test stations. According to the statistics of RMSE and MAE, both methods performed equally well. Similar results have been reported about close performance of ANN and M5 models in rainfall-runoff modeling [43] and river flow forecasting [44]. The main advantage of M5 model tree is that it provides the results in a simple and comprehensible form of regression equations [1], which can be easily used in the calculation of $ET_O$.

It seems generally that $LST_N$ and $LST_D$ values obtained from MODIS sensor, making the use of both ANN and M5 models, can be properly utilized in the estimation of $ET_O$.

## 4. Conclusion

In this study, attempts were made to calculate $ET_O$ using LST values extracted from images of MODIS sensor. For this

TABLE 2: Error statistics in estimating $ET_O$ by ANN and M5 model tree using testing data.

| Station | Model | $R^2$ | RMSE (mm/d) | MAE (mm/d) |
|---------|-------|-------|-------------|------------|
| Khazaei | ANN | 0.79 | 1.54 | 1.14 |
| | M5 | 0.80 | 1.49 | 1.12 |
| Shoeybie | ANN | 0.86 | 1.15 | 0.90 |
| | M5 | 0.85 | 1.17 | 0.90 |

purpose, two methods of ANN and M5 model tree were selected in order to estimate $ET_O$; four parameters including $LST_D$, $LST_N$, $LST_M$, and DOY were considered as model inputs, and $ET_O$ was calculated from FAO-PM equation as model outputs. A comparison of estimated $ET_O$ using each mentioned model with the values obtained from the FAO-PM method revealed high accuracy of both methods. The average amounts of RSME in two test stations for ANN and M5 model tree were, respectively, 1.35 and 1.34 mm/day. According to the error values, it can be observed that both

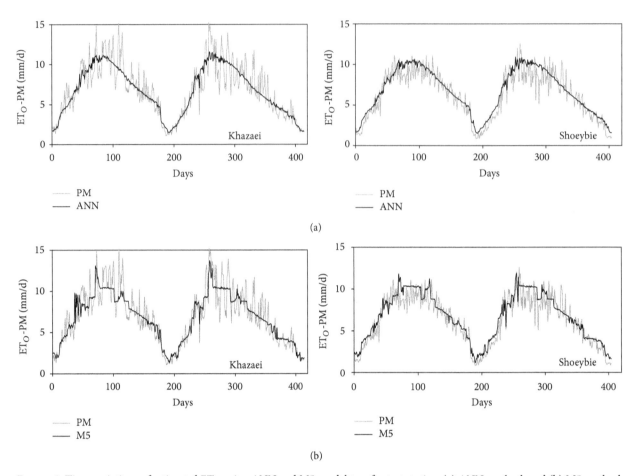

FIGURE 8: Time variations of estimated $ET_O$ using ANN and M5 model tree for test statins, (a) ANN method, and (b) M5 method.

models have shown close performance in determining $ET_O$. It seems that considering other significant variables affecting the ET process such as relative humidity and wind speed (as model inputs) can lead to higher accuracy.

Overall, it can be concluded that, unlike complex ANN method, M5 model tree due to its simple structure and providing simple regression equations, is a more appropriate method in determining $ET_O$ using LST data derived from MODIS sensor.

In this paper A M5 tree model with some simple linear regressions was developed to estimate $ET_O$ using MODIS LST data. Considering simple linear equations and minor data required in the developed M5 tree model, it is expected that this method could be an alternative to calculate $ET_O$ in the study area. However, this model provided acceptable results in the current area but it seems that its applicability in the other region of the world or the region with similar climate condition needs more investigation.

## Conflict of Interests

The authors declare that there is no conflict of interests regarding the publication of this paper.

## References

[1] M. Pal and S. Deswal, "M5 model tree based modelling of reference evapotranspiration," *Hydrological Processes*, vol. 23, no. 10, pp. 1437–1443, 2009.

[2] R. G. Allen, L. S. Pereira, D. Raes, and M. Smith, "Crop evapotranspiration—guidelines for computing crop water requirements," FAO Irrigation and Drainage Paper 56, FAO, Rome, Italy, 1998.

[3] S. Ortega-Farias, S. Irmak, and R. H. Cuenca, "Special issue on evapotranspiration measurement and modeling," *Irrigation Science*, vol. 28, no. 1, pp. 1–3, 2009.

[4] E. E. Maeda, D. A. Wiberg, and P. K. E. Pellikka, "Estimating reference evapotranspiration using remote sensing and empirical models in a region with limited ground data availability in Kenya," *Applied Geography*, vol. 31, no. 1, pp. 251–258, 2011.

[5] A. A. Sabziparvar, H. Tabari, A. Aeini, and M. Ghafouri, "Evaluation of class a pan coefficient models for estimation of reference crop evapotranspiration in cold semi-arid and warm arid climates," *Water Resources Management*, vol. 24, no. 5, pp. 909–920, 2010.

[6] M. E. Jensen, R. D. Burman, and R. G. Allen, "Evapotranspiration and irrigation water requirements," ASCEM Annuals and Reports on Engineering Practice no. 70, ASCE, New York, NY, USA, 1990.

[7] M. Smith, *Report on the Expert Consultation on Revision of FAO Methodologies for Crop Water Requirements*, FAO, Rome, Italy, 1992.

[8] M. Kumar, N. S. Raghuwanshi, R. Singh, W. W. Wallender, and W. O. Pruitt, "Estimating evapotranspiration using artificial neural network," *Journal of Irrigation and Drainage Engineering*, vol. 128, no. 4, pp. 224–233, 2002.

[9] P. Ditthakit and C. Chinnarasri, "Estimation of pan coefficient using M5 model tree," *American Journal of Environmental Sciences*, vol. 8, no. 2, pp. 95–103, 2012.

[10] Z. Huo, S. Feng, S. Kang, and X. Dai, "Artificial neural network models for reference evapotranspiration in an arid area of northwest China," *Journal of Arid Environments*, vol. 82, pp. 81–90, 2012.

[11] A. Rahimikhoob, M. Asadi, and M. Mashal, "A comparison between conventional and M5 model tree methods for converting pan evaporation to reference evapotranspiration for semi-arid region," *Water Resources Management*, vol. 27, no. 14, pp. 4815–4826, 2013.

[12] Y. Yang, S. Shang, and L. Jiang, "Remote sensing temporal and spatial patterns of evapotranspiration and the responses to water management in a large irrigation district of North China," *Agricultural and Forest Meteorology*, vol. 164, pp. 112–122, 2012.

[13] Z. Wan, "New refinements and validation of the MODIS land-surface temperature/emissivity products," *Remote Sensing of Environment*, vol. 112, no. 1, pp. 59–74, 2008.

[14] J. P. Guerschman, A. I. J. M. van Dijk, G. Mattersdorf et al., "Scaling of potential evapotranspiration with MODIS data reproduces flux observations and catchment water balance observations across Australia," *Journal of Hydrology*, vol. 369, no. 1-2, pp. 107–119, 2009.

[15] H. A. Cleugh, R. Leuning, Q. Mu, and S. W. Running, "Regional evaporation estimates from flux tower and MODIS satellite data," *Remote Sensing of Environment*, vol. 106, no. 3, pp. 285–304, 2007.

[16] H. Tian, J. Wen, C.-H. Wang, R. Liu, and D.-R. Lu, "Effect of pixel scale on evapotranspiration estimation by remote sensing over oasis areas in north-western China," *Environmental Earth Sciences*, vol. 67, no. 8, pp. 2301–2313, 2012.

[17] B. Hankerson, J. Kjaersgaard, and C. Hay, "Estimation of evapotranspiration from fields with and without cover crops using remote sensing and in situ methods," *Remote Sensing*, vol. 4, no. 12, pp. 3796–3812, 2012.

[18] X. Li, L. Lua, W. Yangc, and G. Chenga, "Estimation of evapotranspiration in an arid region by remote sensing—a case study in the middle reaches of the Heihe River Basin," *International Journal of Applied Earth Observation and Geoinformation*, vol. 17, no. 1, pp. 85–93, 2012.

[19] S. Emamifar, A. Rahimikhoob, and A. A. Noroozi, "Daily mean air temperature estimation from MODIS land surface temperature products based on M5 model tree," *International Journal of Climatology*, vol. 33, no. 15, pp. 3174–3181, 2013.

[20] B. Bhattacharya and D. P. Solomatine, "Neural networks and M5 model trees in modelling water level-discharge relationship," *Neurocomputing*, vol. 63, pp. 381–396, 2005.

[21] G. Landeras, A. Ortiz-Barredo, and J. J. López, "Comparison of artificial neural network models and empirical and semi-empirical equations for daily reference evapotranspiration estimation in the Basque Country (Northern Spain)," *Agricultural Water Management*, vol. 95, no. 5, pp. 553–565, 2008.

[22] Ö. Kişi and Ö. Öztürk, "Adaptive neurofuzzy computing technique for evapotranspiration estimation," *Journal of Irrigation and Drainage Engineering*, vol. 133, no. 4, pp. 368–379, 2007.

[23] O. Kişi and M. Çimen, "Evapotranspiration modelling using support vector machines," *Hydrological Sciences Journal*, vol. 54, no. 5, pp. 918–928, 2009.

[24] H. Tabari, O. Kisi, A. Ezani, and P. H. Talaee, "SVM, ANFIS, regression and climate based models for reference evapotranspiration modeling using limited climatic data in a semi-arid highland environment," *Journal of Hydrology*, vol. 444-445, pp. 78–89, 2012.

[25] P. B. Shirsath and A. K. Singh, "A comparative study of daily pan evaporation estimation using ANN, regression and climate based models," *Water Resources Management*, vol. 24, no. 8, pp. 1571–1581, 2010.

[26] S. S. Zanetti, E. F. Sousa, V. P. S. Oliveira, F. T. Almeida, and S. Bernardo, "Estimating evapotranspiration using artificial neural network and minimum climatological data," *Journal of Irrigation and Drainage Engineering*, vol. 133, no. 2, pp. 83–89, 2007.

[27] A. Rahimikhoob, "Estimating sunshine duration from other climatic data by artificial neural network for $ET_0$ estimation in an arid environment," *Theoretical and Applied Climatology*, vol. 118, no. 1-2, pp. 1–8, 2014.

[28] I. El-Baroudy, A. Elshorbagy, S. K. Carey, O. Giustolisi, and D. Savic, "Comparison of three data-driven techniques in modelling the evapotranspiration process," *Journal of Hydroinformatics*, vol. 12, no. 4, pp. 365–379, 2010.

[29] M.-B. Aghajanloo, A.-A. Sabziparvar, and P. Hosseinzadeh Talaee, "Artificial neural network-genetic algorithm for estimation of crop evapotranspiration in a semi-arid region of Iran," *Neural Computing and Applications*, vol. 23, no. 5, pp. 1387–1393, 2013.

[30] K. P. Sudheer, A. K. Gosain, and K. S. Ramasastri, "Estimating actual evapotranspiration from limited climatic data using neural computing technique," *Journal of Irrigation and Drainage Engineering*, vol. 129, no. 3, pp. 214–218, 2003.

[31] V. Jothiprakash and A. S. Kote, "Effect of pruning and smoothing while using m5 model tree technique for reservoir inflow prediction," *Journal of Hydrologic Engineering*, vol. 16, no. 7, pp. 563–574, 2011.

[32] M. T. Sattari, M. Pal, K. Yürekli, and A. Ünlükara, "M5 model trees and neural network based modelling of $ET_0$ in Ankara, Turkey," *Turkish Journal of Engineering & Environmental Sciences*, vol. 37, no. 2, pp. 211–219, 2013.

[33] K. Wang, Z. Wan, P. Wang et al., "Estimation of surface long wave radiation and broadband emissivity using moderate resolution imaging spectroradiometer (MODIS) land surface temperature/emissivity products," *Journal of Geophysical Research D: Atmospheres*, vol. 110, no. 11, pp. 1–12, 2005.

[34] C. Vancutsem, P. Ceccato, T. Dinku, and S. J. Connor, "Evaluation of MODIS land surface temperature data to estimate air temperature in different ecosystems over Africa," *Remote Sensing of Environment*, vol. 114, no. 2, pp. 449–465, 2010.

[35] K. K. Singh, M. Pal, and V. P. Singh, "Estimation of mean annual flood in indian catchments using backpropagation neural network and M5 model tree," *Water Resources Management*, vol. 24, no. 10, pp. 2007–2019, 2010.

[36] J. R. Quinlan, "Learning with continuous classes," in *Proceedings of the 5th Australian Joint Conference on Artificial Intelligence (AI '92)*, pp. 343–348, World Scientific, Singapore, 1992.

[37] M. Pal, "M5 model tree for land cover classification," *International Journal of Remote Sensing*, vol. 27, no. 4, pp. 825–831, 2006.

[38] D. P. Solomatine and Y. Xue, "M5 model trees and neural networks: application to flood forecasting in the upper reach of the Huai River in China," *Journal of Hydrologic Engineering*, vol. 9, no. 6, pp. 491–501, 2004.

[39] Y. Wang and I. H. Witten, "Induction of model trees for predicting continuous lasses," in *Proceedings of the Poster Papers of the European Conference on Machine Learning*, University of Economics, Faculty of Informatics and Statistic, Prague, Czech Republic, 1997.

[40] D. Itenfisu, R. L. Elliott, R. G. Allen, and I. A. Walter, "Comparison of reference evapotranspiration calculations as part of the ASCE standardization effort," *Journal of Irrigation and Drainage Engineering*, vol. 129, no. 6, pp. 440–448, 2003.

[41] P. Gavilán, I. J. Lorite, S. Tornero, and J. Berengena, "Regional calibration of Hargreaves equation for estimating reference et in a semiarid environment," *Agricultural Water Management*, vol. 81, no. 3, pp. 257–281, 2006.

[42] I. H. Witten and E. Frank, *Data Mining: Practical Machine Learning Tools and Techniques with Java Implementations*, Morgan Kaufmann, San Francisco, Calif, USA, 2005.

[43] D. P. Solomatine and K. N. Dulal, "Model trees as an alternative to neural networks in rainfall—runoff modelling," *Hydrological Sciences Journal*, vol. 48, no. 3, pp. 399–411, 2003.

[44] S. Londhe and S. Charhate, "Comparison of data-driven modelling techniques for river flow forecasting," *Hydrological Sciences Journal*, vol. 55, no. 7, pp. 1163–1174, 2010.

# 14

# Species Favourability Shift in Europe due to Climate Change: A Case Study for *Fagus sylvatica* L. and *Picea abies* (L.) Karst. Based on an Ensemble of Climate Models

**Wolfgang Falk[1] and Nils Hempelmann[2]**

[1] *Bavarian State Institute of Forestry (LWF), Department Soil and Climate, Hans-Carl-von-Carlowitz-Platz 1, 85354 Freising, Germany*

[2] *Helmholtz-Zentrum Geesthacht Zentrum für Material- und Küstenforschung GmbH, Climate Service Center (CSC), Chilehaus, Fischertwiete 1, 20095 Hamburg, Germany*

Correspondence should be addressed to Wolfgang Falk; wolfgang.falk@lwf.bayern.de

Academic Editors: A. V. Eliseev, T. G. Huntington, and A. Rutgersson

Climate is the main environmental driver determining the spatial distribution of most tree species at the continental scale. We investigated the distribution change of European beech and Norway spruce due to climate change. We applied a species distribution model (SDM), driven by an ensemble of 21 regional climate models in order to study the shift of the favourability distribution of these species. SDMs were parameterized for 1971–2000, as well as 2021–2050 and 2071–2100 using the SRES scenario A1B and three physiological meaningful climate variables. Growing degree sum and precipitation sum were calculated for the growing season on a basis of daily data. Results show a general north-eastern and altitudinal shift in climatological favourability for both species, although the shift is more marked for spruce. The gain of new favourable sites in the north or in the Alps is stronger for beech compared to spruce. Uncertainty is expressed as the variance of the averaged maps and with a density function. Uncertainty in species distribution increases over time. This study demonstrates the importance of data ensembles and shows how to deal with different outcomes in order to improve impact studies by showing uncertainty of the resulting maps.

## 1. Introduction

Climate change affects tree species because trees are sessile, long lived, and comparatively slow in reacting to a changing environment, for example, trees cannot easily colonize new habitats in a strongly fragmented landscape. Because of the slow growth rates of trees to reach maturity, the typical planning horizon in the forestry sector is, for example, for spruce 60 to 80 years in the future. Hence, forestry has to deal with the challenges of a rapidly changing environment in order to avoid breakdown of stands. The same is true for city planners or municipal authorities with regard to tree species. Therefore, forest owners and related professionals need a reliable assessment about shifts of species distribution in order to make management decisions [1].

Even though genetic resources are considered as a key factor of forest ecosystems to adapt to climate change [2], one has to keep in mind that adaptation processes are slow, that niche conservatism is described for some genera [3], and that species will not survive or at least will not have a positive population growth rate if environmental conditions at a site will not meet the physiological requirement of a species.

Species distribution models (SDMs) can be used to gain ecological insights and to predict distributions across landscapes and to extrapolate in space and time [4]. SDMs relate distribution data (e.g., occurrences at known locations) with information on the environmental conditions at these sites [4]. On a continental scale, tree species distribution is mainly caused by climate variables [5] and only on a finer scale by soil or terrain. For example, Coudun et al. showed that soil nutritional factors improve distribution models of species with a high demand of alkaline cations in the soil solution like *Acer campestre* [6]. Additionally, tree species distribution is influenced by management which can result in over- or underdispersion of some species. Reference [7], for example, reviews major effects on beech distribution in north-eastern Europe on the one hand due to conversion of beech stands into agricultural land and later into coniferous forests since the medieval epoch and on the other hand due to planting and promotion of beech near or even outside the reputed distribution range in Prussian times. The SDM approach assumes equilibrium between species distribution and environmental conditions. Problems might arise in some areas where a species is not present due to human impact or when the postglacial migration is not completed (underdispersion). If those areas have comparable environmental conditions as areas with presences, a model prediction for those areas is possible.

SDMs most often use presence-absence data, the model output is a probability of occurrence. This probability can be transformed into binomial data with the help of a threshold value (e.g., [8]). According to Real et al. [9], the probabilities do not only depend on the values of the predictor variables but also on the relative proportion of presences and absences. In order to overcome this effect and to describe the environmental favourability of a site for or against a species, Real et al. [9] suggest the use of a favourability function to receive results that are independent of an unbalanced proportion of presences and absences. Especially beech has more absences than presences in the data base; therefore the favourability transformation is used in this study. Norway spruce (*Picea abies* (L.) Karst.) and European beech (*Fagus sylvatica* L.), two of the most important tree species of the humid zone of Europe referring to forestry, are examined in this study. Spruce is considered to be strongly affected by the climate change, whereas beech is considered to be a possible suitable alternative as sites become dryer and warmer in central Europe [1]. Both species are relatively unspecific to soil and terrain conditions [10, 11]; therefore, their distribution is mainly determined by climatic constraints. Thus, our study concentrates on the relation of climate and distribution. Plant growth is limited amongst others by low temperatures (e.g., [12]) and, therefore, precipitation sum (*PsumVeg*) and temperature sum (*TsumVeg*) are calculated in respect of the start and end of the growing season. Beside these two variables, minimum temperature of the dormancy (*TminDorm*) is used to describe species distribution.

Projected changes in temperature and precipitation patterns differ within seasons in Europe [13, 14]. Furthermore, the annual cycle changes, that is, the vegetation period is enlarging [15, 16] and has to be represented by chosen climate indices (CI). Furthermore, plant growth is limited amongst others by low temperatures (e.g., [12]) and therefore physiological meaningful indices need to be related to a period of higher temperatures. If bud burst is earlier and leaf falls later in the year, the indices forcing the SDM have to take the start and end date of the vegetation period into account. Thus, forcing values are related to length of vegetation period in this study. The terms vegetation period and growing season are used synonymously and refer to a climatological definition.

When using climate data for SDM approaches, the question of appropriate climate data or models has to be solved.

There are a variety of data for present conditions or future climate projections [17–19], calculated with different models, bounding conditions, and assumptions. Many studies on species distribution refer to the WorldClim database [20, 21], because these worldwide data have a very high spatial resolution [17]. WorldClim data are interpolated climate surfaces for global land areas. Future data are calculated with a statistical downscaling of the output of global circulation models (GCM) and these anomalies are applied to the WorldClim data. In general, it is hard to judge which data set is the best [22]. Therefore, the use of an ensemble of data seems to be the state-of-the-art solution to the question of the right data set. This is especially true for future climate data because there are many different data experiments with global circulation models (GCM) which in turn are forcing different dynamic regional circulation models (RCM). The GCMs are set up under defined assumptions, the so-called scenarios (Special Report on Emissions Scenarios, SRES). In this study, only the SRES A1B scenario [23] and one SDM technique has been considered in order to concentrate on differences of circulation models and downscaling methods and to show their effects on SDM output. The A1 storyline and family describe a future world of very rapid economic growth, global population that peaks in mid-century and declines thereafter, and the rapid introduction of new and more efficient technologies [23]. The technological emphasis of A1B is a balance across all sources. That means that similar improvement rates apply to all energy supply and end use technologies [23]. Cumulative emissions of greenhouse gases are in a broad range and in the middle of the different scenarios and families [23]. Due to this, A1B is often used in climate impact studies (e.g., [24]).

The framework from assumptions of boundary conditions for climate change projection to final merged results of spatial favourability distribution for tree species is given by a chain of separate steps:

$$\text{SRES} \longrightarrow \text{GCM} \longrightarrow \text{RCM} \longrightarrow \text{CI} \longrightarrow \text{SDM}. \qquad (1)$$

For every chain member, several different models, approaches, or assumptions can be taken into account and enlarge the number of ensemble members. Due to the usage of data and model ensembles, a corresponding number of SDM results had to be merged. This final step can be taken as an additional chain member.

In this study, we focus on the methodological framework due to the chain of steps and present an approach focusing on ensembles of different GCMs and RCMs. Investigation area is Europe, distribution models are built by means of climate data from the period 1971–2000, and maps are calculated for this period and two future timesteps (2021–2050 and 2071–2100). The study is focusing on the following issues:

(1) shift of climatic favourability for European beech and Norway spruce in Europe;

(2) incorporation of ensembles of gridded and daily resolved climate model variables into a SDM framework;

(3) accentuation of uncertainties of the final results that derive from the use of an ensemble of climate models.

## 2. Data and Data Preprocessing

The presented SDM approach is driven by downscaled regional climate model data. The data are taken from the ENSEMBLE-Project [13, 19] where data experiments with different RCMs, forced by different GCMs, are available. Within the ENSEMBLES-Project 22, data experiments have been realized with regional climate over Europe, all following the assumptions of the IPCC emission scenario A1B [23]. The experiments cover the time period from 1951 to 2100. Seven datasets only completed up to the year 2050; thus only 15 of the 22 realizations are available for the far future. The combination of RCM and appropriate forcing GCM is shown in Table 1. For all ensemble members (single datasets), daily resolved values of near surface air temperature (at 2 m) and surface precipitation were taken. As indicated in the table, we excluded ensemble members, because of implausible values at least in some European regions (e.g., very high precipitation over Spain) which led to physiological implausible SDMs. In case of spruce we left out one realization and in case of beech two (marked in Table 1). The decision to exclude these datasets is based on the shape of the single response curves.

Downloaded ENSEMBLE data were preprocessed in order to receive a uniform file format. To ensure a later on merging, all datasets were remapped to a unique domain from 15.125°W to 39.875°E and 35.125°N to 70.125°N and a grid of 0.25° resolution (according to the E-OBS data grid [25]). Beside these remappings, all units were harmonised for timesteps (in UNIX epoch, seconds after 01.01.1970), daily surface temperature (in Kelvin) and precipitation (in mm/day). All preprocessing of regional climate data were realized with Climate Data Operators (CDO) version 1.5.3 [26]. All in all, we used 21 climate datasets and 3 time slices (see Table 1). Due to too many missing data in nine data experiments for the year 2100, this year was excluded completely for the appropriate realizations.

Beside data describing the climate conditions, observation data of the geographical distribution of present occurrence of tree species is used in this study. The data for beech and spruce were taken from the ICP Forests large-scale forest condition monitoring (Level I) from the period 1987 to 2007. In order to receive a somewhat regular grid, some plots were left out where the density of plots was considerably more than one plot per $225 \text{ km}^2$ (e.g., due to shift of inventory plots within a country or duplicates). This led to more than 7500 inventory plots in a resolution of approximately one plot per $225 \text{ km}^2$ across Europe [27].

The available information for crown condition within the monitoring data were transformed into presence and absence data as required by the SDM approach of this study. Inventory data often comprise fallacious absences [28, 29], for example, due to impact by forest management for long time periods. In case of beech there is fallacious absence in southern Scandinavia because there beech is not surveyed within the monitoring program. This means that the data include absences where a species could live successfully but it is not present due to other reasons than unfavourable climate conditions. The term fallacious absence used by [28] refers to this theoretical concept and does not mean that data

are improper in a technical way. As we aim to describe the climatic favourability, we came up against fallacious absences by using a vegetation map [30] that provides an overview of vegetation units and tree species composition. We attached the information about vegetation units to the Level I plots and transformed it into presence and absence data for beech and spruce. Finally, the two columns were merged. Therefore, a zero means no presence in both of the datasets, whereas a one means a presence in at least one of the datasets. In order to receive a clear signal of the altitudinal alpine distribution limit, we added 56 pseudoabsences above the alpine tree line. The resulting final data set consisted of 7596 plots covering most of Europe with a lower spatial coverage in the eastern European parts in the chosen spatial extent.

We aimed to find independent data for validation and therefore merged data from three different sources: first, the left out Level I data; second, Level II (ICP Forests Intensive Monitoring Programme, http://icp-forests.net/page/level-ii/) data, and, third, data from the European Forest Genetic Resources Programme (http://www.euforgen.org/, resp., http://www.eufgis.org/). All data were transformed into presence and absence data, no treatment for false absences was carried out and the use of the data from the Genetic Resources Programme was restricted to stands with more than one species in order to increase the probability that a full species inventory was behind the data. All in all, a maximum of 2135 stands could be used for validation. The data were unevenly distributed but covered most parts of Europe. Presences and absences of spruce fitted better to rough distribution maps (http://www2.biologie.uni-halle.de/bot/ag_chorologie/areale/index.php?sprache=E/) than presences and absences of beech. Validation data were combined with climate datasets. Because of missing values, for example, close to the shore line in some combinations of GCM with RCM, we left out all those validation sites that did not contain the complete climate information of all 21 ensemble members.

## 3. Methods

The overall method is to relate species occurrence data to an ensemble of climate data and to transfer this relation to climate projections for this ensemble. The distribution modeling, quality assessment, and merging of the final maps is described in this section (Figure 1). The section refers to the last to chain links of the chain mentioned in the introduction.

The SDMs are forced by climate indices which refer to a vegetation period. These indices were calculated for each grid box. A grid box is the smallest spatial unit of the climate datasets which have a resolution of 25 km. As the vegetation period is the base for the indices, the start and end day in the year of this period has to be determined first. The vegetation period, or synonymously the thermal growing season, in this study is defined as a period with daily mean temperatures $\geq 5°C$ for spruce and $\geq 10°C$ for beech. We used a running average over 11 days for the calculation of the start and end day using the daily resolved temperature data. Temperatures had to be above or below the threshold for at least five consecutive

TABLE 1: Ensemble members for distribution calculation and the appropriate GCMs and RCMs. Detailed description of the models is reported in [13].

| GCM | RCM | 1971–2000 | 2021–2050 | 2071–2100 |
|---|---|---|---|---|
| ARPEGE | CNRM-RM4.5 | x | x | — |
| ARPEGE | CNRM-RM5.1 | x | x | x¹ |
| ARPEGE | DMI-HIRHAM5 | x | x | x |
| ARPEGE | DMI-HIRHAM5 | x | x | x |
| BCM | METNOHIRHAM | x | x | — |
| BCM | SMHI-RCA | x | x | x |
| CGCM3 | OURANOSMRCC4.2.1 | x* | x* | — |
| CGCM3 | C4IRCA3 | x | x | — |
| ECHAM5 | DMI-HIRHAM5 | x | x | x¹ |
| ECHAM5 | ICTP-REGCM3 | x | x | x |
| ECHAM5 | KNMI-RACMO2 | x | x | x |
| ECHAM5 | MPI-REMO | x | x | x |
| ECHAM5 | SMHI-RCA | x | x | x |
| ECHAM5 | ETHZ-CLM | x | x | x |
| HadCM3Q0 | METO-HC-HadRM3Q0 | x | x | x |
| HadCM3Q0 | METNOHIRHAM | x | x | — |
| HadCM3Q0 | UCLM-PROMES | x | x | — |
| HadCM3Q0 | VMGO-RCM | x* | x** | — |
| HadCM3Q3 | METO-HC-HadCM3Q3 | x | x | x |
| HadCM3Q3 | SMHI-RCA | x | x | x |
| HadCM3Q16 | METO-HC-HadCM3Q16 | x | x | x |
| IPSL | MPI-REMO | x | x | x |

X: realized SDM; ¹ year 2100 deleted; * skipped for beech; ** skipped for beech and spruce.

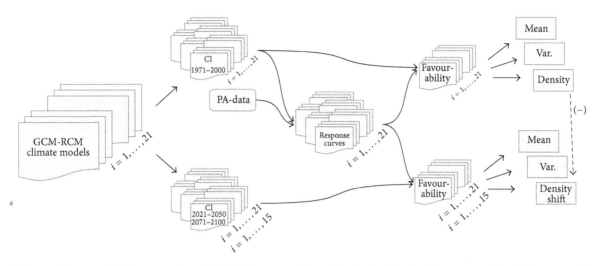

FIGURE 1: General framework of the separate methodical steps to model species distribution based on an ensemble of climate models with 21 respective 15 members.

days. Furthermore, some limitations based on a day length criterion were set as default: the vegetation period was not allowed to start before first of March (60 $y$day) and could not end before first of July (152 $y$day) but had to end latest after first of November (305 $y$day).

If five consecutive days with temperature above, respectively, below 5° (10°)C were detected, the first day of the consecutive timesteps is taken as the start respective end day in the year. The following equations are describing the dedications of vegetations periods begin as well as their ending:

$$V_s = y\text{day if} \left\{ T\text{run}_{(y\text{day}+0,...,4)} \geq T_{\text{base}}, \atop y\text{day} \geq 60 \right\} \text{else } 60 \tag{2}$$

$$V_e = y\text{day if} \left\{ T\text{run}_{(y\text{day}+0,...,4)} \leq T_{\text{base}}, \atop 152 \leq y\text{day} \leq 305 \right\} \text{else } 305, \tag{3}$$

with

$y$day = day in the year

$V_s$ = start of vegetation period ($y$day)

$V_e$ = end of vegetation period ($y$day)

$T$run = running mean with 11 timesteps of daily mean temperature.

The length of the vegetation period (VP) can be defined as

$$VP = V_e - V_s. \tag{4}$$

The following indices are used, which are derived from daily resolved temperature and precipitation values considering the information on the length of the vegetation period.

Precipitation sum in the vegetation period: $P$sumVeg [mm]

$$P\text{sumVeg} = \sum_{y\text{day}=V_s}^{V_e} P_{y\text{day}}, \tag{5}$$

growing degree days in the vegetation period: $T$sumVeg [°C]

$$T\text{sumVeg} = \sum_{y\text{day}=V_s}^{V_e} \left( T_{y\text{day}} - T_{\text{base}} \right), \tag{6}$$

total minimum temperature outside the vegetation period: $T$minDorm [°C]

$$T\text{minDorm} = \min \left( T\text{run}_{(V_s \geq y\text{day} \geq V_e)} \right), \tag{7}$$

with

$P$ = daily precipitation amount (mm)

$T$ = daily mean temperature (°C).

The selection of physiological meaningful variables was done according to our hypotheses on the climate dominated distribution of beech and spruce on the continental scale and according to preliminary work (variable importance of a boosted regression tree analysis; see [4] for details on boosted regression trees).

$T$minDorm is detected based on a smoothed temperature line to reduce misdetection of extreme values and stored for every year. The smooth is again the running average over 11 days. Finally, a mean over the respective 30 year time period was calculated for each index and for each grid box. The climate indices were calculated for each ensemble member and time period separately and joined with the species distribution data.

In order to understand the time dependent trends of the climate data, time series of the indices were calculated as a mean over a small test region for each climate data set. We chose the Bavarian Forest as a test region 12.625° to 13.875°E and 48.875°N to 49.125°N as this landscape is suitable for growing beech and spruce.

We used generalized additive models (GAMs, [31–34]) using the R-package MGVC [33] (R-project version 2.15.0 (2012-03-30)) to calibrate the single SDMs. GAMs are semiparametric extensions of generalized linear models. They allow fitting of response curves with a nonparametric smoothing function instead of parametric terms. This improves the exploration of species responses to environmental gradients. GAMs were used without interaction terms in order to build parsimonious models. An individual GAM was fitted to each climate ensemble member of the present day time slice (1971–2000) for beech and spruce. Response curves were calculated for the climate indices ($T$sumVeg, $P$sumVeg and $T$minDorm) and checked for physiological plausibility according to [35]. We used thin plate regression splines [34]; the degree of smoothness in the package mgcv is in general estimated by a generalized cross validation (GCV) criterion within certain limits set, for example, by the dimension $k$ of the basis used to represent the smooth term. $k$-values were first set to 3 and adapted (maximum $k = 6$) in some cases in order to meet assumptions on physiological plausibility. These values affect the maximum possible degree of freedom for each term of the GAM. One climate data set was excluded due to implausible precipitation data (no decrease of favourability at low precipitation for beech and spruce in Europe, that is, no distribution limit due to drought) and another one for beech (implausible response curve of variable growing degree days) after checking the single response curves for plausibility. All in all, a separate calibration of the GAM for each data set was done.

Different quality criteria were taken from the GAM output and additional parameters like the area under the receiver operating curve (AUC), sensitivity, and specificity were calculated for learning and validation data with the R package PresenceAbsence [36]. Additionally, a tenfold cross validation was run, leaving out one-tenth of the data each time and using these data for validation. The GAMs were used for the calculation of distribution maps. These maps depict the probability of occurrence for *Picea abies* and *Fagus sylvatica* for each ensemble member and time slice (present day and future). In a further step, probabilities were transformed into favourabilities according to [9] in order to allow an easy comparison between the species. This is due to the fact that this transformation corrects the effect of the frequency of presences in the learning data (prevalence) on the probabilities [9]. The output is a floating number between 0 (no) and 1 (maximum favourability) and describes the environmental favourability of a site for a species.

The last step was the merging of the separate results for each ensemble member to a final result. Two strategies were realized: an average over all appropriate ensemble results as well as a pseudodensity calculation. The average favourability (mean) of a grid box was only calculated if grid boxes contain valid values for all ensemble members. If a grid box contains a "Not a Number" (NaN) value, this 25 km rectangle was removed from the averaging procedure (e.g., at coast-lines). Additionally, the variance was calculated. The second way

of integrating the different results was the calculation of a density map. Calculating density functions is a concept of merging ensembles with a large number of members. In this study we received 15 (far future), respectively, 21 (near future) members and therefore called the results due to the low number "pseudodensity function." Favourability output was converted into a binomial response (site is favourable or not) using a threshold of 0.5 (the prevalence transformed to the scale of favourability) according to [8]. For each ensemble result, the favourability of ≥0.5 and ≤1 was taken as a one and otherwise as a zero. The pseudodensity is the proportion of ones at each grid cell. If all ensemble members have a one, then the ratio or pseudodensity is one; if half of the ensemble members have a one and the other half a zero, then the result is 0.5. The distribution shift was determined by subtracting the pseudodensity of the future predictions from that of the present time.

## 4. Results

Due to the chain from future scenario assumption to species distribution projection, several interim data were calculated. The following section presents results and figures that illustrate differences in the climate data ensembles and these differences—as a consequence—can be found again in the distribution models. Single response curves of the GAM approach try to illustrate this as well as averaged distribution maps of beech and spruce. The section ends with the two different averaging methods that are explained in Section 3 and shows the respective distribution shift for future time slices.

*4.1. Climate Indices Ensemble.* The output of the climate model runs is afflicted with high uncertainties, but in this work the focus is on the handling of different climate model output and these uncertainties are not taken into account. Therefore, uncertainties of the final distribution maps within the ensemble approach first of all are due to differences in the climate input data. For example, precipitation data of the 21 ensemble members do have different maximum amounts of precipitation. They agree in high precipitation sums in Alpine regions but differ markedly in the amount of precipitation at the coastlines (not shown). Since the climate data differ, the derived indices differ as well. As an example, Figure 2 shows the length of the vegetation period calculated with a baseline of 10°C for a test region in Eastern Bavaria. While the ensemble mean shows a vegetation period of around 130–150 days in the present time slice, single ensemble members differ with a range of more than 80 days. This variation is decreasing in the future due to the reduction of datasets as only 15 datasets were available for the far future. Additionally, in the case of the 5°C-baseline the fixed limit of 244 days (vegetation period only between April and October) is reached in some years and reduces variation. As climate indices (precipitation sum and growing degree days) base on the length of the vegetation period, these indices inherit this variability. During the century, an extension of the vegetation period of 15 days is described by the climate

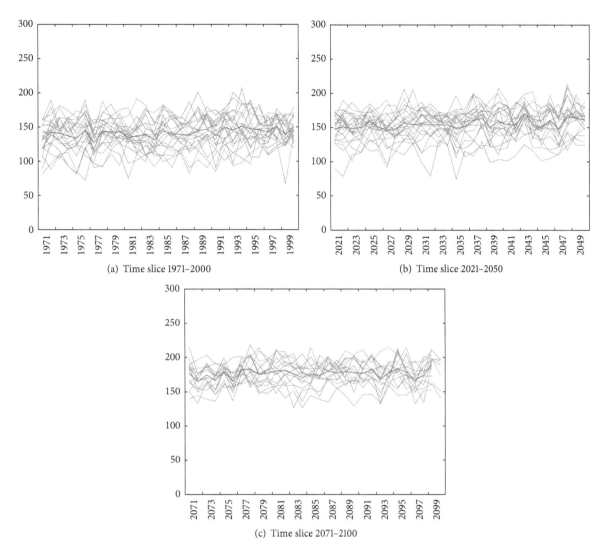

(a) Time slice 1971–2000

(b) Time slice 2021–2050

(c) Time slice 2071–2100

FIGURE 2: Time series of length of vegetation period with the baseline of 10°C for the test region Bavarian Forest for all ensemble members (grey lines). Red line shows the mean.

data within the first half and of up to 35 days by the end of the century. Generally, growing season is longer for spruce as an evergreen conifer than for deciduous beech due to the different baselines (5°C, resp., 10°C) which results in an earlier start and later end day of vegetation period for spruce.

Figure 3 shows a summary of the growing degree days calculated for spruce at the inventory plots with the data of the time slice 1971–2000. This sample illustrates the differences of the climate data ensemble members which in turn have consequences for species distribution modeling. The differences are even stronger for precipitation sum (not shown) because precipitation is harder to model than temperature.

*4.2. GAM Ensemble.* The resulting single response curves of the GAMs of spruce are shown in Figure 4. Even though there is a general accordance of most models in the shape of the response curves, in some cases there are marked deviations due to the different climate data (e.g., Figure 3). The differences are pronounced where the data are scarce, for example, in the range of high precipitation sums which

occur at high altitudes or at some coastlines. Additionally, response curves diverge due to different maxima, for example, of precipitation sum, and therefore there is a shift in the range of values. The marked differences of the input data lead to marked shifts of the response curves but the shape which describes the type of response seems to accord for many SDMs. Basically, GAMs agree in describing a distribution limit at mild winter temperature (3–5°C $T$minDorm), high temperature sums in the vegetation period (>2000–2500°C $T$sumVeg), and low precipitations sums (<200 mm $P$sumVeg). The individual GAMs differ in describing the relationship of low temperature sums and high precipitation sums with the occurrences of spruce. In case of beech (not shown), response curves of $T$minDorm are bell-shaped, $T$sumVeg describes a distribution limit at high temperatures, and $P$sumVeg shows a distribution limit due to drought.

*4.3. Favourability Ensemble.* The 30 years averages of the climate indices are the basis of the SDMs that are used to calculate the spatial distribution of spruce and beech in terms

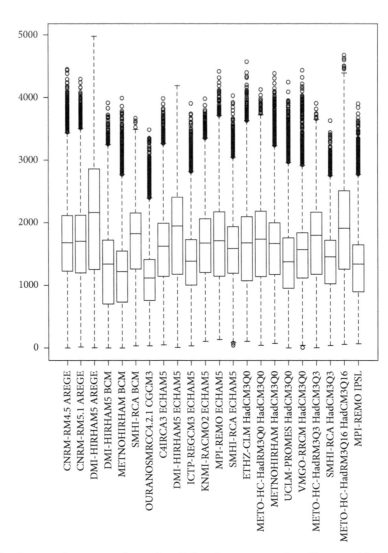

FIGURE 3: Comparison of the climate index growing degree days (˚C) at the inventory plots. The ensemble of climate data is shown for the period 1971–2000. Growing degree days are calculated depending on the length of the vegetation period defined as a period above 5˚C. Base line of the temperature sum is 5˚C (input data for modeling the distribution of *Picea abies*).

of favourability. Accordingly, the resulting maps represent averaged climate conditions for the three periods 1971–2000, 2021–2050, and 2071–2100. As we cannot present all results for each ensemble member, we show the final result as a union of the entire ensemble by averaging the single members of each time period (Figures 5 and 6). Additionally, we present the corresponding variability in the same figures.

The favourability map of beech shows a high accordance with the presences of the training data (white dots in the figure). This is underlined by high quality measures, for example, a mean AUC (area under the receiver operating curve; see [37]) of 0.90 with a minimum of 0.87 for the learning data. Using the prevalence of 0.32 as a threshold, the SDMs on average classify 90 percent of presences and 76 percent of absences correct which leads to 80 percent of the training data correctly classified. The Kappa value (see, e.g., [37]) is determined on average as 0.59 with a minimum of 0.53. Quality measures of the 10-fold cross-validation are in the same order of magnitude with the same mean and

2 to 9 per cent lower minimum values. For the validation data, the figures decrease: a mean AUC of 0.76, 85 percent of presences but only 53 percent of absences are on average correctly classified. The mean Kappa value is 0.35. Besides the present day distribution and quality of favourability maps, the figure characterizes the possible impact of climate change on the species distribution. Beech seems to be stable in middle Europe (e.g., Germany) until 2050 with only moderate shifts from west to north-east but the shift is anticipated to increase by the end of the century (calculation with 15 ensemble members). The changes mostly affect areas with low present day favourabilities of beech like western France.

Overall, the highest variability between the maps for beech is at the edges of distribution and it increases in the future especially in regions with a rising favourability in eastern and north-eastern Europe. Hotspots of variability can be found around the Alps, the Pyrenees as well as certain parts of the Carpathian arc, in the east Ukraine, and in a belt from St. Petersburg, Russia, around the gulf of Bothnia

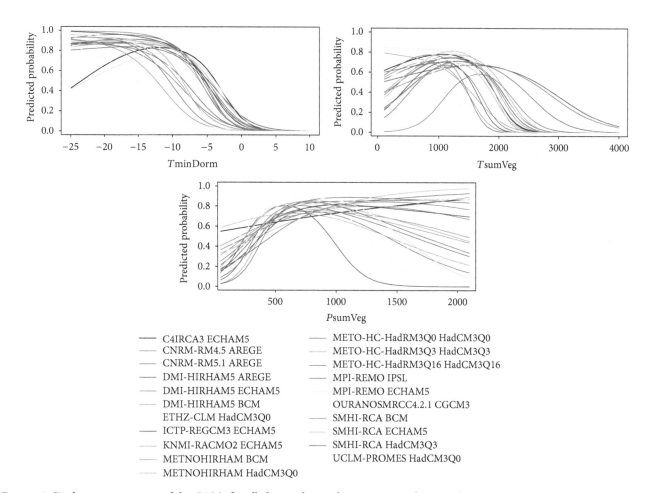

FIGURE 4: Single response curves of the GAMs for all climate data and species *Picea abies*. Single response curves are calculated for each climate index (*T*minDorm, *T*sumVeg, and *P*sumVeg) while keeping the two others constant at the mean value. Single response curves give a hint on model behavior and physiological plausibility. Legend names are combinations of RCM and GCM.

(northernmost arm of the Baltic Sea), via Sweden up to the Norwegian coastline. Generally, regions with a predicted favourability of 0.3 to 0.6 often also show a high variance of more than 0.06. The variance of the present day ensemble is low and thus points to a reliable result.

The situation for spruce is different (Figure 6). First of all, the distribution edge, for example, in Northern Germany or France appears to be not matched as good as in the case of beech (white dots with low favourability). These areas are highly affected by a strong northward shift of spruce's favourability by the end of the century. Second, the predicted distribution shift for spruce is larger than for beech and all resulting maps agree. That means that there is a low variance of the mean. Highest variances are found at the edge of the investigation area (a hotspot in Ukraine stretching into Belarus and Russia marking the southern limit of the distribution). Models have a mean AUC of 0.91 (minimum 0.89), a Kappa value of 0.67 (minimum 0.63), classify 83 percent correctly, a sensitivity of 0.89, and a specificity of 0.77 calculated with the training data and the prevalence as a threshold value (0.51). Again, 10-fold cross-validation results in the same order of magnitude with 1 to 6 per cent lower minimum values. Validation with the independent data

provides the following quality measures: a mean AUC of 0.88, a mean Kappa value of 0.61, and 81 percent correctly classified.

*4.4. Pseudodensity.* Figure 7 presents the calculated pseudo-density functions and the differences between present day and future distribution of beech and spruce. There is a high accordance of the models describing the distribution expressed in a narrow transition belt from high density values to low ones and only a few differing GAM outcomes at the distribution edges (e.g., southeast Europe in case of beech). A few areas with low density in Italy or southwest France are backed as suitable by presences (white dots in Figure 7(a)). Interpreted as a measure of accordance or disaccordance, the pseudodensity helps to illustrate areas where the results are reliable due to high accordance (e.g., areas where 80 percent of the SDM agree). Additionally, the figures show the distribution shift due to climate change in terms of density shifts, allowing a clear picture of where favourable areas for growing beech and spruce decline (values below zero, red areas) and where new areas are gained (values above zero, green areas). Certainly, there is a high accordance between averages and pseudodensity maps.

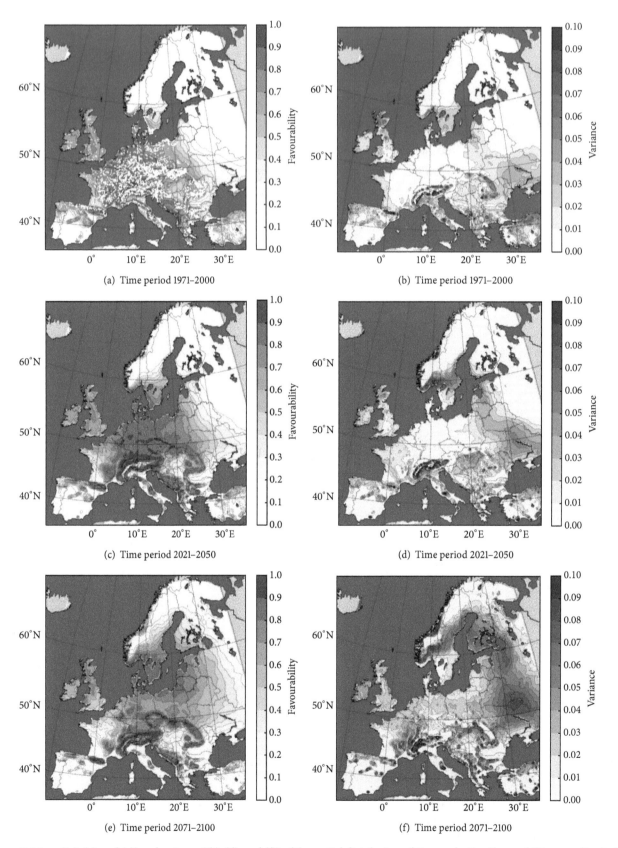

FIGURE 5: Mean ((a), (c), and (e)) and variance ((b), (d), and (f)) of the spatial distribution of *Fagus sylvatica* (favourability according to [9]). White dots in (a) are beech occurrences in the learning data. Mean and variance of (a), (b), (c), and (d) underlie an ensemble of 20 climate data, whereas the ensemble of (e) and (f) consist only of 15 members.

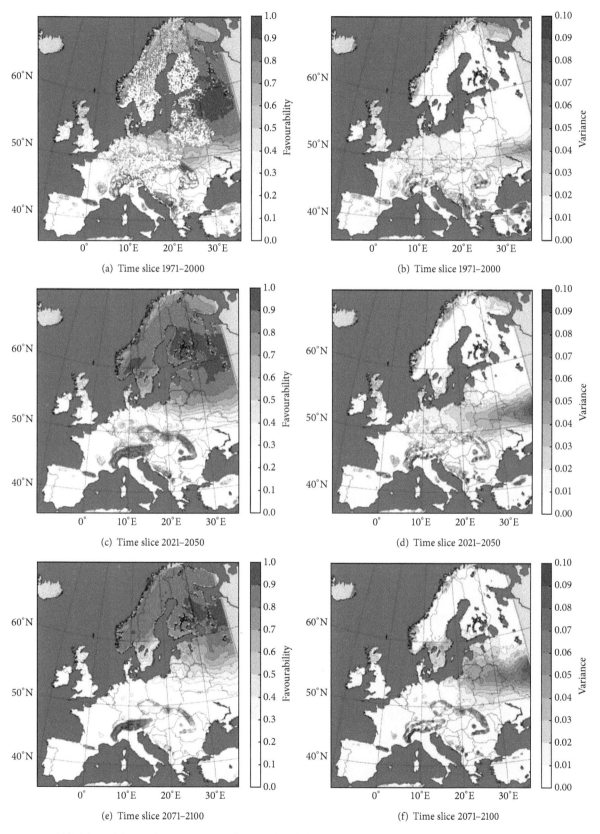

(a) Time slice 1971–2000

(b) Time slice 1971–2000

(c) Time slice 2021–2050

(d) Time slice 2021–2050

(e) Time slice 2071–2100

(f) Time slice 2071–2100

FIGURE 6: Mean ((a), (c), and (e)) and variance ((b), (d), and (f)) of the spatial distribution of *Picea abies* (favourability according to [9]). White dots in (a) are spruce occurrences in the learning data. Mean and variance of (a), (b), (c), and (d) underlie an ensemble of 21 climate data, whereas the ensemble of (e) and (f) consist only of 15 members.

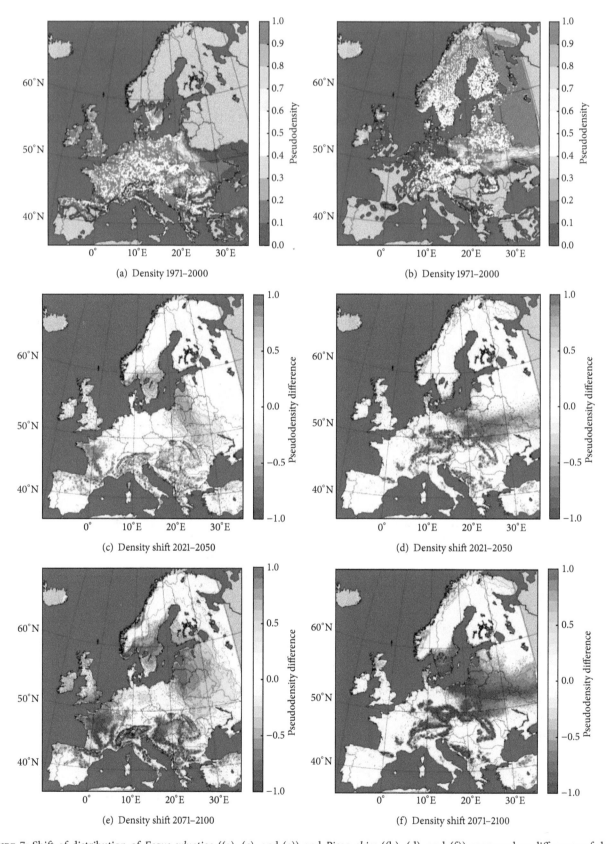

(a) Density 1971–2000

(b) Density 1971–2000

(c) Density shift 2021–2050

(d) Density shift 2021–2050

(e) Density shift 2071–2100

(f) Density shift 2071–2100

FIGURE 7: Shift of distribution of *Fagus sylvatica* ((a), (c), and (e)) and *Picea abies* ((b), (d), and (f)) expressed as difference of density distribution of the SDM ensemble in reference to time slice 1971–2000. White dots in (a) and (b) are beech respective spruce occurrences in the learning data. Density distribution in (c) and (d) underlies an ensemble of 20 respective 21 members, whereas the ensemble of (e) and (f) consists of only 15 members.

The results for beech as shown in Figures 7(a), 7(c), and 7(e) indicate a high accordance of different maps, in the middle of the century a slow north-eastern distribution shift that accelerates by the end of the century. In 2021–2050 beech stands at the western (western France) and southern distribution limit (Apennine Mountains in Italy, Hungary and some parts of southeastern Europe) will be affected first. These areas are widening by the end of the century also affecting the south of England, Ireland, and northern and northeastern Germany. On the other hand, new favourable regions for growing beech are gained in higher altitudes and northeastern Europe. In the case of spruce (Figures 7(b), 7(d), and 7(f)), the main spatial distribution in terms of pseudodensity function is situated around the Baltic Sea. The distribution includes the whole Scandinavian Peninsula except the higher altitudes of the Norwegian Mountains. The southern distribution border is characterized by a relatively large transition zone of lower model agreement along 50°N for the more continental regions and shifting northwards within an oceanic influence. Further south, the agreement of the entire ensemble is large again in the higher altitudes of the Carpathian Arc, the Swiss Alps, and in a few regions of the Pyrenees and the Balkan regions. Beside smaller parts of the Massif Central in France, the higher altitudes of the German low mountain ranges are also pointed out as suitable sites for spruce under present climatic conditions described with the SDMs. Nevertheless, the presences (white dots in Figure 7(b)) in northern Germany, the Benelux countries, and in France emphasizes the problems of many spruce stands in warmer climates today. They are in climatically unfavourable areas outside their natural distribution range and further warming as projected by scenarios indicates a high risk of decline. Sites with a raising favourability for spruce in the future can be found in the Scandinavian mountains and at the actual northern distribution edge. The distribution shift also shows that favourability for spruce declines in Germany (lower mountain ranges and alpine upland), Czech Republic, northeast Poland, Pyrenees and French Massif Central, and southern Sweden.

## 5. Discussion

This section provides a discussion of the major steps of the presented modeling chain including the climate data, the indices driving the SDMs, the SDM approach, and the resulting final maps.

*5.1. Climate Data and Ensemble Modeling.* Using projection ensembles in SDMs is a common technique in order to reduce variance and uncertainty and to raise the accuracy of species distribution models [38–40]. Whereas many studies focus on ensembles of SDM model classes, our focus is on the climate data, an approach becoming more important in the SDM literature [41]. Perhaps due to limited access to climate data or problems in handling huge data volumes, ensembles often are limited to a couple of combinations of GCM and RCM (e.g., [1], but see [24]). Beside the assumptions made on the development of greenhouse gas emissions, the realizations of climate data via GCM and RCM

differ depending on the particular GCM and RCM used. This introduces an additional range of uncertainty into the model chain. Regionalized downscaled climate data differ markedly, even when using only one emission scenario (Figures 2 and 3). We therefore concentrate in this study only on the different realizations of GCMs and RCMs and not on different emission scenarios.

A state-of-the-art ensemble model set is that of the ENSEMBLES Project. Lorenz and Jacob [42] compared the RCMs driven with quasiobserved lateral boundary conditions with gridded observational data [25] for near surface temperatures and concluded an underestimation of the past temperature trends as well as a wide range of the results of the RCMs. Jacob et al. [14] reported an uncertainty of several degrees as well as uncertainties in precipitation among the projected model data. The uncertainties are increasing during the century up to $\Delta$ 2.5 K for running averages and $\Delta$ 30% in case of precipitation. The changes vary during the seasons and different regions. The difficulty of judging the single members of the ensemble leads to an ensemble without weighting their members [22, 43].

Not all ensembles members have a realization over the whole century; in seven cases the future projections were only resized to the year 2050. Therefore, a smaller variance at the end of the century does not mean a lower uncertainty.

*5.2. Climate Indices: Use of Daily Climate Data.* We focused on three climate indices describing the environmental conditions driving species distribution. An alternative for using temperature and precipitation is the use of aridity indices (e.g., [44]). They have the advantage of combining two variables and therefore taking interactions into account. The disadvantage is that they are normally developed for climate data with a monthly or yearly resolution and that they are therefore weaker related to plant physiology. Two of the indices used in this work—temperature and precipitation in the growing season, $P$sumVeg and $T$sumVeg—depend on the length of this period, which can be estimated with daily climate data. Daily resolved data are suitable for the tackled questions but require methods of handling large data amounts.

As shown in Figure 2, the vegetation period is slightly enlarging with rising temperatures in the future projections which is in accordance with extrapolations made from phenological observations [16, 45, 46]. The projected temperature rise for the 21st century has a twofold effect on the calculated indices. On the one hand, there are higher temperatures leading to a higher temperature sum. On the other hand, the reference base for the calculation of the temperature and precipitation sum—the length of the vegetation period—is enlarging too. This can intensify a trend in case of the rising temperature sums as well as compensate decreasing precipitation sum values due to more days which are taken into account to the end of the century.

We use the vegetation period for two of the selected indices because precipitation and temperature during the physiological active phase are most important for water supply and evapotranspiration and therefore tree growth [12] and thus for species distribution (e.g., see [47], a study on drought and mortality of spruce). Due to the large climatic differences within Europe, a fixed time period presumably has a weaker relation with tree growth. We use a simple but straightforward approach to define the growing season (mean temperature threshold and day length criterion). The distribution of beech and spruce are located in climate regions with thermal seasons. The detection defined by five consecutive days higher respective lower than a threshold based on a smoothed temperature line is a very simple but useful approach. There are more sophisticated ones (e.g., [45, 46, 48]), but due to the variances within the ensemble members there is no benefit expected with a more complex approach for vegetation period detection.

The chosen simple definition of the vegetation period may not be suitable for all European regions, that is, in southern Europe where the boundaries 1st of March and 31st of October are reached too often, due to their default exclusion. This happens especially in the case of spruce for which we assumed a temperature limit of 5°C for considerable transpiration and other physiological growth processes (Raspe pers. com., [45, 48–50]). We think our approach still does make sense because the distribution models are developed for species of the temperate to boreal zone and regions with high temperatures almost all over the year are far beyond the distribution range of spruce and beech. An important point is that we opened the door for using regionalized daily climate data. In future works more sophisticated indices concentrating on drought events in the growing season or drought indices can be implemented.

### 5.3. Species Distribution Data and Modeling

#### 5.3.1. Use of Vegetation Map Data for the Treatment of Fallacious Absence.
Species distribution models are only as good as the calibration data. Hence, if environmental data do not explain the distribution well because of unaccounted effects like forest management practice (e.g., preference of one species over another due to cultivation traditions or economic reasons), postglacial migration borders due to competition, or land use effects, models may have high prediction errors. As forest inventory data are affected by human impact, we cannot assume an equilibrium between species traits and distribution leading to errors due to fallacious absences [28, 29]. That means that a species is absent although the site is favourable. The use of expert knowledge is one possibility to treat these fallacious absences (see [44]). So on the one hand, the used vegetation map helps to improve the distribution data. On the other hand, it might introduce some uncertainty due to its scale. In our opinion, the advantages outweigh the disadvantages because the distribution maps now give a general picture where spruce and beech potentially can grow. Whether the data describe the full potential of a species is not clear and it is obvious that our models are influenced by the human impact on species distributions. In case of spruce, we assume that plantations show the distribution limit in continental Europe. Ireland might be an example where our data is not complete and do not properly represent the potential distribution: models describe low to medium favourability for Ireland despite presences in

the learning data. Absences dominates over the few presences in this area. The maps show the potential distribution area [51]. This information is valuable for forest management decisions dealing with adaptation to climate change. Whether species will colonize this potential distribution depends on many factors like, for example, migration speed and barriers, dispersal and establishment limitations, and pests that, for example, can be described with mechanistic models [52].

The independent validation data were not corrected for fallacious absences and therefore reflect the realized distribution and show the differences between potential and realized distribution. Spruce is a tree species that is planted outside its natural distribution range with the help of forest management. Therefore, there are not so much fallacious absences and realized distribution is almost equal to potential distribution. Hence, validation results are in the same magnitude as cross-validation results (AUC, Kappa, percent correctly classified stands). The situation is different in case of beech, where quality measures of the validation with independent data are lower compared to cross-validation results. Beech is missing at sites where vegetation experts judge it to be within its natural distribution range (fallacious absences). It might be missing due to forest management and preference of spruce or oaks, for example, for economic reasons. Fallacious absences lead to higher error rate (percent correctly classified absences, Kappa value) when comparing these absences with a model that describes the potential distribution. That means that there could be a gap between actually realized distribution and potential distribution as depicted by vegetation experts because the cross-validation as the second validation method backs the model quality. Independent validation data could have been corrected for fallacious absences but then there would not be obtained more information of this method than by the cross-validation procedure.

#### 5.3.2. Use of Climate Data as Proxies for Water Supply.
We used climate data to describe the distribution of spruce and beech. Soil influences species distribution, too (e.g., preference of sites with high base saturation in case of *Acer campestre* [6]). However, spruce and beech are highly competitive, widely distributed in Europe, and do not have a special preference to soil conditions (e.g., nutrients) or certain relief positions [10, 11]. Soil is also part of the hydrological cycle and therefore does matter when drought events are the reason for distribution limits. Generally, it is difficult to find adequate European soil data and the influence of soil physics on water availability on a European scale is small compared to the effects of precipitation and temperature. On a local scale it is much more important to incorporate soil data into SDMs. In case of the Level I data, the European demonstration project BioSoil might fill the gap of missing soil data in the future [53]. Nevertheless, climate data as proxies for water supply in the growing season appear to be sufficient to reproduce the general distribution pattern of most tree species in Europe and to risk a glimpse into what climate change might mean for the distribution of these species. The picture will change and distribution models improve when taking into account frequency, intensity, and

duration of drought [54]. Daily climate model data prepare the technical ground to focus on extreme events.

*5.3.3. Species Distribution Modeling Technique—GAM.* Following the argumentation of Wang et al. [41] and referring to the work of Marmion et al. [39], we concentrate on one SDM technique. Reference [39] recommends either a single best method or a consensus approach of different methods like realized within the Biomod framework [40]. Other than [41], we do not use a machine-learning technique like Random Forest or a Boosted Regression Tree technique despite their good performance [4, 52, 55], but use GAM, a well-established semiparametric regression technique [55]. GAM can be handled to model smoothed continuous gradients meeting the continuum theory [56]. Additionally, it can be run in a parsimonious way using only a few dose variables, whereas the strength of data driven machine learning techniques amongst others is to deal with a high number of variables and their interactions [4]. It is especially important to avoid overfitting when balancing between model fit and predictive performance because overfitting may reduce transferability [4, 57]. We calibrated the GAMs in order to give a generalized picture of how the distribution of spruce and beech is determined by precipitation and temperature. Therefore, GAMs were run with low $k$-values (basis dimension of the penalized regression smoother, $k - 1$ is the upper limit of the degrees of freedom and influences computational efficiency) which leads to a high degree of smoothness and therefore results in very general trends (see partial response curves in Figure 4). Thus, we avoid overfitting and the SDMs are as reliable as possible in terms of transferability or more precise forecasting [38, 44].

*5.3.4. Calibration of GAMs and Plausibility of Individual Climate Data Experiments.* One step in the presented species distribution modeling was the individual tuning of the single response curves in order to receive physiological plausible models according to some fundamental hypotheses about the relationship between climate and species distribution. In doing this, we follow [35]. Elith et al. [58] also suggested to deliberately control the fit of models, for example, by using exploratory techniques like checking response curves for their contribution to the final outcome. They even went one step further and successfully integrated information from mechanistic models in order to enhance the reliability of correlative predictions. When calibrating the GAMs individually for each climate data set, we also tried to reach a high sensitivity (correctly modeled presences) because this is a requirement for describing the ecological potential of a species. Nevertheless, response curves differ which is a result of the different climate data and highlights the uncertainty of models in some areas, for example, at the edges of distribution where presences are scarce. As we cannot judge the quality of the different datasets, we use the ensemble technique in order to obtain a more complete picture of the distribution and a possible change with rising temperatures by the end of the century. A comparison of differences in input data (Figure 3) and SDMs (Figure 4) reveals the dependence of the SDMs on the climate data and can be used to cautiously judge the quality of input data for species distribution modeling: if there is no drought limit for the distribution of beech and spruce within a European data base, then this data set seems to be inappropriate, at least for a species distribution modeling approach. Overall, lower levels of temperature or precipitation leads to leftwards shift of single response curves in Figure 4. Thus, the SDMs do not differ markedly but the climate indices.

*5.3.5. Transformation of Probabilities into Favourabilities.* We transformed the probability of occurrence as provided by a GAM into favourabilities according to [9] in order to compare the two species with different presence/absence ratios (prevalence) easily. According to [9], "favourability values can be regarded as a degree of membership of the fuzzy set of sites whose environmental conditions are favourable to the species." As we want to describe the potential distribution area of spruce and beech, favourabilities seem to be an adequate quantity to show where spruce and beech can grow today and in the future.

*5.3.6. Effect of Climate Change on Species Distribution in Europe.* On a European scale it is obvious that spruce as a mountainous species is more affected by climate change than beech. This means that there are many spruce stands that will have a high risk of mortality due to warmer and drier stand conditions. There are different ways of managing risk in managed forests like shortening of rotation period, thinning [59], and in the long term transformation into stable mixed stands. We work on the species level not taking provenances into account due to the European scale and the fact that information on provenances are scarce. It can be expected that on a stand level, changes in vitality will be even more dramatic and adaptation to drier conditions will not keep pace with the projected warming. Therefore, our attempt treating the species as a whole and leaving genetic differences out might lead to an optimistic picture for single stands and a pessimistic one for long term adaptation processes. The latter is because we model the whole distribution range of the species and do not take a closer look on the distribution edges. With the help of forest management, today spruce is already grown beyond its natural distribution range at relatively warm sites. This means that stands that will fall outside the range in the future might survive for a certain time but the mortality risk is increasing strongly. Mortality might not be caused directly by drought but drought leads to lower vitality and subsequent pest attack. It is clearly visible that beech can be an alternative species on stands that will be too dry for spruce in the future because drought is a major discriminating factor at the warm and dry distribution limit between favourabilities for beech and spruce. Beech can withstand drier conditions because of deeper rooting and the chance to reduce leaf area in extreme situations in summer by a premature fall of leaves. Nevertheless, the distribution of beech has a drought limit too. With lower competition (e.g., growth) due to drought [60] beech fails to shade out other species.

*5.3.7. Final Maps: Mean and Density.* Using ensemble modeling techniques entails handling of the multiple output generated. A simple but straightforward approach is to calculate a mean and the variance if there is no evidence of differences in quality of the resulting models or maps [22]. Otherwise, a weighted mean (or median) should be favored [22]. We present this approach (simple mean) as it leads to a map describing the whole range between suitable and unsuitable environments without losing information due to classification. Typically, species distribution models that are built on binomial distribution data lead to binomial output maps with the help of a threshold value. This leads to a clear picture of the final maps but has the disadvantage of losing information. A compromise is the use of multiple thresholds [44], linking the two approaches. We present another way of handling the ensemble output using one threshold, the prevalence [8], and thus combining binomial maps. we call it a "pseudodensity function" as the amount of ensemble members is limited to 20, respectively, 21, or 15, in case of the far future. The maps of the two approaches (mean and pseudodensity function) look similar to a certain degree but the density map has sharper distribution limits due to its binomial input maps.

## 6. Conclusion

Projections of the translocation of climatic favourability for beech and spruce can be done with SDMs forced by climate model and species presence/absence data because the growth of these tree species is mainly depending on climate conditions. Thus, the presented approach can be used for tree species favourability projections where edaphic factors can be neglected on a European scale.

The study shows consequences that can be expected for spruce and beech due to climate change with regard to their distribution in Europe. The general trend is a north-eastward and altitudinal shift of climatological favourability for both species, though stronger in case of spruce. These outcomes are in accordance with other studies [1, 24] and can be used as a basis for risk assessment. The altitudinal shift is only relevant for beech because spruce already is part of the Alpine tree line and the area above this line with sufficient soil substrate for tree growth is limited. In general, the distribution shift will strongly affect the middle European spruce stands and underlines the need for stand conversion of pure spruce stands into mixed forests. Beech can be part of mixed stands and therefore used to stabilize spruce stands but certain beech stands at the warm and dry edge of distribution might need a conversion too. The presented work gives an example on how gridded daily resolved values of climate parameters, like temperature at 2 m and surface precipitation, can be used to describe the climatic site conditions for tree species. With these daily resolved climate model data very specific indices can be calculated. We present a rather simple approach to determine the vegetation period just to give an idea of the potential of temporal highly resolved data experiments.

The use of an ensemble with 21 members reduces the risk of a wrong projection. The study shows the importance and advantages of ensemble projections by pointing out regional differences of the uncertainties for each time slice. The outcome of an impact model like a SDM with climate data as input depends strongly on these differences. Beside a wide spread and common use of an arithmetic average combined with the appropriate variance, pseudodensity functions are another tool to visualize uncertainty originating from the climatic data ensemble. The individual check for physiological plausibility reduces bias and uncertainty deriving from the SDM framework, though this source of uncertainty cannot be eliminated completely although the used GAM technique, separately trained for each ensemble member, minimizes the individual bias of the different climate model members within the ensemble.

## 7. Outlook

The data handling with daily resolved climate model data and the connection to species distribution modeling is done with the presented work. Now that the frame is set, extension into different ways can be made: the focus can be set on variability and extreme events [54] and their influence on species distribution. For example, drought indices or deviations from long term mean might provide a better explanation of species distribution than simple mean values [54]. Certainly, the models could be run for other species, or other SDM techniques.

The Coordinated Regional Downscaling Research Experiment (CORDEX http://wcrp-cordex.ipsl.jussieu.fr/) is the successor of ENSEMBLE-Project and is forced by the RCP scenarios [61] instead of those of the SRES scenario families used in the ENSEMBLE-Project. If these data are available, an ensemble with more members, higher spatial and temporal resolution, and new emission scenarios can be taken to force impact models like SDMs. Amongst other things, the CORDEX data will reduce the data preprocessing due to observing some standards. Future work will be done with these new climate data.

To reduce the workload or invite impact modelers with limited access to climate data, the shown GAM approach will be available in a virtual computing environment (C3Grid-INAD) and therefore can help to process large datasets under the control of the user (e.g., use other climate indices for sensitivity or extreme events studies or different definitions of the vegetation period).

## Acknowledgments

The ENSEMBLES data used in this work was funded by the EU FP6 Integrated Project ENSEMBLES (Contract no. 505539) whose support is gratefully acknowledged. The tree species data are derived from the first grid (Level I) data. These data are surveyed under the International Co-operative Programme on Assessment and Monitoring of Air Pollution Effects on Forests operating under the UNECE Convention on Long-Range Transboundary Air Pollution (ICP Forests, http://icp-forests.net/) and provided by the Programme Co-ordinating Centre (PCC) of ICP Forests in Hamburg, Germany, whose support is gratefully acknowledged.

# References

[1] M. Hanewinkel, S. Hummel, and D. A. Cullmann, "Modelling and economic evaluation of forest biome shifts under climate change in Southwest Germany," *Forest Ecology and Management*, vol. 259, no. 4, pp. 710–719, 2010.

[2] S. Kapeller, M. J. Lexer, T. Geburek, J. Hiebl, and S. Schueler, "Intraspecific variation in climate response of Norway spruce in the eastern Alpine range: selecting appropriate provenances for future climate," *Forest Ecology and Management*, vol. 271, pp. 46–57, 2012.

[3] B. Huntley, P. J. Bartlein, and I. C. Prentice, "Climatic control of the distribution and abundance of beech (Fagus L.) in Europe and North America," *Journal of Biogeography*, vol. 16, no. 6, pp. 551–560, 1989.

[4] J. Elith, J. R. Leathwick, and T. Hastie, "A working guide to boosted regression trees," *Journal of Animal Ecology*, vol. 77, no. 4, pp. 802–813, 2008.

[5] B. Huntley, P. M. Berry, W. Cramer, and A. P. McDonald, "Modelling present and potential future ranges of some European higher plants using climate response surfaces," *Journal of Biogeography*, vol. 22, no. 6, pp. 967–1001, 1995.

[6] C. Coudun, J.-C. Gégout, C. Piedallu, and J.-C. Rameau, "Soil nutritional factors improve models of plant species distribution: an illustration with Acer campestre (L.) in France," *Journal of Biogeography*, vol. 33, no. 10, pp. 1750–1763, 2006.

[7] A. Bolte, T. Czajkowski, and T. Kompa, "The north-eastern distribution range of European beech—a review," *Forestry*, vol. 80, no. 4, pp. 413–429, 2007.

[8] C. Liu, P. M. Berry, T. P. Dawson, and R. G. Pearson, "Selecting thresholds of occurrence in the prediction of species distributions," *Ecography*, vol. 28, no. 3, pp. 385–393, 2005.

[9] R. Real, A. M. Barbosa, and J. M. Vargas, "Obtaining environmental favourability functions from logistic regression," *Environmental and Ecological Statistics*, vol. 13, no. 2, pp. 237–245, 2006.

[10] G. von Wühlisch, "EUFORGEN technical guidelines for genetic conservation and use for European beech (Fagus sylvatica)," Tech. Rep., Bioversity International, Rome, Italy, 2008.

[11] T. Skrøppa, "EURFORGEN technical guidelines for genetic conservation and use for Norway spruce (Picea abies)," Tech. Rep., International Plant Genetic Resources Institute, Rome, Italy, 2003.

[12] F. Went, "The effect of temperature on plant growth," *Annual Review of Plant Physiology and Plant Molecular Biology*, vol. 4, pp. 347–362, 1953.

[13] P. van der Linden and J. Mitchell, "ENSEMBLES: climate change and its impacts: summary of research and results from the ENSEMBLES project," Tech. Rep., Met Office Hadley Centre, London, UK, 2009.

[14] D. Jacob, K. Bulow, L. Kotova, C. Moseley, J. Perterson, and D. Rechid, "CSC—Report 6: Regionale Klimaprojektionen für Europa und Deutschland: Ensembles-Simulationen für die Klimafolgenforschung," Tech. Rep., Climate Service Center (CSC), 2012.

[15] A. Menzel, T. H. Sparks, N. Estrella et al., "European phenological response to climate change matches the warming pattern," *Global Change Biology*, vol. 12, no. 10, pp. 1969–1976, 2006.

[16] N. Estrella, T. H. Sparks, and A. Menzel, "Effects of temperature, phase type and timing, location, and human density on plant phenological responses in Europe," *Climate Research*, vol. 39, no. 3, pp. 235–248, 2009.

[17] R. J. Hijmans, S. E. Cameron, J. L. Parra, P. G. Jones, and A. Jarvis, "Very high resolution interpolated climate surfaces for global land areas," *International Journal of Climatology*, vol. 25, no. 15, pp. 1965–1978, 2005.

[18] P. Brohan, J. J. Kennedy, I. Harris, S. F. B. Tett, and P. D. Jones, "Uncertainty estimates in regional and global observed temperature changes: a new data set from 1850," *Journal of Geophysical Research*, vol. 111, no. D12, Article ID D12106, 2006.

[19] C. D. Hewitt, "The ensembles project: providing ensemble-based predictions of climate changes and their impacts," *The Eggs Newsletter*, vol. 13, pp. 22–25, 2005.

[20] A. Jiménez-Valverde and J. M. Lobo, "Threshold criteria for conversion of probability of species presence to either-or presence-absence," *Acta Oecologica*, vol. 31, no. 3, pp. 361–369, 2007.

[21] R. J. Hijmans and C. H. Graham, "The ability of climate envelope models to predict the effect of climate change on species distributions," *Global Change Biology*, vol. 12, no. 12, pp. 2272–2281, 2006.

[22] A. P. Weigel, R. Knutti, M. A. Liniger, and C. Appenzeller, "Risks of model weighting in multimodel climate projections," *Journal of Climate*, vol. 23, no. 15, pp. 4175–4191, 2010.

[23] IPCC, "IPCC special report emission scenarios: summary for policymakers: a special report of IPCC working group III," Tech. Rep., Intergovernmental Panel on Climate Change, 2000.

[24] M. Hanewinkel, D. A. Cullmann, M. J. Schelhaas, G. J. Nabuurs, and N. E. Zimmermann, "Climate change may cause severe loss in the economic value of European forest land," *Nature Climate Change*, vol. 3, pp. 203–207, 2013.

[25] M. R. Haylock, N. Hofstra, A. M. G. Klein Tank, E. J. Klok, P. D. Jones, and M. New, "A European daily high-resolution gridded data set of surface temperature and precipitation for 1950–2006," *Journal of Geophysical Research D*, vol. 113, no. 20, Article ID D20119, 2008.

[26] U. Schulzweida, L. Kornblueh, and R. Quast, *Climate Data Operators. User's Guide*, MPI for Meteorology Brockmann Consult, 2009.

[27] M. Lorenz, "International co-operative programme on assessment and monitoring of air pollution effects on forests—ICP forests," *Water, Air, and Soil Pollution*, vol. 85, no. 3, pp. 1221–1226, 1995.

[28] A. H. Hirzel and G. Le Lay, "Habitat suitability modelling and niche theory," *Journal of Applied Ecology*, vol. 45, no. 5, pp. 1372–1381, 2008.

[29] P. L. Zarnetske, T. C. Edwards Jr., and G. G. Moisen, "Habitat classification modeling with incomplete data: pushing the habitat envelope," *Ecological Applications*, vol. 17, no. 6, pp. 1714–1726, 2007.

[30] U. Bohn and R. Neuhäusl, "Interaktive CD-ROM zur Karte der natürlichen Vegetation Europas/Interactive CD-Rom Map of the Natural Vegetation of Europe.Mafistab/Scale 1:2 500 000. Teil 1: Erläuterungstext mit CD-ROM, Teil 2: Legende, Teil 3: Karten," Tech. Rep., Landwirtschaftsverlag, Münster, Germany, 2003.

[31] T. Hastie and R. Tibshirani, *Generalized Additive Models*, Chapman and Hall, 1990.

[32] S. N. Wood and N. H. Augustin, "GAMs with integrated model selection using penalized regression splines and applications to environmental modelling," *Ecological Modelling*, vol. 157, no. 2-3, pp. 157–177, 2002.

[33] S. Wood, "Mixed GAM computation vehicle with GCV/AIC/REML smoothness estimation," 2012.

[34] S. Wood, *Generalized Additive Models: An Introduction With R*, vol. 66, CRC Press, 2006.

[35] K. H. Mellert, V. Fensterer, H. Küchenhoff et al., "Hypothesis-driven species distribution models for tree species in the Bavarian Alps," *Journal of Vegetation Science*, vol. 22, no. 4, pp. 635–646, 2011.

[36] E. A. Freeman and G. Moisen, "PresenceAbsence: an R package for PresenceAbsence analysis," *Journal of Statistical Software*, vol. 23, no. 11, pp. 1–31, 2008.

[37] A. H. Fielding and J. F. Bell, "A review of methods for the assessment of prediction errors in conservation presence/absence models," *Environmental Conservation*, vol. 24, no. 1, pp. 38–49, 1997.

[38] M. B. Araújo and M. New, "Ensemble forecasting of species distributions," *Trends in Ecology & Evolution*, vol. 22, no. 1, pp. 42–47, 2007.

[39] M. Marmion, M. Parviainen, M. Luoto, R. K. Heikkinen, and W. Thuiller, "Evaluation of consensus methods in predictive species distribution modelling," *Diversity and Distributions*, vol. 15, no. 1, p. 5969, 2009.

[40] W. Thuiller, B. Lafourcade, R. Engler, and M. B. Araújo, "BIOMOD-a platform for ensemble forecasting of species distributions," *Ecography*, vol. 32, no. 3, pp. 369–373, 2009.

[41] T. Wang, E. M. Campbell, G. A. O'Neill, and S. N. Aitken, "Projecting future distributions of ecosystem climate niches: uncertainties and management applications," *Forest Ecology and Management*, vol. 279, pp. 128–140, 2012.

[42] P. Lorenz and D. Jacob, "Validation of temperature trends in the ENSEMBLES regional climate model runs driven by ERA40," *Climate Research*, vol. 44, no. 2-3, pp. 167–177, 2010.

[43] M. Marmion, M. Luoto, R. K. Heikkinen, and W. Thuiller, "The performance of state-of-the-art modelling techniques depends on geographical distribution of species," *Ecological Modelling*, vol. 220, no. 24, pp. 3512–3520, 2009.

[44] W. Falk and K. H. Mellert, "Species distribution models as a tool for forest management planning under climate change: risk evaluation of Abies alba in Bavaria," *Journal of Vegetation Science*, vol. 22, no. 4, pp. 621–634, 2011.

[45] A. Menzel, *Phänologie von Waldbäumen unter sich ändernden Klimabedingungen: Auswertung der Beobachtungen in den internationalen phänologischen Gärten und Möglichkeiten der Modellierung von Phänodaten [Ph.D. thesis]*, Technical University München, München, Germany, 1997.

[46] A. Menzel and P. Fabian, "Growing season extended in Europe," *Nature*, vol. 397, no. 6721, p. 659, 1999.

[47] S. Solberg, "Summer drought: a driver for crown condition and mortality of Norway spruce in Norway," *Forest Pathology*, vol. 34, no. 2, pp. 93–104, 2004.

[48] K. von Wilpert, *Die Jahrringstruktur von Fichten in Abhängigkeit vom Bodenwasserhaushalt auf Pseudogley und Parabraunerde. Ein Methodenkonzept zur Erfassung standortsspezifischer Wasserstreßdisposition [Ph.D. thesis]*, Institut für Bodenkunde und Waldernährungslehre der Albert-Ludwigs-Universität, Freiburg im Breisgau, Germany, 1990.

[49] R. Man and P. Lu, "Effects of thermal model and base temperature on estimates of thermal time to bud break in white spruce seedlings," *Canadian Journal of Forest Research*, vol. 40, no. 9, pp. 1815–1820, 2010.

[50] M. Hannerz, "Evaluation of temperature models for predicting bud burst in Norway spruce," *Canadian Journal of Forest Research*, vol. 29, no. 1, pp. 9–19, 1999.

[51] M. B. Araújo and A. T. Peterson, "Uses and misuses of bioclimatic envelope modeling," *Ecology*, vol. 93, no. 7, pp. 1527–1539, 2012.

[52] L. Iverson, A. Prasad, S. Matthews, and M. Peters, "Lessons learned while integrating habitat, dispersal, disturbance, and life-history traits into species habitat models under climate change," *Ecosystems*, vol. 14, pp. 1005–1020, 2011.

[53] R. Hiederer, E. Micheli, and T. Durrant, *Evaluation of Biosoil. Demonstration Project. Soil Data Analysis*, European Union, 2011.

[54] N. E. Zimmermann, N. G. Yoccoz, T. C. Edwards Jr. et al., "Climatic extremes improve predictions of spatial patterns of tree species," *Proceedings of the National Academy of Sciences of the United States of America*, vol. 106, no. 2, pp. 19723–19728, 2009.

[55] A. Guisan, N. E. Zimmermann, J. Elith, C. H. Graham, S. Phillips, and A. T. Peterson, "What matters for predicting the occurrences of trees: techniques, data, or species' characteristics?" *Ecological Monographs*, vol. 77, no. 4, pp. 615–630, 2007.

[56] M. Austin, "Spatial prediction of species distribution: an interface between ecological theory and statistical modelling," *Ecological Modelling*, vol. 157, no. 2-3, pp. 101–118, 2002.

[57] C. F. Randin, T. Dirnböck, S. Dullinger, N. E. Zimmermann, M. Zappa, and A. Guisan, "Are niche-based species distribution models transferable in space?" *Journal of Biogeography*, vol. 33, no. 10, pp. 1527–1539, 2006.

[58] J. Elith, M. Kearney, and S. Phillips, "The art of modelling range-shifting species," *Methods in Ecology and Evolution*, vol. 1, no. 4, pp. 330–342, 2010.

[59] M. Kohler, J. Sohn, G. Nägele, and J. Bauhus, "Can drought tolerance of Norway spruce (Picea abies (L.) Karst.) be increased through thinning?" *European Journal of Forest Research*, vol. 129, pp. 1109–1118, 2010.

[60] A. S. Jump, J. M. Hunt, and J. Penuelas, "Rapid climate change-related growth decline at the southern range edge of Fagus sylvatica," *Global Change Biology*, vol. 12, no. 11, pp. 2163–2174, 2006.

[61] R. H. Moss, J. A. Edmonds, K. A. Hibbard et al., "The next generation of scenarios for climate change research and assessment," *Nature*, vol. 463, no. 7282, pp. 747–756, 2010.

# 15

# Recent Extreme Precipitation and Temperature Changes in Djibouti City (1966–2011)

## Pierre Ozer[1] and Ayan Mahamoud[2]

[1] *Department of Environmental Sciences and Management, University of Liège, Avenue de Longwy 185, 6700 Arlon, Belgium*
[2] *Research Center, Department of Geomatics and Environmental Sciences, University of Djibouti, Avenue Georges Clemenceau, BP 1904, Djibouti*

Correspondence should be addressed to Pierre Ozer; pierre.ozer@gmail.com

Academic Editors: A. V. Eliseev, S. Feng, P. Gober, and S. Gualdi

A dataset of 23 derived indicators has been compiled to clarify whether the frequency of rainfall and temperature extremes has changed over the last decades in Djibouti City, eastern Africa. Results show that all precipitation indices have declined over the last decades, although only the very wet day frequency and the very wet day proportion present a significant decline. Annual total precipitation has decreased by 17.4% per decade from 1980 to 2011 and recent mean yearly rainfall (44 mm on average from 2007 to 2011) meets a 73% deficit compared to the 30-year (1981–2010) average (164 mm). The average temperature increase is +0.28°C per decade. Extremely warm days (maximum temperature ≥45.0°C) have become 15 times more frequent than in the past while extremely cool nights (minimum temperature ≤8.6°C) have almost disappeared. Current rainfall shortages and increasing temperature extremes are impacting local people who urgently need adaptation strategies.

## 1. Introduction

Warming of the climate system is unequivocal, as is now evident from observations of increases in global average air and ocean temperatures, widespread melting of snow and ice and rising global average sea level [1].

According to the Climate Research Unit of the University of East Anglia [2], global air temperature during the period 2001–2010 (0.48°C above 1961–1990 mean) was 0.22°C warmer than that in the 1991–2000 decade (0.27°C above 1961–1990 mean). The warmest year of the entire series (1850–2012) was 2010, with a temperature of 0.54°C above the 1961–1990 mean. After 2010, the next ten warmest years in the series are all in the period 1998–2012. Most of the observed increase in global average temperatures since the mid-20th century is very likely due to the observed increase in anthropogenic greenhouse gases (GHGs) concentrations [1].

Some extreme weather events have received increased attention in the last few years within the perspective of climate change. Studies show that they have changed in frequency and/or intensity over the last 50 years. Yet, it is very likely that

cold days, cold nights, and frost have become less frequent over most land areas, while hot days and hot nights have become more frequent. It is also likely that heatwaves have become more frequent and that the frequency of heavy precipitation events (or proportion of total rainfall from heavy falls) has increased over most areas [1].

In addition to the climate change impacts, population and infrastructure continue to develop in areas that are vulnerable to extremes such as flooding, storm damage, and extreme temperatures. Furthermore, land use change can often further increase vulnerability by creating more potential for catastrophic impacts from climate extremes, such as flooding due to extreme precipitation events.

Although the frequency, the intensity, and the impacts of extreme weather events are well documented in most parts of the world, there has been paucity of information on trends in daily extreme rainfall events in Africa [3]. The lack of long-term daily climate data suitable for analysis of extremes is the biggest obstacle to quantifying whether extreme events have changed over the last decades, either continental or on a more regional basis [4].

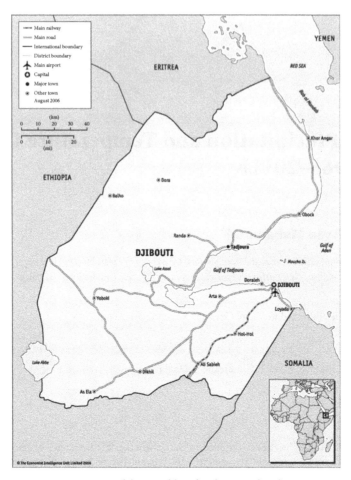

FIGURE 1: Location of the Republic of Djibouti and Djibouti City.

In this paper, evidence for changes in the intensity of extreme daily rainfall and temperature events during the last decades is assessed in the Republic of Djibouti, the smallest country in the Horn of Africa bordered by Eritrea in the north, Ethiopia in the west and south, Somalia in the southeast, and facing the Red Sea and the Gulf of Aden at the east. The country occupies a total area of about 23,200 km$^2$ (Figure 1).

According to the Köppen-Geiger world climate classification, the climate of the Republic of Djibouti is defined as a hot desert (BWh) [5]. Hot conditions prevail year-round with monthly average temperatures ranging from 25.4°C in January to 36.8°C in July, while rainfall are very scarce all year-long with monthly total precipitation below 10 mm from May to August and between 10 and 28 mm the rest of the year (Figure 2). As in many arid regions, annual rainfall is significantly irregular and extreme rainfall can be observed in a number of years [6]. We use here meteorological data from the unique synoptic station of the country, Djibouti airport, located south of the capital city of Djibouti City (Figure 1).

## 2. Data and Methods

*2.1. Database.* For the analysis of recent trends of extreme precipitation and temperatures, data were made available for

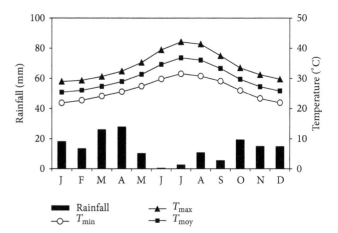

FIGURE 2: Monthly distribution of rainfall and temperature in Djibouti City (average 1981–2010).

the synoptic station of Djibouti City (Lat: 11.55° N; Long: 43.15° E; altitude: 13 m asl, WMO station code: 63125) located near the International Airport of Djibouti from the "Agence Nationale de la Météorologie (ANM) de Djibouti." The airport is located outside and south of the capital city. The database includes daily precipitation data from January 1980

TABLE 1: Rainfall and temperature indices with their definitions and units.

| ID | Indicator name | Definition | Unit |
|---|---|---|---|
| PTOT | Precipitation total | Annual total precipitation | mm |
| Rd | Rainfall days | Annual total of wet days (rainfall ≥1 mm) | days |
| SDII | Simple day intensity index | Average rainfall from wet days | mm/day |
| Rx1d | Maximum 1-day rainfall | Annual maximum 1-day rainfall | mm |
| R10 mm | Number of heavy precipitation days | Annual count of days when rainfall ≥10 mm | days |
| R20 mm | Number of very heavy precipitation days | Annual count of days when rainfall ≥20 mm | days |
| R95p | Very wet day frequency | Annual count of days when rainfall ≥95th percentile of 1981–2010 | days |
| R99p | Extreme rainfall frequency | Annual count of days when rainfall ≥99th percentile of 1981–2010 | days |
| R95pSUM | Very wet day intensity | Annual precipitation from days when rainfall ≥95th percentile of 1981–2010 | mm |
| R99pSUM | Extreme rainfall intensity | Annual precipitation from days when rainfall ≥99th percentile of 1981–2010 | mm |
| R95pTOT | Very wet day proportion | Percentage of annual precipitation from days when rainfall ≥95th percentile of 1981–2010 | % |
| R99pTOT | Extreme rainfall proportion | Percentage of annual precipitation from days when rainfall ≥99th percentile of 1981–2010 | % |
| ATN | Average annual $T_{min}$ | Annual average value of daily minimum temperature | °C |
| ATX | Average annual $T_{max}$ | Annual average value of daily maximum temperature | °C |
| ATM | Average annual $T_{mean}$ | Annual average value of daily mean temperature | °C |
| TN1p | Extreme cool night | Annual count of days when $T_{min}$ ≤1th percentile of 1971–2000 (18.6°C) | days |
| TN5p | Cool night | Annual count of days when $T_{min}$ ≤5th percentile of 1971–2000 (20.2°C) | days |
| TN95p | Warm night | Annual count of days when $T_{min}$ ≥95th percentile of 1971–2000 (32.2°C) | days |
| TN99p | Extreme warm night | Annual count of days when $T_{min}$ ≥99th percentile of 1971–2000 (33.6°C) | days |
| TX1p | Extreme cool day | Annual count of days when $T_{max}$ ≤1th percentile of 1971–2000 (27.5°C) | days |
| TX5p | Cool day | Annual count of days when $T_{max}$ ≤5th percentile of 1971–2000 (28.5°C) | days |
| TX95p | Warm day | Annual count of days when $T_{max}$ ≥95th percentile of 1971–2000 (43.9°C) | days |
| TX99p | Extreme warm day | Annual count of days when $T_{max}$ ≥99th percentile of 1971–2000 (45.0°C) | days |

to December 2011 and maximum, minimum, and mean daily temperatures from January 1966 to December 2011. There were no missing data, so the database can be used for trend analysis [7].

### 2.2. Extreme Rainfall Indices.

In this study, 12 rainfall indices were calculated over the January to December period and are listed in Table 1. These are the annual total precipitation (PTOT), the annual total of wet days (with daily rainfall ≥1 mm, Rd), the simple day intensity index (SDII) that was calculated as the average rainfall from wet days, the annual maximum rainfall recorded during 1 day (Rx1d), and the number of heavy precipitation (rainfall ≥10 mm, R10 mm) and very heavy precipitation (rainfall ≥20 mm, R20 mm) days. Other six indices are based on the 95th and 99th

percentiles which define a very wet day and an extreme rainfall event, respectively [7–13]. These percentile values were calculated from daily rainfall data over the 1981–2010 period. For the station of Djibouti City, the thresholds calculated from percentiles are 46.9 mm and 108.1 mm to define a very wet day and an extreme rainfall event, respectively. Based on these percentiles, two extreme precipitation indices were chosen. Very wet day and extreme rainfall frequency are based on the annual count of days when rainfall ≥95th and 99th percentiles of 1981–2010 (R95p and R99p). Very wet day and extreme rainfall intensity correspond to the annual total precipitation recorded from days when rainfall ≥95th and 99th percentiles of 1981–2010 (R95pSUM and R99pSUM) and give an indication on the rain received from very wet or extreme rainfall. Very wet day and extreme rainfall proportion are the percentage of the annual total

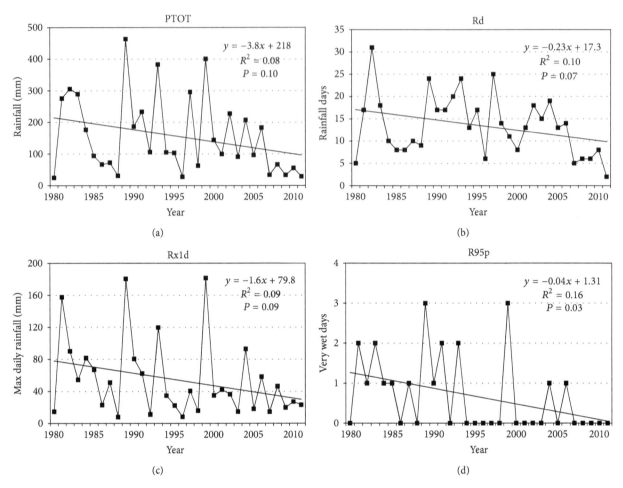

FIGURE 3: Evolution and trends of PTOT, Rd, Rx1d, and R95p in Djibouti City (1980–2011).

precipitation recorded from days when rainfall ≥95th and 99th percentiles of 1981–2010 (R95pTOT and R99pTOT) and measure how much total rain comes from very wet or extreme events.

*2.3. Extreme Temperature Indices.* The analysis of minimum, maximum, and mean temperatures is based on 11 indices calculated over the January to December period (Table 1). Three of them are based on the annual average value of daily minimum, maximum, and mean temperatures (ATN, ATX, and ATM) in order to analyse the global trends in temperatures. All other indices are based on the 1st, 5th, 95th, and 99th percentiles which define "extremely cool," "cool," "warm," and "extremely warm" nights (using $T_{\min}$) and days (using $T_{\max}$), respectively, (adapted from [10–13]). These percentile values were calculated from daily rainfall data over the 1971–2000 period. Threshold temperature values calculated from percentiles are presented in Table 1. In Djibouti, an extremely cool night is characterized by minimum temperatures ≤8.6°C while an extremely warm day is defined when maximum temperatures are ≥45.0°C.

*2.4. Trend Analysis.* In the analysis, trend coefficients are determined using linear regression modelling, which represent the increasing or decreasing rate of the given index during 1980–2011 period for precipitation and 1966–2011 period for temperatures. This method is now extensively used worldwide [7–13]. Each slope (positive or negative) was categorized in six classes indicating very significant, significant, or nonsignificant trends. The regression procedure supplies a Student $t$-test and its resulting significance $P$ level to analyse the hypothesis that the slope is equal to 0. This $P$-level was used as a criterion to define the class boundaries. The trends, for each index, were labelled as "very significant" if the $P$-level exceeded 0.01 for the one-tailed $t$-test, "significant" if the $P$-level ranged between 0.01 and 0.05, and otherwise "nonsignificant" if the $P$-level is up to 0.05.

## 3. Results

*3.1. Precipitation.* Time series of several precipitation indices can be seen in Figure 3. All indices show decreasing trends although most of them are statistically nonsignificant (Table 2). Yet, the annual total precipitation (PTOT), the rainfall days (Rd), the annual maximum rainfall recorded during 1 day (Rx1d), and the very wet day rainfall intensity (R95pSUM) decrease in a moderate way. Only the very wet day frequency (R95p) and the very wet day proportion (R95pTOT) present a significant decline. It is likely that

TABLE 2: Rainfall trends analysis from 1980 to 2011.

| ID | Unit | Average 1981–2010 | Trend units/decade | Significance (P level) | Trend %/decade |
|----|------|-------------------|--------------------|------------------------|----------------|
| PTOT | mm | 164 | −38.0 | 0.10 | −17.4 |
| Rd | days | 14.1 | −2.3 | 0.07 | −13.5 |
| SDII | mm/day | 11.1 | −0.5 | 0.70 | −4.2 |
| Rx1d | mm | 56.4 | −15.6 | 0.09 | −19.5 |
| R10 mm | days | 3.9 | −0.5 | 0.38 | −10.7 |
| R20 mm | days | 2.1 | −0.5 | 0.27 | −16.7 |
| R95p | days | 0.70 | −0.39 | 0.03 | −30.2 |
| R99p | days | 0.13 | −0.07 | 0.26 | −29.7 |
| R95pSUM | mm | 59.7 | −31.4 | 0.07 | −29.1 |
| R99pSUM | mm | 21.3 | −11.5 | 0.28 | −29.5 |
| R95pTOT | % | 23.4 | −14.2 | <0.01 | −31.3 |
| R99pTOT | % | 5.8 | −3.7 | 0.20 | −32.1 |

the extreme rainfall events (R99p) decline may be significant but it is difficult to make statistics on such a limited sample.

The most significant rainfall shortage is found over the last five years (see Figure 3). Yet, since 2007, the yearly rainfall average has been 44 mm, which is an extreme rainfall deficit of near 75% when compared to the 1981–2010 average (164 mm). During this same recent period, the synoptic station of Djibouti City did not record any extreme rainfall nor very wet day events. The maximum rainfall recorded during 1 day was 46 mm over the last five years.

These data are of the highest importance in terms of vulnerability evolution as well as future adaptation strategies to natural hazard reduction, especially concerning floods. This aspect will be discussed afterwards.

*3.2. Temperature.* The analysis of the annual time series of the temperature indices indicates that changes in temperature extremes over the 1966–2011 period reflect warming for the Djibouti City area. The trends in annual average minimum, maximum, and mean temperatures (ATN, ATX, and ATM) presented in Figure 4 and given in Table 3 show a very significant increase of the three indices. Mean temperature increased by 1.24°C (0.28°C per decade) during the 1966–2011 period. The warmest year of the entire series was 2010 with mean temperature of 31.3°C, which is 1.18°C above the 1971–2000 mean. The ten warmest years of the whole record have been registered since 1998. The period 2001–2011 was 0.66°C warmer than the 1971–2000 mean (Table 3).

The annual number of warm and extremely warm days and nights, analyzed through the TX95p, TX99p, TN95p, and TN99p indices, has significantly increased. Extremely warm days and nights were 11.5 and 13.3 on average during 2001–2011, a much higher figure than the averages of 4.7 and 3.8 recorded during the 1971–2000 reference period (Table 3). Conversely, the number of cool and extremely cool nights and days, analyzed through TN5p, TN1p, TX5p, and TX1p,

has decreased significantly. Trends of extremes are shown in Figure 4.

## 4. Discussion

Lack of long-term climate data suitable for analysis of extremes is the biggest obstacle to quantifying whether extreme events have changed over the last decades in Africa. This paper presents, for the first time, the analysis of extreme precipitation and temperatures in Djibouti City.

Results show that all rainfall indices have declined over the last 32 years, although only the very wet day frequency (R95p) and the very wet day proportion (R95pTOT) present a significant decline. Such recent decrease in annual total precipitation (PTOT) is in line with recent findings in the Arab region [12], over the Great Horn of Africa region [13], in the Arabian Peninsula [14], or in Ethiopia [15] and is likely to be increasingly influenced by the Indian Ocean sea surface temperature [16]. In terms of all other rainfall indices, a mainly nonsignificant decreasing trend is observed that may suggest a consistent change towards drier conditions as emphasized in the northern part of the Great Horn of Africa [13] and in the eastern part of the Arab region [12].

All trends of temperature indices indicate serious significant warming in Djibouti City. Mean temperature increased by 1.24°C during the 1966–2011 period and the period 2001–2011 was 0.66°C warmer than the 1971–2000 mean. This increase in temperature is much higher than the global warming [2] and is fully consistent with other studies carried out in the Arab region [12], over the Great Horn of Africa region [13], in the Arabian Peninsula [14], or in Sudan [17]. Extremely warm days characterized by daily maximum temperatures ≥45°C have become 15 times more frequent than in the past (comparing the decades 1966–75 and 2002–2011 periods) while extremely cool nights (<18.7°C) have almost disappeared. These impressive changes are observed at the global level [1] and are very significant in our region of interest [12, 13].

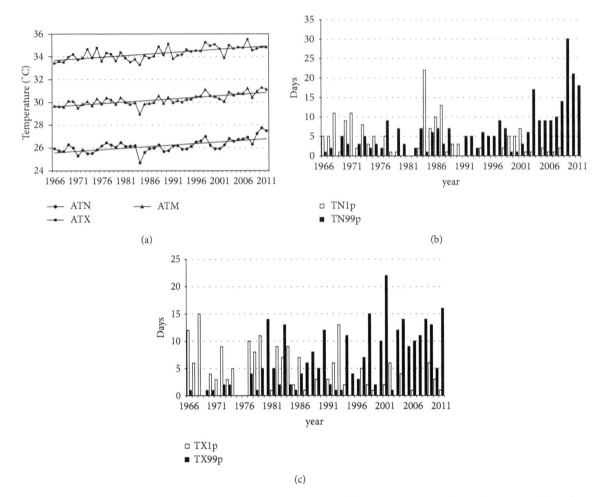

FIGURE 4: Evolution of (a) ATN, ATX, and ATM; (b) extremely cool (TN1p) and warm (TN99p) nights; and (c) extremely cool (TX1p) and warm (TX99p) days in Djibouti City (1966–2011).

TABLE 3: Temperature trends analysis from 1966 to 2011 and averages calculated for the 1971–2000 and 2001–2011 periods.

| ID | Unit | Average 1971–2000 | Trend (1966–2011) units/decade | Significance (P level) | Average 2001–2011 |
|---|---|---|---|---|---|
| ATN | °C | 26.0 | +0.27 | <0.001 | 26.8 |
| ATX | °C | 34.2 | +0.28 | <0.001 | 34.8 |
| ATM | °C | 30.1 | +0.28 | <0.001 | 30.8 |
| TN1p | days | 3.7 | −1.2 | 0.014 | 1.4 |
| TN5p | days | 19.1 | −4.8 | <0.001 | 9.2 |
| TN95p | days | 18.8 | +6.3 | <0.001 | 38.3 |
| TN99p | days | 3.8 | +2.8 | <0.001 | 13.3 |
| TX1p | days | 4.0 | −1.2 | 0.005 | 2.1 |
| TX5p | days | 21.9 | −9.0 | <0.001 | 6.6 |
| TX95p | days | 19.0 | +3.8 | <0.001 | 28.5 |
| TX99p | days | 4.7 | +2.7 | <0.001 | 11.5 |

## 5. Impact of Extreme Weather in Recent Years

Since 2007, mean yearly rainfall (44 mm) has met a 73% deficit when compared to the 30-year average (164 mm), a situation that is much worse than what was observed in the early 1980s. Yet, it has been shown that 2009 was a year of exceptionally widespread drought in Ethiopia, the second driest year ever recorded after the historic year 1984 [15]. In 2012, this period of consecutive dry year continued [18]. This impacted the well-being and, in some cases, the survival of the inhabitants of the Republic of Djibouti, especially rural population whose migration towards Djibouti City has increased in recent years. By the end of 2011, nearly 19 million people were declared food insecure in East Africa, among which 206,000 people are in the Republic of Djibouti: 120,000 rural, 26,000 Somalian and Ethiopian refugees, and 60,000 urban [19, 20].

Drought is one of the main factors of migration to Djibouti City [21]. These new informal settlements continue to grow in the south and west of the city avoiding the restrictions put by the state and may be at risk [21]. Limited access to risk knowledge may lead to settlements in risk-prone areas. Yet, the maximum daily rainfall recorded over those last five years was 46 mm in 2008. But we showed here that extreme daily rainfall events are characterized by an amount over 108 mm. In addition, it has been reported that the maximum rainfall recorded during 1 day reached 211 mm in Djibouti City [22]. Such rainfall occurred in the past, and they will occur in the future. In addition, the fourth IPCC report states that precipitation extremes are projected to increase worldwide independently of the trend in annual total precipitation [1]. Uncontrolled urbanization process may turn into catastrophic hazard in case of heavy rainfall as it has been seen elsewhere in arid zones of Africa [23–26]. The latest dramatic flood that impacted Djibouti occurred in 2004. It has been estimated that approximately 300 people died; 600 houses were destroyed and other 100 ones were inundated; 3,000 persons were made homeless; and the lives of a total of 100,000 persons were affected [27]. But the country also experienced killing floods in 1981, 1989, and 1994, systematically affecting over 100,000 people [28]. The risk of flooding is therefore real. Moreover, since the rainfall decline in the Greater Horn of Africa is influenced by the warming of the southern tropical Indian Ocean sea surface temperature [16], it is likely that migration from rural to urban areas due to rainfall shortages will not stop nor reverse in the coming decades. Local authorities should pay special attention to this specific hazard within urban planning policies.

Large augmentation of temperature indices has been highlighted. Such increase in heatwaves clearly has an impact on human health [29]. The greater absolute burden of adverse health impact from heatwaves is in the general community, but workers in various heat exposed workplaces are particularly vulnerable. The impact is therefore also economical. Considering this increasing threat, the health sector of the Republic of Djibouti should play a central role to communicate the health risks of heatwave, to initiate studies on the real impact of such high temperatures on mortality and morbidity, and to promote, lead, and evaluate a range of adaptive strategies. On its side, the National Agency of Meteorology of Djibouti should develop a heatwave early warning system in order to alert the population when weather conditions pose risks to health as it has been done elsewhere [30].

## 6. Conclusion

Although the daily rainfall and temperature datasets over Djibouti should be extended, the analysis of climatic features over the last decades justifies national and international concerns about the recent drought and heatwaves.

Yet, with yearly precipitation ranging between 29 and 66 mm between 2007 and 2011, the last 5-year period (average rainfall of 44 mm) has been the driest ever recorded since 1980 with precipitation amount as low as 27% of the long-term mean (1981–2010). Such large rainfall shortages were recorded elsewhere in the neighbouring countries of the Great Horn of Africa [15, 20] and partly explain recent food insecurity in the country and the large migration fluxes towards Djibouti City [19–21]. The conjunction of a very dry period with the absence of extreme daily rainfall events and limited access to risk knowledge increases the settlements in flood-prone areas and contribute to flood vulnerability. Local authorities have now to challenge two rainfall related hazards: the current drought impact and the effects of future exceptional rainfalls with probable large damages in recent settlements.

With an average mean temperature of 31.0°C, the 2007–2011 period has been the hottest 5-year period ever recorded since 1966, 0.9°C warmer than the long-term mean (1971–2000). Extremely warm nights and days have become much more frequent than in the past while extremely cool nights and days tend to disappear. These trends observed in the bordering countries [12–14] clearly have an impact on human health that needs to be managed by national authorities.

## Acknowledgments

This paper would not be possible without the collaboration of Mr. Osman Saad and Mr. Abdourahman Youssouf from the "Agence Nationale de la Météorologie (ANM) de Djibouti" who provided meteorological data.

## References

[1] IPCC, "Climate change 2007: synthesis report," in *Contribution of Working Groups I, II and III To the Fourth Assessment Report of the Intergovernmental Panel on Climate Change*, K. Pachauri and A. Reisinger, Eds., IPCC, Geneva, Switzerland, 2007.

[2] P. Jones, "Global temperature record," Climate Research Unit, University of East Anglia, 2013, http://www.cru.uea.ac.uk/cru/info/warming/.

[3] M. New, B. Hewitson, D. B. Stephenson et al., "Evidence of trends in daily climate extremes over southern and West Africa," *Journal of Geophysical Research*, vol. 111, Article ID D14102, 11 pages, 2006.

[4] D. R. Easterling, H. F. Diaz, A. V. Douglas et al., "Long-term observations for monitoring extremes in the Americas," *Climatic Change*, vol. 42, no. 1, pp. 285–308, 1999.

[5] M. C. Peel, B. L. Finlayson, and T. A. McMahon, "Updated world map of the Köppen-Geiger climate classification," *Hydrology and Earth System Sciences*, vol. 11, no. 5, pp. 1633–1644, 2007.

[6] J. F. Griffiths, "Climates of Africa," in *World Survey of Climatology*, vol. 10, Elsevier Publishing, New York, NY, USA, 1972.

[7] A. M. G. Klein Tank, J. B. Wijngaard, and A. van Engelen, *Climate of Europe: Assessment of Observed Daily Temperature and Precipitation Extremes*, KNMI, De Bilt, Netherlands, 2002.

[8] G. M. Griffiths, M. J. Salinger, and I. Leleu, "Trends in extreme daily rainfall across the South Pacific and relationship to the South Pacific convergence zone," *International Journal of Climatology*, vol. 23, no. 8, pp. 847–869, 2003.

[9] M. R. Haylock, T. C. Peterson, L. M. Alves et al., "Trends in total and extreme South American rainfall in 1960-2000 and links with sea surface temperature," *Journal of Climate*, vol. 19, no. 8, pp. 1490–1512, 2006.

[10] M. J. Manton, P. M. Della-Marta, M. R. Haylock et al., "Trends in extreme daily rainfall and temperature in southeast Asia and the south Pacific: 1961–1998," *International Journal of Climatology*, vol. 21, no. 3, pp. 269–284, 2001.

[11] E. Aguilar, T. C. Peterson, P. R. Obando et al., "Changes in precipitation and temperature extremes in Central America and northern South America, 1961–2003," *Journal of Geophysical Research D*, vol. 110, no. 23, Article ID D23107, pp. 1–15, 2005.

[12] M. G. Donat, T. C. Peterson, M. Brunet et al., "Changes in extreme temperature and precipitation in the Arab region: long-term trends and variability related to ENSO and NAO," *International Journal of Climatology*, 2013.

[13] P. A. Omondi, J. L. Awange, E. Forootan et al., "Changes in temperature and precipitation extremes over the Great Horn of Africa region from 1961 to 2010," *International Journal of Climatology*, 2013.

[14] M. Almazroui, M. Nazrul Islam, H. Athar, P. D. Jones, and M. A. Rahman, "Recent climate change in the Arabian Peninsula: annual rainfall and temperature analysis of Saudi Arabia for 1978–2009," *International Journal of Climatology*, vol. 32, no. 6, pp. 953–966, 2012.

[15] E. Viste, D. Korecha, and A. Sorteberg, "Recent drought and precipitation tendencies in Ethiopia," *Theoretical and Applied Climatology*, vol. 112, pp. 535–551, 2013.

[16] A. P. Williams, C. Funk, J. Michaelsen et al., "Recent summer precipitation trends in the Greater Horn of Africa and the emerging role of Indian Ocean sea surface temperature," *Climate Dynamics*, vol. 39, pp. 2307–2328, 2012.

[17] N. A. Elagib, "Trends in intra-and inter-annual temperature variabilities across Sudan," *Ambio*, vol. 39, no. 6, pp. 413–429, 2010.

[18] ICPAC, "IGAD Climate Prediction and Applications Centre Monthly Bulletin, April 2012," Report ICPAC/02/242, 11 pages, 2012.

[19] OCHA, in *Consolidated Appeal for Djibouti 2012* , Office for the Coordination of Humanitarian Affairs (OCHA), United Nations, New York, NY, USA, 2011.

[20] M. V. K. Sivakumar, "Current droughts: Context and need for national drought policies," in *Towards a Compendium on National Drought Policy. Proceedings of an Expert Meeting on the Preparation of a Compendium on National Drought Policy, July 14-15, 2011, Washington, DC, USA*, V. K. . Sivakumar, R. P. Motha, D. A. Wilhite, and J. J. Qu, Eds., pp. 2–12, World Meteorological Organization, Geneva, Switzerland, 2011.

[21] A. S. Chiré, *Le Nomade et la Ville a Djibouti: Stratégies D'Insertion Urbaine et production de Territoire*, Karthala, Paris, France, 2012.

[22] M. Shahin, *Water Resources and Hydrometeorology of the Arab Region*, Springer, Dordrecht, The Netherlands, 2007.

[23] S. Sene and P. Ozer, "Evolution pluviométrique et relation inondations: événements pluvieux au Sénégal," *Bulletin de la Société Géographique De Liège*, vol. 42, pp. 27–33, 2002.

[24] A. Tarhule, "Damaging rainfall and flooding: the other Sahel hazards," *Climatic Change*, vol. 72, no. 3, pp. 355–377, 2005.

[25] M. A. Ould Sidi Cheikh, P. Ozer, and A. Ozer, "Flood risks in the city of Nouakchott (Mauritania)," *Geo-Eco-Trop*, vol. 31, no. 1, pp. 19–42, 2007.

[26] A. Maheu, "Urbanization and flood vulnerability in a peri-urban neighborhoodof Dakar, Senegal: How can participatory GIS contribute to flood management?" in *Climate Change and the Sustainable Use of Water Resources. Climate Change Management*, W. L. Filho, Ed., pp. 185–207, Springer, Berlin, Germany, 2012.

[27] W. H. O. ], *Assessment Report on the April 2004 Floods*, World Health Organization, Republic of Djibouti, 2004.

[28] "Djibouti: disasters statistics," 2013, http://www.preventionweb .net/.

[29] T. McMichael, H. Montgomery, and A. Costello, "Health risk, present and future, from global climate change," *British Medical Journal*, vol. 344, Article ID e1359, 5 pages, 2012.

[30] K. L. Ebi, T. J. Teisberg, L. S. Kalkstein, L. Robinson, and R. F. Weiher, "Heat watch/warning systems save lives: estimated costs and benefits for Philadelphia 1995–98," *Bulletin of the American Meteorological Society*, vol. 85, no. 8, pp. 1067–1073, 2004.

# 16

# Trends of Dust Transport Episodes in Cyprus Using a Classification of Synoptic Types Established with Artificial Neural Networks

**S. Michaelides, F. Tymvios, S. Athanasatos, and M. Papadakis**

*Cyprus Meteorological Service, 28 Nikis Avenue, 1086 Nicosia, Cyprus*

Correspondence should be addressed to S. Michaelides; silas@ucy.ac.cy

Academic Editors: E. Paoletti and A. Rutgersson

The relationship between dust episodes over Cyprus and specific synoptic patterns has long been considered but also further supported in recent studies by the authors. Having defined a dust episode as a day when the average PM10 measurement exceeds the threshold of 50 mg/(m$^3$ day), the authors have utilized Artificial Neural Networks and synoptic charts, together with satellite and ground measurements, in order to establish a scheme which links specific synoptic patterns with the appearance of dust transport over Cyprus. In an effort to understand better these complicated synoptic-scale phenomena and their associations with dust transport episodes, the authors attempt in the present paper a followup of the previous tasks with the objective to further investigate dust episodes from the point of view of their time trends. The results have shown a tendency for the synoptic situations favoring dust events to increase in the last decades, whereas, the synoptic situations not favoring such events tend to decrease with time.

## 1. Introduction

It is common knowledge that dust transport from desert areas is not confined to the regions adjacent to the desert source itself, but dust can be transported extensively, and the final deposition can be thousands of kilometers away from the originating source (see [1]). According to Marelli [2], this natural contribution to Particulate Matter (PM) may range from 5% to 50% in different European countries. The Mediterranean Basin is generally recognized as a major recipient of desert dust originating from the Sahara and Saudi Arabia deserts; several t/km$^2$ are deposited each year in the Mediterranean Sea [3] profoundly affecting its coastal regions (see [4–15]).

Dust transport can be defined as a three-stage process: initially, dust is lifted and suspended in the atmospheric air, resulting in the occurrence of high level concentrations of dust in desert areas and then transported to great distances from the initial source where it finally settles on the ground. This is a quite complex phenomenon since a large number of factors contribute to its intensity and frequency. Particularly, in countries of the southeastern Mediterranean region which is particularly affected, there is an acute interest in understanding the phenomenon itself as well as the various mechanisms which govern important attributes like the severity, duration, and time trends of dust transportation (see [16]). The present study focuses on the third stage of dust transportation, namely the deposition of desert dust at ground level.

The importance of such a task is undoubted since dust transport episodes affect a number of human activities ranging from public health to ground and air transportation safety and renewable energy efficiency (to name but a few). Restricting the exemplars of dust implications to these three areas, it is worth elaborating that implications of high dust levels to human health include respiratory disorders, eye inflammations, heart disease, and lung cancer (see [17, 18]). A deep understanding of dust events and their associated processes is essential for the application of renewable energy systems that utilize solar radiation (e.g., photovoltaic systems for the generation of electricity). These systems are affected profoundly by dust deposition (wet or dry) which can lead to serious problems, such as system degradation or inefficiency due to frequent cleaning requirements [19]. Also, aviation safety can be severely affected in some circumstances, as the abrasive effect of dust on aircraft hull and engines can have an accumulating adverse impact (see [20]). Dust events can lead to a considerable reduction of visibility at airports, thus resulting in delays and cancellations (see [21]). Under severe conditions, road transportation can also be affected by reduced visibility [22].

A novel approach has recently been proposed contributing to the diagnosis and prediction of dust events in the eastern Mediterranean. The proposed methodology incorporates the adoption of Artificial Neural Networks (ANN) in order to diagnose and predict atmospheric pollutant levels in the eastern Mediterranean resulting from the transportation of dust from the originating sources located in adjacent regions (see [23]). In that study, synoptic circulation types and surface measurements were employed, along with neural methodologies [24] and satellite measurements. In the present paper, the association between synoptic patterns that was established by using the ANN methodology and dust events is presented. Thus, the first objective of the research is to identify favorable synoptic situations that lead to dust events over Cyprus, in order to justify objectively the long existing empirical knowledge of preference of occurrence of dust events to synoptic types.

Having performed this, a second objective of this paper is to investigate whether the frequency of appearance of such favorable or unfavorable synoptic patterns exhibits any time tendencies which can subsequently be used to explain the increase with time of dust episodes; this increase has been (qualitatively and subjectively) registered, at least during the last decade, at the weather observing station network maintained by the Meteorological Service of Cyprus.

Section 2 summarizes the methodologies and data used in this study; also, a brief account of the known relationship between dust episodes and prevailing weather conditions is given. The results section (Section 3) is separated into two parts: in the first, the relationship between synoptic patterns and dust events is established (first objective, above), and in the second the trends with respect to time of the synoptic patterns are investigated (second objective, above). Conclusions are presented in Section 4.

## 2. Methodologies and Data

Michaelides et al. [23] employed digital meteorological charts, satellite data, and surface measurements of dust in order to determine which synoptic circulation types are most likely to produce conditions which favor dust transport episodes over the island of Cyprus. For that study, a number of integral factors were investigated and discussed, namely, the atmospheric conditions leading to dust transportation and the meteorological conditions associated with the phenomenon, along with a presentation of the surface measurements of PM10 (particles that are less than 10 $\mu$m in aerodynamic diameter) used for the definition of a dust event. Also, the methodology for using ANN in order to construct a classification of synoptic patterns as well as the identification of those selected types favoring dust transportation was discussed. Furthermore, the exploitation of satellite technology

in estimating dust load in the atmosphere (using the Atmospheric Optical Thickness determined by the Moderate Resolution Imaging Spectrometer (MODIS) sensor onboard the Aqua-Terra satellites) was presented. The application of multiple regression in combination with the synoptic classification for the prediction of dust episodes was discussed, as well as a neural network prediction methodology. Finally, an integrated approach for the prediction of dust episodes that makes use of either the multiple regression or the neural approaches was considered.

In the following, a brief outline of the ANN methodology adopted is given. The reader is referred to Michaelides et al. [23, 25] for more details on this methodology, the mathematical formulations, and the results obtained for the synoptic classification.

The neural network architecture used in this research is Kohonen's Self Organizing Maps (SOM) [26, 27]. These networks provide a way of representing multidimensional data in much lower dimensional spaces, usually one or two dimensions. An advantage of the SOM networks over other neural network classification techniques is that the Kohonen technique creates a network that stores information in such a way that any topological relationships within the training set are maintained; for example, even if the Kohonen network associates weather patterns with dust events inaccurately, the error obtained will not be of great amplitude, since the result will be a class with similar characteristics. For a recent review of the advantages in using SOM as a tool in synoptic climatology, the reader is referred to Sheridan and Lee [28]. Also, the work by Tsakovski et al. [29] stresses the importance and usefulness of Kohonen's Self Organizing Maps in reaching such classification goals, as well as their role in decision-making processes.

Kohonen's SOM algorithm in its unsupervised mode was chosen for building the neural network models, because neither the number of output classes nor the desired output is known a priori. This is a typical example where unsupervised learning is more appropriate, since the domain expert will be given the chance to see the results and decide which model gives the best results. The expert's guidance can help to decide the number of output classes that better represent the system [30].

In unsupervised learning, there are no target values, as in the case of other methods of ANN. Given a training set of data, the objective is to discover significant features or regularities in the training data (input data). The neural network attempts to map the input feature vectors onto an array of neurons (usually one- or two-dimensional), thereby, it compresses information while preserving the most important topological and metric relationships of the primary data items on the display. By doing so, the input feature vectors can be clustered into $n$ clusters, where $n$ is less or equal to the number of neurons used. Input vectors are presented sequentially in time without specifying the desired output (see [31, 42]). The two-dimensional rectangular grid architecture of Kohonen's SOM was adopted in the present research.

The input vector is connected with each unit of the network through weights whose number is equal to the number of grid points, for the area in study. The training procedure utilizes competitive learning. When a training example is fed to the network, its Euclidean distance to all weight vectors is computed. The neuron whose weight vector is closest to the input vector (in terms of Euclidean distance) is the winner. The weight vectors of the winner neuron, as well as its neighborhood neurons, are updated in such a way that they become closer to the input pattern. The radius of the neighborhood around the winner unit is relatively large to start with, in order to include all neurons. As the learning process continues, the neighborhood is consecutively shrunk down to the point where only the winner unit is updated [32]. As more input vectors are represented to the network, the size of the neighborhood decreases until it includes only the winning unit or the winning unit and some of its neighbors. Initially, the values of the weights are selected at random.

The number of outputs is not *a priori* determined, but an "optimum" can be adopted by experimentation, in relation to the specific application under study. For several applications, it appears that the optimum number of outputs is around 30, as this exhibits the level of discretization required for the synoptic scale phenomena examined (see [33]).

Local weather forecasters have long identified specific atmospheric circulations as favoring the overall process for dust transportation over the area of eastern Mediterranean. The type of synoptic-scale atmospheric circulation which favors wet or dry dust deposits over the eastern Mediterranean is comprehensively described as one yielding a southerly to south-westerly flow throughout the entire troposphere (see [34]). The first step in the process is the elevation and suspension of dust in the atmosphere which starts with the development of a north African low pressure system, giving rise to a dust storm. This low pressure is primarily initiated in two ways: either by an upper-level trough which occurs on the polar front jet, when it overlies a heat low, or by the presence of a low level frontal system southeast of the Atlas Mountains [35, 36]. Both "initial conditions" are more frequent in late winter and spring (see [37]); indeed, this is the time of the year when dust events are most frequent over the eastern Mediterranean [1]. The second step, the transition of the dust "cloud" from its point of origin to its final destination (which can span a considerable distance), is supported by a south-westerly tropospheric flow. Finally, the third step of dust deposition can occur under dry conditions (gradual sedimentation due to gravity) or under conditions of increased humidity (where dust particles mix with rain droplets and fall on the ground as colored precipitation). Spring and autumn are the two seasonal periods favoring dust episodes, whereas summer appears to be suppressing these events. Dust episodes are also present in winter, although not as much as during spring or fall.

The ability of ANN to group synoptic patterns into seasonally dependent clusters was originally noted by Michaelides et al. [25]. Since the number of distinctive synoptic patterns used in order to classify synoptic patters over any particular geographical region is by no means fixed, it was decided to run a number of experiments and build classification models with different numbers of output nodes (i.e., classes). For the present analysis, 35 output nodes were used. For the construction of the synoptic patterns, data from

the National Centers for Environmental Prediction (NCEP) were retrieved. The data consist of the 500 hPa isobaric level values distributed on a $2.5° \times 2.5°$ grid and valid at 1200 UTC of each day. Such data for the 25-year period 1980–2005 were exploited in order to establish the synoptic classification which is based on ANN, as explained above.

A dust transport episode is considered as a day when the average PM10 measurement exceeds the threshold of $50 \, mg/(m^3 \, day)$, (which is the threshold used by Cyprus' Ministry of Labor and Social Insurance). For the needs of this study, ground measurements of PM10 were used, obtained from the Background Representative Station at Ayia Marina Xyliatou in Cyprus ($35 \, 02' \, 17''N, \, 33 \, 03' \, 28''E$), and operated by the Cyprus Ministry of Labor and Social Insurance. The location has been selected because it has negligible local particulate sources, so the pollution levels registered are ascribed, to a large extent, to external sources. The data retrieved for this station cover the three-year period 2003–2005, which consists of the available data provided by the ministry.

Having established a relationship between synoptic patterns and dust events (by using the available data for the three-year period, as explained above), the synoptic classification over the entire 25-year period (1980–2005) was exploited in an attempt to identify trends in those synoptic conditions which favor or not dust transport episodes for this period. For each year, the total of days belonging to each of the 35 classes was counted, and the trends were exposed by using linear regression. This task is performed assuming that the three-year period used for the identification of classes that favor or not dust events is representative enough to establish such a relationship.

## 3. Results

*3.1. Synoptic Patterns and Dust Events.* In the three-year period 2003–2005, 85 dust deposition events (as defined above) were recorded (out of a total of 1096 days). Figure 1 displays the monthly distribution of these episodes during the three-year period 2003–2005. It is evident from this figure that there is a seasonal preference for dust events to occur.

It has long been realized that there is a strong association between large scale atmospheric circulation patterns and regional meteorological phenomena that are observed at Earth's surface, making synoptic charts a valuable tool for the operational weather forecaster to predict qualitatively occurrences of certain weather phenomena over particular areas (see [33, 38]). One such typical example is the close association between the atmospheric circulation and the onset and maintenance of desert dust transport episodes.

Several techniques for weather type classification can be found in the literature: automatic, objective, and consistent methodologies, each of which being applicable to different circumstances and different problems (e.g., [39–41]). Each method has its strong and weak points. For the present study, a new methodology with the adoption of ANN for the classification of synoptic circulations, as it was originally proposed by Michaelides et al. [25], is utilized. The methodology employs Kohonen's Self Organizing Maps (SOM) architecture

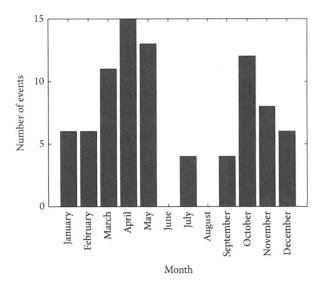

FIGURE 1: Monthly totals of dust events in the three-year period 2003–2005.

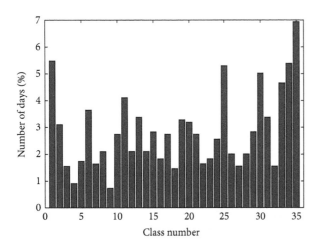

FIGURE 2: Percentage of days per class, for the 35 classes in the synoptic classification, during 2003–2005.

[27]; this is a neural network method with unsupervised learning (see also [25, 26, 42]) for the classification of distribution of isobaric patterns on 500 hPa charts. The result of this classification is the formulation of 35 synoptic prototypes (classes). Using these results, a correspondence was made between dust events and particular synoptic classes, as explained below.

Figure 2 shows the frequency of appearance of the 35 synoptic patterns for the three-year period 2003–2005, with class 35 being the most frequently encountered, followed by classes 1, 25, and 34.

Figure 3 shows the way in which the 85 dust transport events are distributed among the 35 classes in the synoptic classification adopted above. There appears to be a certain preference of classes associated with these dust events: most prone to dust events is class 1, followed by class 31. For the sake

TABLE 1: Categorization of classes according to the association between dust events.

| Category | Classes |
|---|---|
| Classes associated with dust events (ordered in diminishing importance) | 1, 31, 11, 15, 20, 23, 6, 7, 10, 32, 16, 17, 35, 13, 14, 26, 2, 8, 18, 19, 22, 24, 29, 30, 33, and 34 |
| Classes with no association to dust events | 3, 4, 5, 9, 12, 21, 25, 27, and 28 |

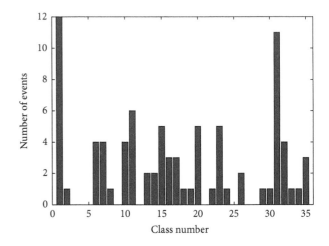

FIGURE 3: Distribution of dust deposition events per class, for the 35 classes in the synoptic classification, during 2003–2005.

of brevity, representative synoptic situations for these two classes are selectively shown in Figure 4. Figure 4(a) refers to 1200 UTC February 1 2003, and Figure 4(b) at 1200 UTC May 10 2004; in the former, a central Mediterranean upper trough extends well into the north African desert; in the later, the trough axis extends southwards from the Iberian Peninsula. In both cases, typical patterns are identified favoring dust raising and its transfer eastwards with the prevailing southwesterly airflow over the eastern Mediterranean.

Table 1 summarizes the findings of the above analysis: 26 classes appear to be associated with dust events, whereas 9 classes appear not to be associated with any dust events at all.

*3.2. Trends in Synoptic Classes and Dust Deposition Events.* Figure 5 shows the frequency of appearance of the synoptic patterns in the classification adopted here, for the 25-year period 1980–2005. Class 35 is again the most frequently encountered, followed by classes 30 and 25.

Figure 6 presents the time evolution of Classes 1, 31 and 11 in the 25-year period considered here; the first two (i.e., classes 1 and 31) exhibit an overall negative trend, while the third (i.e., class 11) exhibits a positive one. These classes were found above to represent the synoptic conditions that mostly favor dust deposition in the 3-year period.

Table 2 summarizes the trends for each of the 35 classes in the classification. It is evident from this table that while

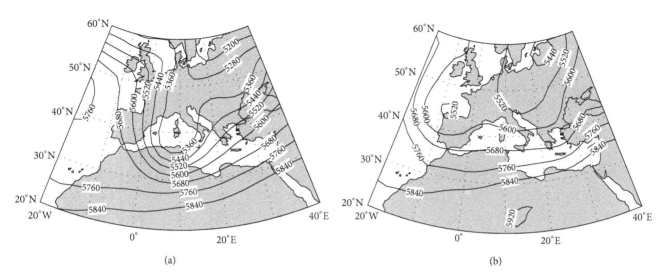

(a)                                                              (b)

FIGURE 4: Representative synoptic situations at 500 hPa corresponding to (a) class 1, (b) class 31. Isolines are drawn for every 60 geopotential meters.

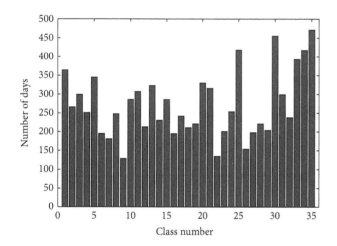

FIGURE 5: Number of days per class, for the 35 classes in the synoptic classification, during 1980–2005.

for both categories (dust and no dust), an overall prevalence of negative trends was found; this is more notable in the no dust Classes. Indeed, 8 out of the 9 no dust classes exhibit a negative trend; regarding the dust related cases, for 14 out of 26 classes the trend is negative, too. Furthermore, for the latter category, if we consider only the classes that are associated with more than two dust events, then 6 out of 13 classes exhibit negative trends. Bearing the above in mind, we can postulate that there is an increased presence (with time) of Classes that favor the presence of dust; however, this postulation surely needs to be investigated further by involving a larger period than the one used for this research work.

## 4. Concluding Remarks

As mentioned in the introduction, local weather forecasters have traditionally identified specific atmospheric circulations

as favoring the overall process for dust transportation over the area of eastern Mediterranean. However, such a traditional association of atmospheric circulations and dust events is purely qualitative and subjective. The present research presents an objective methodology which may be applied in establishing an association between the prevailing synoptic circulation and the occurrence of dust events. However, because of the limited time span of the ground dust collection data used, further research is required in order to reach a firm conclusion.

In this respect, an objective classification of synoptic types as they are portrayed by the 500 hPa isobaric analyses was performed. Adopting this classification, an association between dust transportation and prevailing weather conditions in the middle troposphere is established. These results can subsequently be utilized in studying dust transport trends for the island of Cyprus. Despite the fact that the results are site specific and the ANN methodologies used are generally highly data demanding, the methodology has some merit since it allows expanding our knowledge on the phenomenon. Indeed, by utilizing the results of this study, one can achieve some idea of how these synoptic types are associated with dust episodes. The findings can also be used to note the temporal evolution of the synoptic patterns established and also reveal their trends.

The findings of the research performed can be summarized as follows.

(i) There seems to be a preference of synoptic patterns favoring dust event occurrence over Cyprus; also, it is revealed that there are some other synoptic situations that do not favor such events at all.

(ii) There seems to be a tendency for the synoptic situations favoring dust events to increase with time, whereas, the synoptic situations not favoring such events tend to decrease with time.

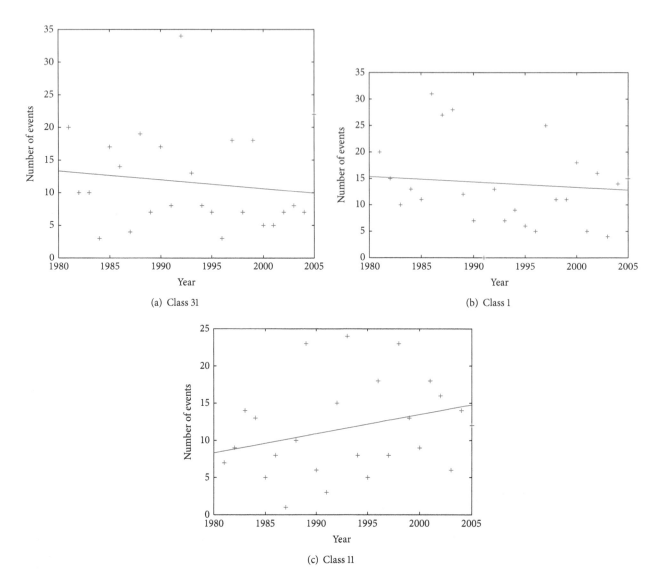

FIGURE 6: The time evolution of classes 31 (a), 1 (b), and 11 (c) in the 25-year period (1980–2005).

The fundamental mechanism behind the initiation and further evolution of a dust episode over the study area is one of synoptic scale; bearing this in mind, the above outcomes of this research can be used to explain the observed increase in dust episodes over the area, as revealed from synoptic observations at weather observing stations in the country (independently recorded from the data used in this study).

Although, as mentioned above, the findings in this study are based on a limited time span of dust ground measurements, the above conclusions are qualitatively justified, using the opinion, knowledge, and experience of operational weather forecasters working in the area for many decades. This qualitative validation of the findings focuses mainly on the justification of the relationship between dust events and specific synoptic situations and the trend for increasing number of dust events over the area over the past decade, as mentioned in the introduction.

The use of an objective methodology for synoptic pattern classification is appropriate. Any subjective classification method can only be used as a first guess approach, as it can assist to identify important favorable patterns but cannot provide an objective quantitative background for an investigation such as the one presented here (see [43]). ANN appears to be appropriate for the purpose as it comprises a nonlinear approach to tackling a very complex and definitely non-linear problem (see [44]).

Plans for further research by the group include the expansion of the ground dust measurements once it becomes available and experimentation with other synoptic classification methodologies (see [41, 45]).

## Acknowledgments

Parts of this work have been carried out within the framework of the Extreme Weather impacts on European Networks

TABLE 2: Synoptic classes and their trends for the 25-year period 1980–2005. Classes associated with dust events are in bold, whereas classes with no association to dust events are not in bold; positive trends are shown as POS and negative trends as NEG. The $R^2$ for each class is also shown.

| Class | Slope | Trend | $R^2$ |
|---|---|---|---|
| **1** | **−0.1019** | **NEG** | **0.0083** |
| **2** | **−0.0588** | **NEG** | **0.0085** |
| 3 | −0.0537 | NEG | 0.006 |
| 4 | −0.1255 | NEG | 0.033 |
| 5 | −0.3484 | NEG | 0.148 |
| **6** | **0.1036** | **POS** | **0.036** |
| **7** | **−0.0803** | **NEG** | **0.025** |
| **8** | **0.0581** | **POS** | **0.01** |
| 9 | −0.0058 | NEG | 0.001 |
| **10** | **0.147** | **POS** | **0.052** |
| **11** | **0.2574** | **POS** | **0.1** |
| 12 | −0.1015 | NEG | 0.018 |
| **13** | **−0.0147** | **NEG** | **0.001** |
| **14** | **−0.0291** | **NEG** | **0.001** |
| **15** | **−0.0472** | **NEG** | **0.004** |
| **16** | **−0.0817** | **NEG** | **0.022** |
| **17** | **−0.0574** | **NEG** | **0.009** |
| **18** | **−0.0229** | **NEG** | **0.002** |
| **19** | **0.0131** | **POS** | **0.045** |
| **20** | **0.0499** | **POS** | **0.006** |
| 21 | −0.0048 | NEG | 0.001 |
| **22** | **0.012** | **POS** | **0.001** |
| **23** | **0.0961** | **POS** | **0.025** |
| **24** | **−0.121** | **NEG** | **0.022** |
| 25 | 0.0834 | POS | 0.012 |
| **26** | **0.0602** | **POS** | **0.021** |
| 27 | −0.0048 | NEG | 0.001 |
| **28** | **−0.1959** | **NEG** | **0.092** |
| **29** | **−0.0178** | **NEG** | **0.001** |
| **30** | **0.1084** | **POS** | **0.013** |
| **31** | **−0.1364** | **NEG** | **0.02** |
| **32** | **0.026** | **POS** | **0.001** |
| **33** | **−0.1159** | **NEG** | **0.013** |
| **34** | **−0.1338** | **NEG** | **0.002** |
| **35** | **0.6376** | **POS** | **0.26** |

of Transport (EWENT) project that was funded by the European Commission under its 7th Framework Programme (Transport, Horizontal Activities). The authors wish to thank the anonymous reviewers for their constructive comments that have led to important improvement of the paper.

# References

[1] U. Dayan, J. Heffter, J. Miller, and G. Gutman, "Dust intrusion events into the Mediterranean basin," *Journal of Applied Meteorology*, vol. 30, no. 8, pp. 1185–1199, 1991.

[2] L. Marelli, "Contribution of natural sources to air pollution levels in the EU a technical basis for the development of guidance for the member states," in *Post Workshop Report From "Contribution of Natural Sources to PM Levels in Europe"*, EUR 22779 EN, JRC, Ispra, 2007.

[3] E. Ganor and Y. Mamane, "Transport of Saharan dust across the eastern Mediterranean," *Atmospheric Environment*, vol. 16, no. 3, pp. 581–587, 1982.

[4] G. Bergametti, L. Gomes, M. N. Le Coustumer, G. Coude-Gaussen, and P. Rognon, "African dust observed over Canary Islands: source-regions identification and transport pattern for some summer situations," *Journal of Geophysical Research*, vol. 94, no. 12, pp. 14–864, 1989.

[5] S. Guerzoni and R. Chester, *The Impact of the Desert Dust Across the Mediterranean*, Kluwer Academic Publishers, Norwell, Mass, USA, 1996.

[6] X. Querol, A. Alastuey, J. A. Puicercus et al., "Seasonal evolution of suspended particles around a large coal-fired power station: particulate levels and sources," *Atmospheric Environment*, vol. 32, no. 11, pp. 1963–1978, 1998.

[7] S. Rodríguez, X. Querol, A. Alastuey, G. Kallos, and O. Kakaliagou, "Saharan dust contributions to PM10 and TSP levels in Southern and Eastern Spain," *Atmospheric Environment*, vol. 35, no. 14, pp. 2433–2447, 2001.

[8] J. Barkan, P. Alpert, H. Kutiel, and P. Kishcha, "Synoptics of dust transportation days from Africa toward Italy and central Europe," *Journal of Geophysical Research D*, vol. 110, no. 7, Article ID D07208, pp. 1–14, 2005.

[9] M. Escudero, S. Castillo, X. Querol et al., "Wet and dry African dust episodes over eastern Spain," *Journal of Geophysical Research D*, vol. 110, no. 18, Article ID D18S08, pp. 1–15, 2005.

[10] E. Gerasopoulos, G. Kouvarakis, P. Babasakalis, M. Vrekoussis, J. P. Putaud, and N. Mihalopoulos, "Origin and variability of particulate matter (PM10) mass concentrations over the Eastern Mediterranean," *Atmospheric Environment*, vol. 40, no. 25, pp. 4679–4690, 2006.

[11] G. Kallos, A. Papadopoulos, P. Katsafados, and S. Nickovic, "Transatlantic Saharan dust transport: model simulation and results," *Journal of Geophysical Research D*, vol. 111, no. 9, Article ID D09204, 2006.

[12] G. Kallos, M. Astitha, P. Katsafados, and C. Spyrou, "Long-range transport of anthropogenically and naturally produced particulate matter in the Mediterranean and North Atlantic: current state of knowledge," *Journal of Applied Meteorology and Climatology*, vol. 46, no. 8, pp. 1230–1251, 2007.

[13] M. Koçak, N. Mihalopoulos, and N. Kubilay, "Contributions of natural sources to high PM10 and PM2.5 events in the eastern Mediterranean," *Atmospheric Environment*, vol. 41, no. 18, pp. 3806–3818, 2007.

[14] C. Mitsakou, G. Kallos, N. Papantoniou et al., "Saharan dust levels in Greece and received inhalation doses," *Atmospheric Chemistry and Physics*, vol. 8, no. 23, pp. 7181–7192, 2008.

[15] C. D. Papadimas, N. Hatzianastassiou, N. Mihalopoulos, X. Querol, and I. Vardavas, "Spatial and temporal variability in aerosol properties over the Mediterranean basin based on 6-year (2000–2006) MODIS data," *Journal of Geophysical Research D*, vol. 113, no. 11, Article ID D11205, 2008.

[16] S. Michaelides, P. Evripidou, and G. Kallos, "Monitoring and predicting Saharan desert dust events in the eastern Mediterranean," *Weather*, vol. 54, no. 11, pp. 359–365, 1999.

[17] D. W. Griffin, V. H. Garrison, J. R. Herman, and E. A. Shinn, "African desert dust in the Caribbean atmosphere: microbiology and public health," *Aerobiologia*, vol. 17, no. 3, pp. 203–213, 2001.

[18] T. Sandstrom and B. Forsberg, "Desert dust: an unrecognized source of dangerous air pollution?" *Epidemiology*, vol. 19, no. 6, pp. 808–809, 2008.

[19] H. K. Elminir, A. E. Ghitas, R. H. Hamid, F. El-Hussainy, M. M. Beheary, and K. M. Abdel-Moneim, "Effect of dust on the transparent cover of solar collectors," *Energy Conversion and Management*, vol. 47, no. 18-19, pp. 3192–3203, 2006.

[20] J. J. Simpson, G. L. Hufford, R. Servranckx, J. Berg, and D. Pieri, "Airborne Asian dust: case study of long-range transport and implications for the detection of volcanic ash," *Weather & Forecasting*, vol. 18, no. 2, pp. 121–141, 2003.

[21] A. Sunnu, F. Resch, and G. Afetia, "Back-trajectory model of the Saharan dust flux and particle mass distribution in West Africa," *Aeolian Research*, vol. 9, pp. 125–132, 2013.

[22] L. Mona, Z. Liu, D. Müller et al., "Lidar measurements for desert dust characterization: an overview," *Advances in Meteorology*, vol. 2012, Article ID 356265, 36 pages, 2012.

[23] S. Michaelides, F. Tymvios, D. Paronis, and A. Retalis, "Artificial neural networks for the diagnosis and prediction of desert dust transport episodes," *Studies in Fuzziness and Soft Computing*, vol. 269, pp. 285–304, 2011.

[24] M. H. Beale, M. T. Hagan, and H. B. Demuth, Neural Network Toolbo User's Guide. The MathWorks, 2010, http://www.mathworks.com/help/pdf_doc/nnet/nnet.pdf.

[25] S. C. Michaelides, F. Liassidou, and C. N. Schizas, "Synoptic classification and establishment of analogues with artificial neural networks," *Pure and Applied Geophysics*, vol. 164, no. 6-7, pp. 1347–1364, 2007.

[26] T. Kohonen, "The self-organizing map," *Proceedings of the IEEE*, vol. 78, no. 9, pp. 1464–1480, 1990.

[27] T. Kohonen, *Self-Organizing Maps*, vol. 30 of *Series in Information Sciences*, Springer, Heidelberg, Germany, 2nd edition, 1997.

[28] S. C. Sheridan and C. C. Lee, "The self-organizing map in synoptic climatological research," *Progress in Physical Geography*, vol. 35, no. 1, pp. 109–119, 2011.

[29] S. Tsakovski, P. Simeonova, V. Simeonov, M. C. Freitas, I. Dionísio, and A. M. G. Pacheco, "Air-quality assessment of Pico-mountain environment (Azores) by using chemometric and trajectory analyses," *Journal of Radioanalytical and Nuclear Chemistry*, vol. 281, no. 1, pp. 17–22, 2009.

[30] C. S. Pattichis, C. N. Schizas, and L. T. Middleton, "Neural network models in EMG diagnosis," *IEEE Transactions on Biomedical Engineering*, vol. 42, no. 5, pp. 486–496, 1995.

[31] F. Schnorrenberg, C. S. Pattichis, C. N. Schizas, K. Kyriacou, and M. Vassiliou, "Computer-aided classification of breast cancer nuclei," *Technology and Health Care*, vol. 4, no. 2, pp. 147–161, 1996.

[32] D. W. Patterson, *Artificial Neural Networks, Theory and Applications*, Prentice Hall, 1995.

[33] F. Tymvios, K. Savvidou, and S. C. Michaelides, "Association of geopotential height patterns with heavy rainfall events in Cyprus," *Advances in Geosciences*, vol. 23, pp. 73–78, 2010.

[34] N. G. Prezerakos, A. G. Paliatsos, and K. V. Koukouletsos, "Diagnosis of the relationship between dust storms over the Sahara desert and dust deposit or coloured rain in the South Balkans," *Advances in Meteorology*, vol. 2010, Article ID 760546, 14 pages, 2010.

[35] N. G. Prezerakos, "Synoptic flow patterns leading to the generation of north-west African depressions," *International Journal of Climatology*, vol. 10, no. 1, pp. 33–48, 1990.

[36] N. G. Prezerakos, S. C. Michaelides, and A. S. Vlassi, "Atmospheric synoptic conditions associated with the initiation of north-west African depressions," *International Journal of Climatology*, vol. 10, no. 7, pp. 711–729, 1990.

[37] N. Kubilay, S. Nickovic, C. Moulin, and F. Dulac, "An illustration of the transport and deposition of mineral dust onto the eastern Mediterranean," *Atmospheric Environment*, vol. 34, no. 8, pp. 1293–1303, 2000.

[38] S. Michaelides, F. Tymvios, and D. Charalambous, "Investigation of trends in synoptic patterns over Europe with artificial neural networks," *Advances in Geosciences*, vol. 23, pp. 107–112, 2010.

[39] A. K. A. El-Kadi and P. A. Smithson, "Atmospheric classifications and synoptic climatology," *Progress in Physical Geography*, vol. 16, no. 4, pp. 432–455, 1992.

[40] P. Maheras, I. Patrikas, T. Karacostas, and C. Anagnostopoulou, "Automatic classification of circulation types in Greece: methodology, description, frequency, variability and trend analysis," *Theoretical and Applied Climatology*, vol. 67, no. 3-4, pp. 205–223, 2000.

[41] A. J. Cannon, P. H. Whitfield, and E. R. Lord, "Synoptic map-pattern classification using recursive partitioning and principal component analysis," *Monthly Weather Review*, vol. 130, no. 5, pp. 1187–1206, 2002.

[42] S. C. Michaelides, C. S. Pattichis, and G. Kleovoulou, "Classification of rainfall variability by using artificial neural networks," *International Journal of Climatology*, vol. 21, no. 11, pp. 1401–1414, 2001.

[43] H. M. van den Dool, "A new look at weather forecasting through analogues," *Monthly Weather Review*, vol. 117, no. 10, pp. 2230–2247, 1989.

[44] E. N. Lorenz, "Atmospheric predictability as revealed by naturally occurring analogues," *Journal of the Atmospheric Sciences*, vol. 26, no. 4, pp. 636–646, 1969.

[45] A. Philipp, J. Bartholy, C. Beck et al., "Cost733cat—a database of weather and circulation type classifications," *Physics and Chemistry of the Earth*, vol. 35, no. 9–12, pp. 360–373, 2010.

# 17

# Interglacials, Milankovitch Cycles, Solar Activity, and Carbon Dioxide

**Gerald E. Marsh**

*Argonne National Laboratory, 5433 East View Park, Chicago, IL 60615, USA*

Correspondence should be addressed to Gerald E. Marsh; gemarsh@uchicago.edu

Academic Editor: Maxim Ogurtsov

The existing understanding of interglacial periods is that they are initiated by Milankovitch cycles enhanced by rising atmospheric carbon dioxide concentrations. During interglacials, global temperature is also believed to be primarily controlled by carbon dioxide concentrations, modulated by internal processes such as the Pacific Decadal Oscillation and the North Atlantic Oscillation. Recent work challenges the fundamental basis of these conceptions.

## 1. Introduction

The history [1] of the role of carbon dioxide in climate begins with the work of Tyndall [2] in 1861 and later in 1896 by Arrhenius [3]. The conception that carbon dioxide controlled climate fell into disfavor for a variety of reasons until it was revived by Callendar [4] in 1938. It came into full favor after the work of Plass in the mid-1950s. Unlike what was believed then, it is known today that, for Earth's present climate, water vapor is the principal greenhouse gas with carbon dioxide playing a secondary role.

Climate models nevertheless use carbon dioxide as the principal variable while water vapor is treated as a feedback. This is consistent with, but not mandated by, the assumption that—except for internal processes—the temperature during interglacials is dependent on atmospheric carbon dioxide concentrations. It now appears that this is not the case: interglacials can have far higher global temperatures than at present with no increase in the concentration of this gas.

## 2. Glacial Terminations and Carbon Dioxide

Even a casual perusal of the data from the Vostok ice core shown in Figure 1 gives an appreciation of how temperature and carbon dioxide concentration change synchronously. (Temperature is given in terms of relative deviations of a ratio of oxygen isotopes from a standard. $\delta^{18}O$ means $\delta^{18}O = [[(^{18}O/^{16}O)_{SAMPLE} - (^{18}O/^{16}O)_{STANDARD}]/(^{18}O/^{16}O)_{STANDARD})] \times 10^3$‰, measured in parts per thousand (‰). See R. S. Bradley, *Paleoclimatology* (Harcourt Academic Press, New York, 1999).) The role of carbon dioxide concentration in the initiation of interglacials, during the transition to an interglacial, and its control of temperature during the interglacial is not yet entirely clear.

Between glacial and interglacial periods, the concentration of atmospheric carbon dioxide varies between about 200 and 280 p.p.m.v., being at ~280 p.p.m.v. during interglacials. The details of the source of these variations is still somewhat controversial, but it is clear that carbon dioxide concentrations are coupled and in equilibrium with oceanic changes [5]. The cause of the glacial to interglacial increase in atmospheric carbon dioxide is now thought to be due to changes in ventilation of deep water at the ocean surface around Antarctica and the resulting effect on the global efficiency of the "biological pump" [6]. If this is indeed true, then a perusal of the interglacial carbon dioxide concentrations shown in Figure 1 tells us that the process of increased ventilation coupled with an increasingly productive biological pump appears to be self-limiting during interglacials, rising little above ~280 p.p.m.v., despite warmer temperatures in past interglacials. This will be discussed more extensively below.

The above mechanism for glacial to interglacial variation in carbon dioxide concentration is supported by the observation that the rise in carbon dioxide lags the temperature

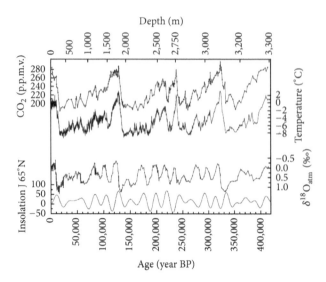

FIGURE 1: Time series from the Vostok ice core showing $CO_2$ concentration, temperature, $\delta^{18}O_{atm}$, and mid-June insolation at 85°N in $Wm^{-2}$. Based on Figure 3 of Petit et al. [22].

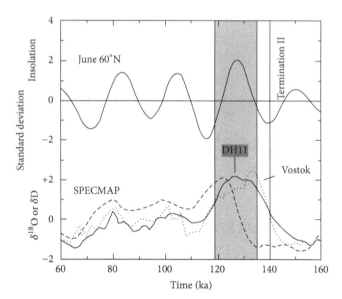

FIGURE 2: Devils Hole (DH) is an open fault zone in south-central Nevada. This figure shows a superposition of DH-11, Vostok, and SPECMAP curves for the period 160 to 60 ka in comparison with June 60°N insolation. The shading corresponds to sea levels at or above modern levels [Figure 4 from [8]].

increase by some 800–1000 years—ruling out the possibility that rising carbon dioxide concentrations were responsible for terminating glacial periods. As a consequence, it is now generally believed that glacial periods are terminated by increased insolation in polar regions due to quasiperiodic variations in the Earth's orbital parameters. And it is true that paleoclimatic archives show spectral components that match the frequencies of Earth's orbital modulation. This Milankovitch insolation theory has a number of problems associated with it [7], and the one to be discussed here is the so called "causality problem"; that is, what came first—increased insolation or the shift to an interglacial. This would seem to be the most serious objection, since if the warming of the Earth preceded the increased insolation, it could not be caused by it. This is not to say that Milankovitch variations in solar insolation do not play a role in changing climate, but they could not be the principal cause of glacial terminations.

Figure 2 shows the timing of the termination of the penultimate ice age (Termination II) some 140 thousand years ago. The data shown is from Devils Hole (DH), Vostok, and the $\delta^{18}O$ SPECMAP record (The acronym stands for Spectral Mapping Project.). The DH-11 record shows that Termination II occurred at 140 ± 3 ka; the Vostok record gives 140 ± 15 ka; and the SPECMAP gives 128 ± 3 ka [8]. The latter is clearly not consistent with the first two. The reason has to do with the origin of the SPECMAP time scale.

The SPECMAP record was constructed by averaging $\delta^{18}O$ data from five deep-sea sediment cores. The result was then correlated with the calculated insolation cycles over the last 800,000 years. This procedure serves well for many purposes, but, as can be seen from Figure 2, sea levels were at or above modern levels before the rise in solar insolation often thought to initiate Termination II. The SPECMAP chronology must therefore be adjusted when comparisons are made with records not being dependent on the SPECMAP timescale.

The above considerations imply that Termination II was not initiated by an increase in carbon dioxide concentration or increased insolation. The question then remains what did initiate Termination II?

A higher time-resolved view of the timing of glacial Termination II is shown in Figure 3 from Kirkby et al. [9]. What these authors found was that "the warming at the end of the penultimate ice age was underway at the minimum of 65°N June insolation, and essentially complete about 8 kyr prior to the insolation maximum." In this figure, the Visser et al. data and the galactic cosmic ray rate are shifted to an 8 kyr earlier time to correct for the SPECMAP time scale upon which they are based. As discussed above, the SPECMAP timescale is tuned to the insolation cycles.

The galactic cosmic ray flux—using an inverted scale—is also shown in the figure. The most striking feature is that the data strongly imply that Termination II was initiated by a reduction in cosmic ray flux. Such a reduction would lead to a reduction in the amount of low-altitude cloud cover, as discussed below, thereby reducing the Earth's albedo with a consequent rise in global temperature [10].

There is another compelling argument that can be given to support this hypothesis. Sime et al. [11] have found that past interglacial climates were much warmer than previously thought. Their analysis of the data shows that the maximum interglacial temperatures over the past 340 kyr were between 6°C and 10°C above present day values. From Figure 1, it can be seen that past interglacial carbon dioxide concentrations were not higher than those of the current interglacial, and therefore carbon dioxide could not have been responsible for this warming. In fact, the concentration of carbon dioxide that would be needed to produce a 6–10°C rise in temperature above present day values exceeds the maximum

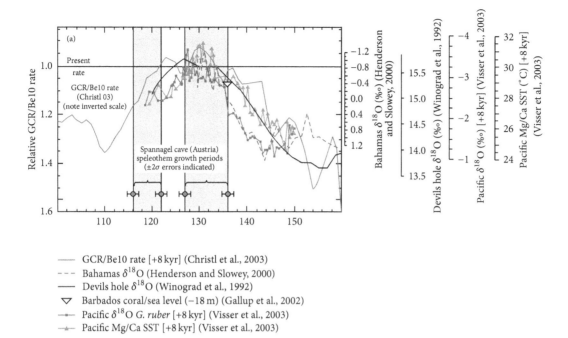

FIGURE 3: The timing of glacial Termination II. The figure shows the corrected galactic cosmic ray (GCR) rate [23] along with the Bahamian $\delta^{18}O$ record [24]; the date when the Barbados sea level was within 18 m of its present value [25] (shown by the single inverted triangle at 136 kyr); the $\delta^{18}O$ temperature record from the Devils Hole cave in Nevada [8]; and the corrected Visser et al. measurements of the Indo-Pacific Ocean surface temperature and $\delta^{18}O$ records [26] from Kirkby et al. [9].

(1000 p.p.m.v.) for the range of validity of the usual formula $[\Delta F = \alpha \ln(C/C_0)]$ used to calculate the forcing in response to such an increase.

In addition, it should be noted that the fact that carbon dioxide concentrations were not higher during periods of much warmer temperatures confirms the self-limiting nature of the process driving the rise of carbon dioxide concentration during the transition to interglacials; that is, where an increase in the ventilation of deep water at the surface of the Antarctic ocean and the resulting effect on the efficiency of the biological pump cause the glacial to interglacial rise carbon dioxide.

## 3. Past Interglacials, Albedo Variations, and Cosmic Ray Modulation

If it is assumed that solar irradiance during past interglacials was comparable to today's value as is assumed in the Milankovitch theory, it would seem that the only factor left—after excluding increases in insolation or carbon dioxide concentrations—that could be responsible for the glacial to interglacial transition is a change in the Earth's albedo. During glacial periods, the snow and ice cover could not melt without an increase in the energy entering the climate system. This could occur if there was a decrease in albedo caused by a decrease in cloud cover.

The Earth's albedo is known to be correlated with galactic cosmic-ray flux. This relationship is clearly seen over the eleven-year cycle of the sun as shown in Figure 4, which shows a very strong correlation between galactic cosmic rays,

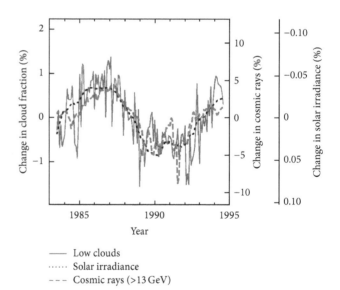

FIGURE 4: Variations of low-altitude cloud cover (less than about 3 km), cosmic rays, and total solar irradiance between 1984 and 1994 from [10]. Note the inverted scale for solar irradiance.

solar irradiance, and low cloud cover. Note that increased lower cloud cover (implying an increased albedo) closely follows cosmic ray intensity.

It is generally understood that the variation in galactic cosmic ray flux is due to changes in the solar wind associated with solar activity. The sun emits electromagnetic radiation and energetic particles known as the solar wind. A rise in solar activity—as measured by the sun spot cycle—affects the solar wind and the interplanetary magnetic field by driving matter and magnetic flux trapped in the plasma of the local interplanetary medium outward, thereby creating what is called the heliosphere and partially shielding this volume, which includes the earth, from galactic cosmic rays—a term used to distinguish them from solar cosmic rays, which have much less energy.

When solar activity decreases, with a consequent small decrease in irradiance, the number of galactic cosmic rays entering the Earth's atmosphere increases as does the amount of low cloud cover. This increase in cloud cover results in an increase in the Earth's albedo, thereby lowering the average temperature. The sun's 11-year cycle is therefore not only associated with small changes in irradiance but also with changes in the solar wind, which in turn affect cloud cover by modulating the cosmic ray flux. This, it is argued, constitutes a strong positive feedback needed to explain the significant impact of small changes in solar activity on climate. Long-term changes in cloud albedo would be associated with long-term changes in the intensity of galactic cosmic rays.

The great sensitivity of climate to small changes in solar activity is corroborated by the work of Bond et al., who have shown a strong correlation between the cosmogenic nuclides $^{14}C$ and $^{10}Be$ and centennial to millennial changes in proxies for drift ice as measured in deep-sea sediment cores covering the Holocene time period [12]. The production of these nuclides is related to the modulation of galactic cosmic rays, as described above. The increase in the concentration of the drift ice proxies increases with colder climates [13]. These authors conclude that Earth's climate system is highly sensitive to changes in solar activity.

For cosmic ray driven variations in albedo to be a viable candidate for initiating glacial terminations, cosmic ray variations must show periodicities comparable to those of the glacial/interglacial cycles. The periodicities are shown in Figure 5 taken (a) from Kirkby et al. [9] and (b) from Schulz and Zeebe [14]. Figure 5(a) is derived from the galactic cosmic ray flux—shown in Figure 6—over the last 220 kyr.

The mechanism for the modulation of cosmic ray flux discussed above was tied to solar activity, but the 41 kyr and 100 kyr cycles seen in Figure 5(a) correspond to the small quasiperiodic changes in the Earth's orbital parameters underlying the Milankovitch theory. For these same variations, to affect cosmic ray flux, they would have to modulate the geomagnetic field or the shielding due to the heliosphere. Although the existence of these periodicities and the underlying mechanism are still somewhat controversial, the lack of a clear understanding of the underlying theory does not negate the fact that these periodicities do occur in galactic cosmic ray flux.

If cosmic ray driven albedo change is responsible for Termination II and a lower albedo was also responsible for the warmer climate of past interglacials, rather than higher carbon dioxide concentrations, the galactic cosmic ray flux would have had to be lower during past interglacials than it is during the present one. That this appears to be the case is suggested by the record in Figure 6 (note inverted scale).

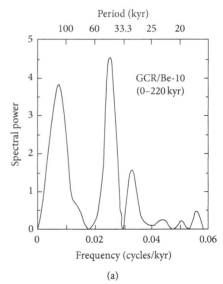

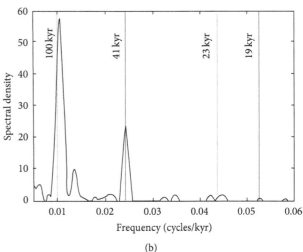

FIGURE 5: (a) Spectral power of the galactic cosmic ray flux for the past 220 ky as shown by $^{10}Be$ in ocean sediments—from Figure 6 below; (b) the detrended and normalized $\delta^{18}O$ power spectral densities versus frequency over the last 900 kyr [(a) from Kirkby et al. [9] and (b) from Schulz and Zeebe [14]]. The resolution in (a) is not as good as in (b) because of the relatively short record.

Reconstructions of solar activity on the temporal scale of Figure 6 are based on records of the cosmogenic radionuclides $^{14}C$ and $^{10}Be$ and must be corrected for geomagnetic variations. Using physics-based models [15], solar activity can now be reconstructed over many millennia. The longer the time period, the greater the uncertainty due to systematic errors.

The appearance of the 100 ky and 41 ky periods in the galactic cosmic ray flux of Figure 5(a) is somewhat surprising. With regard to orbital forcing of the geomagnetic field intensity, Frank [16] has stated that "Despite some indications from spectral analysis, there is no clear evidence for a significant orbital forcing of the paleointensity signals," although there were some caveats. In terms of the galactic cosmic ray flux, Kirkby et al. [9] maintain that "...previous conclusions that

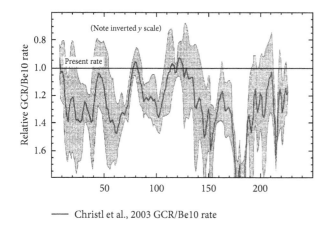

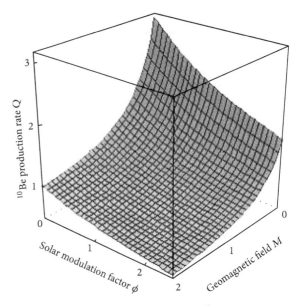

FIGURE 6: Galactic cosmic ray (GCR) flux over the last 220 kyr from combined $^{10}$Be deep-sea sediments, based on the SPECMAP timescale. The shaded region represents the $1\sigma$ confidence interval for the vertical scale [from Kirkby et al. [9]].

FIGURE 7: Atmospheric production rate of $^{10}$Be as a function of the solar modulation factor $\phi$ and the geomagnetic field intensity $M$. The plot corresponds to (2). The variables are normalized to present values of $Q$, $M$, and $\phi$. Given the $^{10}$Be production rate and geomagnetic field from independent records, the normalized solar modulation factor may be calculated for any time in the past.

orbital frequencies are absent were premature." An extensive discussion of "Interstellar-Terrestrial Issues" has also been given by Scherer et al. [17].

Using the fact that the galactic cosmic ray flux incident on the heliosphere boundary is known to have remained close to constant over the last 200 kyr and that there exist independent records of geomagnetic variations over this period, Sharma [18] was able to use a functional relation reflecting the existing data to give a good estimate of solar activity over this 200 kyr period. The atmospheric production rate of $^{10}$Be depends on the geomagnetic field intensity and the solar modulation factor—the energy lost by cosmic ray particles traversing the heliosphere to reach the Earth's orbit (this is also known as the "heliocentric potential," an electric potential centered on the sun, which is introduced to simplify calculations by substituting electrostatic repulsion for the interaction of cosmic rays with the solar wind).

If $Q$ is the average production rate of $^{10}$Be, $M$ the geomagnetic field strength, and $\phi$ the solar modulation factor, the functional relation between the data used by Sharma is

$$Q\left(M_t, \phi_t\right)$$
$$= Q\left(M_0, \phi_0\right)\left[\frac{1}{a + b\left(\phi_t/\phi_0\right)}\right]\left[\frac{c + \left(M_t/M_0\right)}{d + e\left(M_t/M_0\right)}\right], \quad (1)$$

where the subscript "0" corresponds to present day values. By setting these equal to unity, the normalized expression simplifies to

$$Q\left(M_t, \phi_t\right) = \left[\frac{1}{a + b\phi_t}\right]\left[\frac{c + M_t}{d + eM_t}\right]. \quad (2)$$

The values of the constants [19] are $a = 0.7476$, $b = 0.2458$, $c = 2.347$, $d = 1.077$, and $e = 2.274$. $Q$ may now be plotted as a function of both $M$ and $\phi$, as shown in Figure 7.

Using (2) and a globally stacked record of relative geomagnetic field intensity from marine sediments, as well as a globally stacked, $^{230}$Th-normalized $^{10}$Be record from

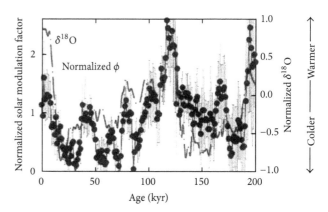

FIGURE 8: Normalized solar modulation factor and $\delta^{18}$O record over the last 200 kyr [from Sharma [18]]. The heavy dots correspond to values of the normalized solar modulation factor over the last 200 kyr. Each point is shown with an uncertainty of $1\sigma$ standard deviation. Only the $y$-direction uncertainties are large enough to be seen in the plot.

deep marine sediments, Sharma was able to calculate the normalized solar modulation factor over the last 200 kyr. ("Stacked" means aligned with respect to important stratigraphic features but without a timescale.) The result is shown in Figure 8.

The 100 kyr periodicity is readily apparent in Figure 8. It is also seen that the $\delta^{18}$O record and solar modulation are coherent and in phase. Sharma concludes from this that "… variations in solar surface magnetic activity cause changes in the Earth's climate on a 100 ka timescale."

Variations of solar activity over tens to hundreds of thousands of years do not seem to be a feature of the standard solar model. Ehrlich, [20] however, has developed a modification of the model based on resonant thermal diffusion waves which shows many of the details of the paleotemperature record over the last 5 million years. This includes the transition from glacial cycles having a 41 kyr period to a ~100 kyr period about 1 Myr ago. While the model has some problems—as noted by the author—the work shows that a reasonable addition to the standard solar model could help explain the long-term paleotemperature record. However, the fact that ice ages did not begin until some 2.75 million years ago, when atmospheric carbon dioxide concentration levels fell to values comparable to today, tells us that solar variations consistent with Ehrlich's model did not significantly affect climate before this oscillatory phase of climate began [21].

## 4. Summary

It has been shown above that low-altitude cloud cover closely follows cosmic ray flux; that the galactic cosmic ray flux has the periodicities of the glacial/interglacial cycles; that a decrease in galactic cosmic ray flux was coincident with Termination II; and that the most likely initiator for Termination II was a consequent decrease in Earth's albedo.

The temperature of past interglacials was higher than today most likely as a consequence of a lower global albedo due to a decrease in galactic cosmic ray flux reaching the Earth's atmosphere. In addition, the galactic cosmic ray intensity exhibits a 100 kyr periodicity over the last 200 kyr which is in phase with the glacial terminations of this period. Carbon dioxide appears to play a very limited role in setting interglacial temperature.

## Conflict of Interests

The author declares that there is no conflict of interests regarding the publication of this paper.

## References

[1] An excellent summary of this history is available in the January-February 2010 issue of the American Scientist. This issue reprints portions of Gilbert Plass' 1956 article in the magazine along with two commentaries. See this article for the references to scientific and more popular articles by Plass from the period 1953-1955.

[2] J. Tyndall, "On the absorption and radiation of heat by gases and vapours; on the physical connection of radiation, absorption, and conduction," *Philosophical Magazine, Series 4*, vol. 22, no. 169–194, pp. 273–285, 1861.

[3] S. Arrhenius, "On the influence of carbonic acid in the air upon the temperature of the ground," *Philosophical Magazine*, vol. 4, pp. 237–276, 1896.

[4] G. S. Callendar, "The artificial production of carbon dioxide and its influence on temperature," *Quarterly Journal of the Royal Meteorological Society*, vol. 64, pp. 223–240, 1938.

[5] W. S. Broecker, "Glacial to interglacial changes in ocean chemistry," *Progress in Oceanography*, vol. 11, no. 2, pp. 151–197, 1982.

[6] D. M. Sigman and E. A. Boyle, "Glacial/interglacial variations in atmospheric carbon dioxide," *Nature*, vol. 407, no. 6806, pp. 859–869, 2000.

[7] D. B. Karner and R. A. Muller, "Paleoclimate: a causality problem for Milankovitch," *Science*, vol. 288, no. 5474, pp. 2143–2144, 2000.

[8] I. J. Winograd, T. B. Coplen, J. M. Landwehr et al., "Continuous 500,000-year climate record from vein calcite in Devils Hole, Nevada," *Science*, vol. 258, no. 5080, pp. 255–260, 1992.

[9] J. Kirkby, A. Mangini, and R. A. Muller, "The glacial cycles and cosmic rays," Tech. Rep. CERN-PH-EP/2004-027, http://arxiv.org/abs/physics/0407005.

[10] K. S. Carslaw, R. G. Harrison, and J. Kirkby, "For a discussion of the phenomenology behind galactic cosmic ray intensity and cloud formation," *Science*, vol. 298, p. 1732, 2002.

[11] L. C. Sime, E. W. Wolff, K. I. C. Oliver, and J. C. Tindall, "Evidence for warmer interglacials in East Antarctic ice cores," *Nature*, vol. 462, no. 7271, pp. 342–345, 2009.

[12] G. Bond, B. Kromer, J. Beer et al., "Persistent solar influence on north atlantic climate during the Holocene," *Science*, vol. 294, no. 5549, pp. 2130–2136, 2001.

[13] G. Bond, W. Showers, M. Cheseby et al., "A pervasive millennial-scale cycle in North Atlantic Holocene and glacial climates," *Science*, vol. 278, no. 5341, pp. 1257–1266, 1997.

[14] K. G. Schulz and R. E. Zeebe, "Pleistocene glacial terminations triggered by synchronous changes in Southern and Northern Hemisphere insolation: the insolation canon hypothesis," *Earth and Planetary Science Letters*, vol. 249, no. 3-4, pp. 326–336, 2006.

[15] I. G. Usoskin, "A history of solar activity over millennia," *Living Reviews in Solar Physics*, vol. 5, no. 3, 2008.

[16] M. Frank, "Comparison of cosmogenic radionuclide production and geomagnetic field intensity over the last 200 000 years," *Philosophical Transactions of the Royal Society A*, vol. 358, no. 1768, pp. 1089–1107, 2000.

[17] K. Scherer, H. Fichtner, T. Borrmann et al., "Interstellar-terrestrial relations: variable cosmic environments, the dynamic heliosphere, and their imprints on terrestrial archives and climate," *Space Science Reviews*, vol. 127, no. 1–4, pp. 327–465, 2006.

[18] M. Sharma, "Variations in solar magnetic activity during the last 200 000 years: is there a Sun-climate connection?" *Earth and Planetary Science Letters*, vol. 199, no. 3-4, pp. 459–472, 2002.

[19] J. Masarik and J. Beer, "Simulation of particle fluxes and cosmogenic nuclide production in the Earth's atmosphere," *Journal of Geophysical Research D*, vol. 104, no. 10, pp. 12099–12111, 1999.

[20] R. Ehrlich, "Solar resonant diffusion waves as a driver of terrestrial climate change," *Journal of Atmospheric and Solar-Terrestrial Physics*, vol. 69, pp. 759–766, 2007.

[21] G. E. Marsh, "Climate stability and policy: a synthesis," *Energy & Environment*, vol. 22, no. 8, pp. 1085–1090, 2008.

[22] J. R. Petit, J. Jouzel, D. Raynaud et al., "Climate and atmospheric history of the past 420,000 years from the Vostok ice core, Antarctica," *Nature*, vol. 399, no. 6735, pp. 429–436, 1999.

[23] M. Christl, C. Strobl, and A. Mangini, "Beryllium-10 in deep-sea sediments: a tracer for the Earth's magnetic field intensity during the last 200,000 years," *Quaternary Science Reviews*, vol. 22, no. 5–7, pp. 725–739, 2003.

[24] G. M. Henderson and N. C. Slowey, "Evidence from U-Th dating against Northern Hemisphere forcing of the penultimate deglaciation," *Nature*, vol. 404, no. 6773, pp. 61–66, 2000.

[25] C. D. Gallup, H. Cheng, F. W. Taylor, and R. L. Edwards, "Direct determination of the timing of sea level change during Termination II," *Science*, vol. 295, pp. 310–313, 2002.

[26] K. Visser, R. Thunell, and L. Stott, "Magnitude and timing of temperature change in the Indo-Pacific warm pool during deglaciation," *Nature*, vol. 421, no. 6919, pp. 152–155, 2003.

# Permissions

All chapters in this book were first published in JCLI, by Hindawi Publishing Corporation; hereby published with permission under the Creative Commons Attribution License or equivalent. Every chapter published in this book has been scrutinized by our experts. Their significance has been extensively debated. The topics covered herein carry significant findings which will fuel the growth of the discipline. They may even be implemented as practical applications or may be referred to as a beginning point for another development.

The contributors of this book come from diverse backgrounds, making this book a truly international effort. This book will bring forth new frontiers with its revolutionizing research information and detailed analysis of the nascent developments around the world.

We would like to thank all the contributing authors for lending their expertise to make the book truly unique. They have played a crucial role in the development of this book. Without their invaluable contributions this book wouldn't have been possible. They have made vital efforts to compile up to date information on the varied aspects of this subject to make this book a valuable addition to the collection of many professionals and students.

This book was conceptualized with the vision of imparting up-to-date information and advanced data in this field. To ensure the same, a matchless editorial board was set up. Every individual on the board went through rigorous rounds of assessment to prove their worth. After which they invested a large part of their time researching and compiling the most relevant data for our readers.

The editorial board has been involved in producing this book since its inception. They have spent rigorous hours researching and exploring the diverse topics which have resulted in the successful publishing of this book. They have passed on their knowledge of decades through this book. To expedite this challenging task, the publisher supported the team at every step. A small team of assistant editors was also appointed to further simplify the editing procedure and attain best results for the readers.

Apart from the editorial board, the designing team has also invested a significant amount of their time in understanding the subject and creating the most relevant covers. They scrutinized every image to scout for the most suitable representation of the subject and create an appropriate cover for the book.

The publishing team has been an ardent support to the editorial, designing and production team. Their endless efforts to recruit the best for this project, has resulted in the accomplishment of this book. They are a veteran in the field of academics and their pool of knowledge is as vast as their experience in printing. Their expertise and guidance has proved useful at every step. Their uncompromising quality standards have made this book an exceptional effort. Their encouragement from time to time has been an inspiration for everyone.

The publisher and the editorial board hope that this book will prove to be a valuable piece of knowledge for researchers, students, practitioners and scholars across the globe.

# List of Contributors

**Thomas Hede, Caroline Leck and Jonas Claesson**
Department of Meteorology, Stockholm University, 106 91 Stockholm, Sweden

**Piia Post**
Institute of Physics, University of Tartu, 50090 Tartu, Estonia

**Olavi Kärner**
Tartu Observatory, 61602 Tõravere, Estonia

**Igor G. Zurbenko**
Department of Epidemiology & Biostatistics, University at Albany, 1 University Place, Rensselaer, NY 12144, USA

**Amy L. Potrzeba-Macrina**
Department of Mathematics, Northern Virginia Community College, Annandale Campus, 8333 Little River Turnpike, Annandale, VA 22003, USA

**Katerina G. Tsakiri**
School of Computing, Engineering and Mathematics, University of Brighton, Brighton BN2 4GJ, UK
Division of Math, Science and Technology, Nova Southeastern University, 3301 College Avenue, Fort Lauderdale-Davie, FL 33314, USA

**Antonios E. Marsellos**
School of Environment and Technology, University of Brighton, Brighton BN2 4GJ, UK
Department of Geological Sciences, University of Florida, 241 Williamson Hall, P.O. Box 112120, Gainesville, FL 32611, USA

**Igor G. Zurbenko**
Department of Biometry and Statistics, State University of New York at Albany, One University Place, Rensselaer, NY 12144, USA

**Nicola Telcik and Charitha Pattiaratchi**
School of Civil, Environmental and Mining Engineering, The UWA Oceans Institute, The University of Western Australia, 35 Stirling Highway, Crawley, WA 6009, Australia

**Ali S. Alghamdi**
Department of Geography, King Saud University, P.O. BOX 2456, Riyadh 11451, Saudi Arabia
Department of Geography and Environmental Planning, Towson University, 8000 York Road, Towson, MD 21252, USA

**ToddW. Moore**
Department of Geography, King Saud University, P.O. BOX 2456, Riyadh 11451, Saudi Arabia
Department of Geography and Environmental Planning, Towson University, 8000 York Road, Towson, MD 21252, USA

**K. V. S. Namboodiri, Dileep Puthillam Krishnan, Rahul Karuna karan Nileshwar and Koshy Mammen**
Meteorology Facility, Thumba Equatorial Rocket Launching Station (TERLS), Vikram Sarabhai Space Centre (VSSC), Indian Space Research Organisation, Thiruvananthapuram 695 022, India

**Nadimpally Kiran kumar**
Space Physics Laboratory, Vikram Sarabhai Space Centre (VSSC), Indian Space Research Organization, Thiruvananthapuram 695 022, India

**Alain Tchakoutio Sandjon**
Laboratory of Industrial Systems and Environmental Engineering, Fotso Victor Technology Institute, University of Dschang, Cameroon
Laboratory for Environmental Modeling and Atmospheric Physics, Department of Physics, Faculty of Sciences, University of Yaoundé 1, P.O. Box 812, Yaoundé, Cameroon

**Armand Nzeukou**
Laboratory of Industrial Systems and Environmental Engineering, Fotso Victor Technology Institute, University of Dschang, Cameroon

**Clément Tchawoua**
Laboratory of Mechanics, Department of Physics, Faculty of Science, University of Yaoundé 1, Cameroon

**Tengeleng Siddi**
Laboratory of Industrial Systems and Environmental Engineering, Fotso Victor Technology Institute, University of Dschang, Cameroon
Higher Institute of Sahel, University of Maroua, Cameroon

**K. V. S. Namboodiri, P. K. Dileep and Koshy Mammen**
Meteorology Facility, Thumba Equatorial Rocket Launching Station (TERLS), Vikram Sarabhai Space Centre (VSSC), Indian Space Research Organization, Thiruvananthapuram 695 022, India

**Markus Lindholm, Risto Jalkanen and Tarmo Aalto**
Metla, Rovaniemi Research Unit, P.O. Box 16, 96301 Rovaniemi, Finland

**Maxim G. Ogurtsov**
A.F. Ioffe Physico-Technical Institute, St. Petersburg 194 021, Russia

**Björn E. Gunnarson**
Bolin Centre for Climate Research, Department of Physical Geography and Quaternary Geology, Stockholm University, 106 91 Stockholm, Sweden

**Peter Domonkos**
Centre for Climate Change, University of Rovira i Virgili, Campus Terres de l'Ebre, Avenue Remolins 13-15, 43500 Tortosa Tarragona, Spain

**Agu Eensaar**
Centre of Real Sciences, Tallinn University of Applied Sciences, Pärnu Maantee 62, 10135 Tallinn, Estonia

**Armin Alipour, Jalal Yarahmadi and Maryam Mahdavi**
Department of Irrigation and Drainage Engineering, College of Aburaihan, University of Tehran, P.O. Box 33955-159, Pakdasht, Tehran, Iran

**Wolfgang Falk**
Bavarian State Institute of Forestry (LWF), Department Soil and Climate, Hans-Carl-von-Carlowitz-Platz 1, 85354 Freising, Germany

**Nils Hempelmann**
Helmholtz-Zentrum Geesthacht Zentrum für Material- und Küstenforschung GmbH, Climate Service Center (CSC), Chilehaus, Fischertwiete 1, 20095 Hamburg, Germany

**Pierre Ozer**
Department of Environmental Sciences and Management, University of Li`ege, Avenue de Longwy 185, 6700 Arlon, Belgium

**Ayan Mahamoud**
Research Center, Department of Geomatics and Environmental Sciences, University of Djibouti, Avenue Georges Clemenceau, BP 1904, Djibouti

**S. Michaelides, F. Tymvios, S. Athanasatos and M. Papadakis**
Cyprus Meteorological Service, 28 Nikis Avenue, 1086 Nicosia, Cyprus

**Gerald E. Marsh**
Argonne National Laboratory, 5433 East View Park, Chicago, IL 60615, USA

Printed in the USA
CPSIA information can be obtained
at www.ICGtesting.com
JSHW051442221024
72173JS00006B/1558

9 781632 396327